Hemopoietic Cellular Proliferation

SYMPOSIUM ON

Hemopoietic Cellular Proliferation

ST. ELIZABETH'S HOSPITAL CENTENNIAL
1869–1969
Boston, Massachusetts, November 5–6, 1969

Edited by

FREDERICK STOHLMAN, JR., M.D.

Director of Medicine and Research, St. Elizabeth's Hospital, Professor of Medicine, Tufts University Medical School, Boston, Massachusetts

GRUNE & STRATTON **New York and London**

International Standard Book Number 0–8089–0656–9

Library of Congress Catalog Card Number 79–114689

757 Third Avenue
New York, New York 10017

Printed in U.S.A.

Contents

SECTION I

General Considerations of Hemopoiesis and the Pluripotential Stem Cell

Chairman: Eugene P. Cronkite, M.D.

SECTION II

The Relationship of Bone Marrow Structure to Hemopoiesis and Myeloproliferative Syndromes

Chairman: Victor P. Bond, M.D., Ph.D.

SECTION III

The Control of Red Cell Production

Chairman: William B. Castle, M.D.

SECTION IV

Granulopoiesis and Leukemia

Chairman: William C. Moloney, M.D.

SECTION V

The Control of Platelet Production

Chairman: George Brecher, M.D.

Contributors

ADAMSON, JOHN W., M.D., Assistant Professor of Medicine, University of Washington School of Medicine, Seattle, Washington.

ATHENS, JOHN W., M.D., Professor and Head, Division of Hematology, Department of Medicine, University of Utah School of Medicine, Salt Lake City, Utah.

BALDINI, MARIO G., M.D., Director, Division of Hematologic Research, The Memorial Hospital, Pawtucket, and Professor of Medicine, Division of Biological and Medical Sciences, Brown University, Providence, Rhode Island.

BISHOP, CARTER, R., M.D., Assistant Professor of Medicine, University of Utah, College of Medicine, Salt Lake City, Utah.

BOHNE, FELICITAS, M.D., Institute of Radiation Biology, Nuclear Research Center, Karlsruhe, Germany.

BOND, VICTOR P., M.D., Associate Director, Brookhaven National Laboratory, Upton, New York.

BOYER, SAMUEL H., M.D., Associate Professor of Medicine and Biology, Division of Medical Genetics and the Clayton Laboratories, Departments of Medicine and Biology, The Johns Hopkins University School of Medicine, Baltimore, Maryland.

BRECHER, GEORGE, M.D., Professor of Clinical Pathology, University of California School of Medicine, San Francisco, California.

CALVO, WENCESLAO, M.D., Scientist of the European Atomic Energy Community, EURATOM, Center for Basic Clinical Research, University of Ulm, Ulm/Donau, Germany.

CARMEL, RICHARD, Harvard Medical School, Boston, Massachusetts, U.S.A.

CARNEVALI, C., M.D., Instituto di Semeiotica Medica, Universita' di Parma, Italy.

CASTLE, WILLIAM B., M.D., Distinguished Physician, Veterans Administration, Francis W. Peabody Faculty Professor of Medicine, Emeritus, Harvard Medical School, Boston, Massachusetts.

CHAFFEY, JOHN T., M.D., Instructor in Radiation Therapy, Harvard Medical School, Boston, Massachusetts.

CHERNOFF, AMOZ I., M.D., Director and Research Professor, University of Tennessee Memorial Research Center and Hospital, Knoxville, Tennessee.

CRONKITE, EUGENE, P., M.D., Director, Medical Division, Brookhaven National Laboratory, Upton, New York.

CROSBY, WILLIAM H., M.D., Professor of Medicine, Tufts University School of Medicine, Chief of Hematology, Tufts-New England Medical Center, Boston, Massachusetts.

DELSIGNORE, R., M.D., Instituto di Semeiotica Medica, Universita' di Parma, Italy.

DUPLAN, J. F., M.D., L'Institut du Radium, Paris, France.

EBBE, SHIRLEY N., M.D., Associate Director of Research, St. Elizabeth's Hospital, Boston, Massachusetts, Associate Professor of Medicine, Tufts University School of Medicine, Boston, Massachusetts.

FESSLER, A., Rutgers Medical School, New Brunswick, New Jersey.

FERRARI, L., M.D., Instituto di Semeiotica Medica, Universita' di Parma, Italy.

FLIEDNER, THEODOR M., M.D., Professor of Physiology, Director of the Department of Clinical Physiology and Chairman of the Task Group Hematology, Center for Basic Clinical Research, University of Ulm, Ulm/Donau, Germany.

FORTEZA-VILA, JERONIMO, M.D., Center for Basic Clinical Research, University of Ulm, Ulm/Donau, Germany.

FRIED, W., M.D., Associate Professor, Department of Medicine, University of Illinois, Chief, Clinical Hematology Laboratory, Rush-Presbyterian St. Lukes Medical Center, Chicago, Illinois.

GOODMAN, JOAN WRIGHT, Ph.D., Research Staff Member, Biology Division, Oak Ridge National Laboratory, Oak Ridge, Tennessee.

GORDON, A. S., Ph.D., Professor of Biology, Department of Biology, New York University, New York.

GRATE, HELEN E., Senior Research Assistant, Harvard Medical School, Boston, Massachusetts.

GRUBBS, CHAREEN G., M.S., Research Assistant, Biology Division, Oak Ridge National Laboratory, Oak Ridge, Tennessee.

GURNEY, CLIFFORD W., M.D., Professor and Chairman, Department of Medicine, University of Kansas Medical Center, Kansas City, Kansas.

HAAS, RAINER J., M.D., Department of Clinical Physiology, Center for Basic Clinical Research, University of Ulm, Ulm/Donau, Germany.

HELLMAN, SAMUEL, M.D., Associate Professor of Radiation Therapy, Harvard Medical School, Boston, Massachusetts.

HOWARD, DONALD, Research Associate, Department of Medicine, St. Elizabeth's Hospital, Boston, Massachusetts.

JACKSON, C. W., B.A., Biology Division, Oak Ridge National Laboratory, Oak Ridge, Tennessee.

JORDON, TOMMYE ANN, M.T., Research Assistant, University of Tennessee Memorial Research Center and Hospital, Knoxville, Tennessee.

KILLMANN, SVEN-AAGE, M.D., Ph.D., Head, Division of Hematology, Department of Medicine A, Rigshospitalet, University Hospital of Copenhagen, Copenhagen, Denmark.

KING-SMITH, ERIC A., Ph.D., National Institutes of Health Special Fellow in Bioengineering, University of Michigan, Ann Arbor, Michigan.

KRETCHMAR, A. L., M.D., Ph.D., Professor of Research, University of Tennessee Memorial Research Center and Hospital, Knoxville, Tennessee.

KUBANEK, BERNHARD, M.D., Department für Innere Medizin, Universität Ulm, Ulm/Donau, Germany.

LAMPKIN, BEATRICE C., M.D., Associate Professor of Pediatrics, University of Cincinnati College of Medicine, Cincinnati, Ohio.

LANGE, ROBERT D., M.D., Assistant Director and Professor of Research, University of Tennessee Memorial Research Center and Hospital, Knoxville, Tennessee.

LEVIN, JACK, M.D., Assistant Professor of Medicine, The Johns Hopkins University School of Medicine and Hospital, Baltimore, Maryland, and John and Mary R. Markle Scholar in Academic Medicine.

LUCARELLI, GUIDO, Director of Hematology, Instituto di Semeiotica Medica, Universita' di Parma, Italy.

MCCULLOCH, E. A., M.D., Professor, Department of Medical Biophysics and Associate Professor, Department of Medicine, University of Toronto Faculty of Medicine, Toronto, Ontario, Canada.

MCDONALD, T. P., Ph.D., Assistant Professor of Research, University of Tennessee Memorial Research Center and Hospital, Knoxville, Tennessee.

MALONEY, MARY A., B.A., Associate Biologist, Laboratory of Radiobiology, University of California, San Francisco, California.

MAUER, ALVIN M., M.D., Professor of Pediatrics, University of Cincinnati College of Medicine, Cincinnati, Ohio.

MIRAND, E. A., Roswell Park Memorial Institute, Buffalo, New York.

MOLONEY, WILLIAM C., M.D., Professor of Medicine, Harvard Medical School, Chief of Hematology, Peter Bent Brigham Hospital, Boston, Massachusetts.

MORLEY, ALEC, M.D., Instructor in Medicine, Tufts University School of Medicine, Fellow in Hematology, St. Elizabeth's Hospital, Boston, Massachusetts.

MORSE, BERNARD, M.D., Associate Professor of Medicine, New Jersey College of Medicine, Newark, New Jersey.

NAGAO, TAKESHI, Director, Division of Hematology, Department of Pediatrics, University of Tokyo, Japan.

ODELL, T. T., JR., Ph.D., Biology Divison, Oak Ridge National Laboratory, Oak Ridge, Tennessee.

PATT, HARVEY M., Ph.D., Professor and Director, Laboratory of Radiobiology, University of California, San Francisco, California.

PIKE, BEVERLEY L., A.A.I.M.L.T., Research Assistant, Department of Medicine, University of Colorado Medical Center, Denver, Colorado.

PORCELLINI, ADOLFO, M.D., Instituto di Semeiotica Medica, Universita' di Parma, Italy.

RENCRICCA, NICHOLAS J., Ph.D., Fellow in Hematology, St. Elizabeth's Hospital, Boston, Massachusetts.

RICHARD, KEVIN, M.D., Fellow in Hematology, St. Elizabeth's Hospital, Boston, Massachusetts.

RIZZOLI, VITTORIO, Instituto di Semeiotica Medica, Universita' di Parma, Italy.

ROBINSON, STEPHEN H., M.D., Associate Professor of Medicine, Harvard Medical School and Beth Israel Hospital, Boston, Massachusetts.

ROBINSON, WILLIAM A., M.D., Ph.D., Assistant Professor of Medicine, University of Colorado Medical Center, Denver, Colorado.

ROSSE, CORNELIUS, M.D., Assistant Professor, Department of Anatomy, University of Washington School of Medicine, Seattle, Washington.

SCHOOLEY, JOHN C., M.D., Donner Laboratory of Lawrence Radiation Laboratory, University of California, Berkeley, California.

SHADDUCK, R. K., M.D., Instructor in Medicine, University of Pittsburgh School of Medicine, Pittsburgh, Pennsylvania.

SMITH, WILLIE W., Ph.D., Laboratory of Physiology, National Cancer Institute, National Institutes of Health, Bethesda, Maryland.

STOHLMAN, FREDERICK, JR., M.D., Director of Medicine and Research, St. Elizabeth's Hospital, Professor of Medicine, Tufts University School of Medicine, Boston, Massachusetts.

TILL, J. E., Ph.D., Professor, Department of Medical Biophysics, University of Toronto, and Head, Division of Biological Research, Ontario Cancer Institute, Toronto, Canada.

VINCENT, PAUL C., M.D., Clinical Hematologist, Medical Research Department, Sydney Hospital, Sydney, N.S.W., Australia.

WILSON, SHIRLEY M., B.S., Laboratory of Physiology, National Cancer Institute, National Institutes of Health, Bethesda, Maryland.

ZANJANI, ESMAIL D., Assistant Professor of Biology, Department of Biology, New York University, New York.

TROBAUGH, FRANK E., JR., M.D., Professor of Medicine, University of Illinois College of Medicine, Chicago, Illinois.

Hemopoietic Cellular Proliferation

Opening Remarks

Frederick Stohlman, Jr.

It is a great pleasure for me to welcome my friends to this Conference on Hemopoietic Cellular Proliferation which is being held as a part of the Centennial Celebration of St. Elizabeth's.

I am sure that you as well as I are saddened by the absence of Bill Dameshek who had kindly agreed to serve as chairman of the second session of this symposium. Quite frankly it is hard for me to grasp the reality of his sudden and tragic death. I had looked forward to his participation in this symposium not only as a friend but also to the comments he would certainly have made in discussion of those areas of research in which he held such an active and lively interest. Perhaps his greatest attribute was his ability to take seemingly unrelated experimental observations, together with his great clinical experience of many years, and synthesize new concepts, many of which have stood the tests of time and critical challenge. We all recognize that his ideas, while frequently controversial, were always provocative and never dull. We will miss him not only today but in the years to come.

We have also recently been deprived by death of one of the pioneers in hemopoietic cellular proliferation, Doctor Edwin E. Osgood. Oz in many respects was ahead of his time and for years his ideas received limited acceptance. As we gather here today and review the literature of the past decade many of Oz's ideas clearly show through: the alpha cell and its limited growth potential, the importance of microenvironment, and ecology both in vivo and in vitro. Indeed we have lost a great experimental hematologist.

I would be remiss if I were not to extend to our European colleagues here today our deepest sympathies on the death of the renowned German hematologist Ludwig Heilmeyer. Professor Heilmeyer's contributions to hematology are widely acknowledged as was attested by his election to the presidency of the International Congress of Hematology which is to be held in Munich next year. Despite his many and varied contributions to our knowledge of blood, perhaps a more lasting monument will be the introduction of new ideas and approaches to medical education in Germany embodied in the new school which he helped to develop in Ulm.

I would like to suggest that before proceeding with the symposium that we stand for a moment in silent tribute to these three great pioneers in hematology, Bill Dameshek, Edwin Osgood, and Ludwig Heilmeyer. They have served us and their fellowman well.

At this time I should like to thank Vic Bond, no stranger to cell proliferation, for accepting chairmanship of the second session in Dr. Dameshek's place. I would also like to express my appreciation to the other chairman, Bill Castle, Gene Cronkite, Bill Moloney and George Brecher, for accepting what may prove to be an arduous task.

A decade age a conference on cellular proliferation was held in Salt Lake

City in which many of us here present were privileged to participate. At that time Dr. M. M. Wintrobe * stated, "We know a great deal about the morphologic aspects of cell growth and we have learned much about the material needed for the formation of cells as well as the chemical changes which the products of their breakdown undergo. In quantitative terms, however, our information is comparatively meager." Accordingly, much of the conference was devoted to presentation and discussion of morphologic data and their interrelationship with the burgeoning information which had been developed with various isotopic labels. Mathematical approaches to interpretation of various data received special emphasis. Physiologic studies on the regulation of hemopoietic cells, while discussed, received rather restricted attention and there was little focus on clinical correlation.

In significant measure as a result of that conference, efforts were focused on the application of these technics to an understanding of hemopoietic cell proliferation. New technics have since been developed and new models brought forward. Whereas a decade ago we spoke of a single precursor cell, mounting evidence today points to the existence of two functional, if not morphologic, stem cells—one pluripotential, the other committed. New technics for study of the stem cell have been developed by Till and McCulloch, and Bradley and Metcalf, which will be considered today. The importance of bone marrow structure in the production of hemopoietic cells, which in Salt Lake City received only passing mention, to experiments by Fliedner has gained new and much deserved attention. Finally, even the concept of the steady state has been challenged as being a state of sustained oscillation rather than a true "steady" state.

It is my hope that not only will some of the more important physiologic facts and concepts of hemopoiesis of the past decade be presented and discussed but that also we can begin to consider them in the light of the many diseases of hemopoiesis which are seen in the clinic. Some of these will come out formally, others will be developed in discussion.

To derive the maximum benefits from the conference, it is my feeling that the symposium should be published and as promptly as possible. In this respect I should like to thank the participants for their cooperation in providing their manuscripts at the time of the conference. Henry Stratton who will publish the conference has assured me every cooperation in obtaining an early publication date and from past experience I am sure that we will be able to get the proceedings out within the next 6–9 months.

Finally I would like to express my deep appreciation to Mr. Wallace Coulter of Coulter Electronics, Merck, Sharpe and Dohme Co., and the Bristol Laboratories for their generous support of this Symposium.

* Wintrobe, M. M.—Opening Remarks.In Stohlman, F., Jr. (Ed.): Kinetics of Cellular Proliferation. New York, Grune & Stratton, 1969.

The Oscillatory Nature of Hemopoiesis

By Alec Morley, Eric A. King-Smith, and Frederick Stohlman, Jr.

Introduction

It is a truism that in the body any biologic system exists as a unified entity. Nevertheless, the most commonly used approach to investigating a system involves isolating one of its components and studying its response to manipulation. Although providing valuable information, this method of analysis has an important inherent limitation since the behavior of the whole system is determined not only by the behavior of each of its parts but also by the manner in which they are organized. A particularly important form of biologic organization is control by negative feedback.

Increasing evidence suggests that negative feedback loops control erythropoiesis, granulopoiesis, and thrombopoiesis. It is the thesis of this paper that, owing to the presence of these loops, hemopoiesis is basically an oscillatory system and that the resultant oscillation may be manifest in the normal steady state, in disease, and following experimental manipulation.

Organization of Hemopoiesis

Figure 1 shows a very simple type of negative feedback system, that containing an absolute time-delay. It can readily be shown that this system will always show some oscillation following a brief perturbation and that the oscillation has the following features[1]:

1. The period is equal or approximately equal to twice the time-delay.
2. The ultimate amplitude of oscillation depends on the loop gain.

The term "gain" is defined as the slope of the input-output curve of the correcting mechanism at the point at which the system is operating. It is also the factor by which a small change in input is multiplied as it traverses the loop to emerge as a small change in output. If the loop gain is greater than one, the amplitude of successive oscillations will increase, if equal to one the amplitude will remain constant, and if less than one the amplitude will decrease, i.e., the oscillation will be damped.

The physiologic counterpart of the term "gain" is the term "sensitivity." A system of high gain has a high sensitivity since it shows a large response to a given disturbance but it also shows a correspondingly high tendency to oscillate in the absence of disturbance.

Figure 2 shows a simplified model for hemopoiesis which, although involving many assumptions, has been able to account for and predict much of the oscillatory behavior shown by hemopoiesis. Erythropoiesis, granulopoiesis

From the Department of Medicine, St. Elizabeth's Hospital, Tufts Medical School, Boston, Massachusetts, and the Department of Aerospace Engineering, University of Michigan, Ann Arbor, Michigan.

The study was supported by Grants HE7542 and 5600 from the National Heart Institute to St. Elizabeth's Hospital.

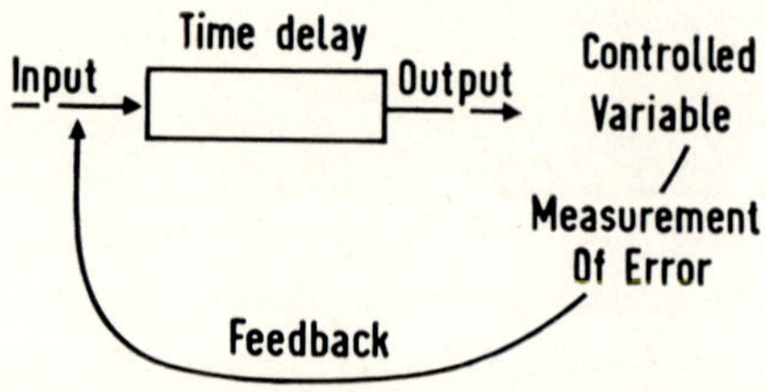

FIG. 1—Model of a negative feedback system containing a time-delay.

and thrombopoiesis each are regarded as being controlled by 2 feedback loops, one controlling the production rate and the other the release rate of cells.

The Production Rate Loop

For erythropoiesis the components of the production rate loop are the circulating red cell, the kidney, erythropoietin, and the bone marrow.[2] Erythropoietin controls erythropoiesis primarily by regulating the rate at which stem cells differentiate into erythroid precursors.[3] Since the latter must proliferate and mature before entering the blood, the feedback loop for erythropoiesis is dominated by a substantial time-delay, the marrow transit time T_M, and is thus prone to oscillate. For the rat the relationship between the circulating red cell level and the rate of erythropoiesis has been reported [4] to be a sigmoid curve with the normal operating point of the system being just on the lower

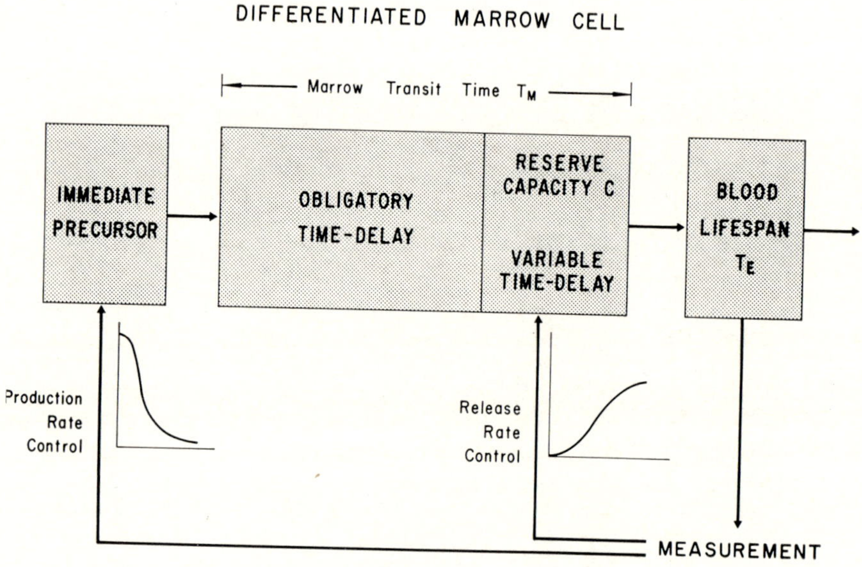

FIG. 2—Proposed model for hemopoiesis.

inflexion of the curve, i.e., at a point of moderately steep slope and hence of relatively high gain.

Less is known about the loops controlling granulopoiesis and thrombopoiesis. Both loops seem to control production by regulating the marrow at an early stage,[5,6] possibly at the stem-cell stage and, if so, both are feedback loops containing a substantial time-delay. In a series of experiments we studied the response to neutropenia by irradiating mice while shielding one leg.[6] The first change in the proliferative granulocyte compartment in the shielded leg was an increase in the size of the myeloblast-promyelocyte compartment, which was interpreted as being due to an increase in the rate of stem-cell differentiation. Figure 3 shows data from an experiment in which groups of mice were irradiated at various doses in order to produce different degrees of neutropenia. The relationship between the neutrophil count and the size of the myeloblast-promyelocyte compartment suggests that as the neutrophil count falls the rate of stem-cell differentiation, and hence ultimately the rate of granulocyte production, increases exponentially. Since there must be a limit above which the rate of production cannot increase, the relationship between the neutrophil count and granulocyte production rate may well approximate a sigmoid curve.

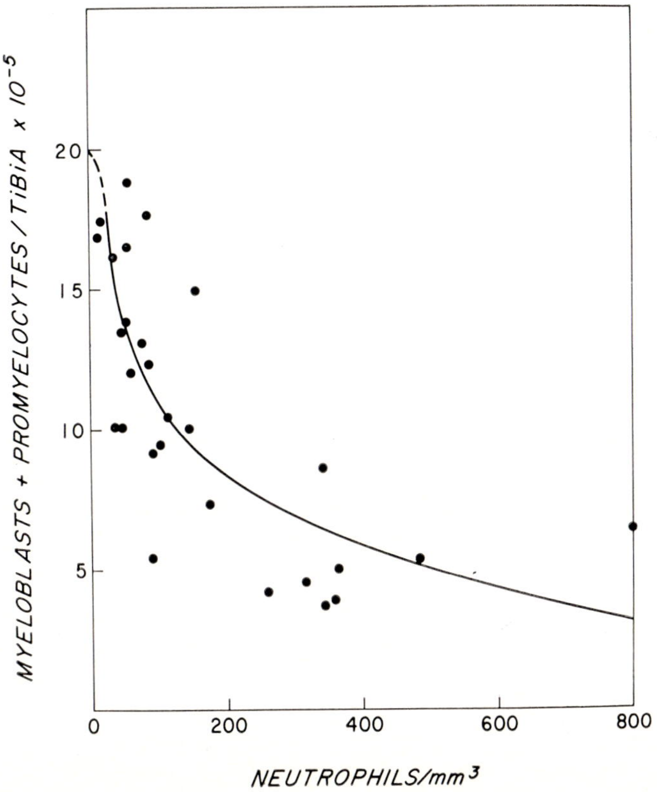

Fig. 3—Relationship between the size of the myeloblast-promyelocyte compartment and the circulating neutrophil count in irradiated leg-shielded mice.

The Release Rate Loop

A change in the fractional release rate of cells results in a change in the marrow transit time, T_M, and a control over release rate may equally well be regarded as a control over T_M. This form of control certainly exists for erythropoiesis, since an erythropoietin stimulus leads to premature release of reticulocytes,[7] and for granulopoiesis, since neutropenia leads to accelerated release of band and segmented neutrophils.[8] Whether it exists for thrombopoiesis is uncertain. There is at present no evidence for immediate release of platelets in response to thrombocytopenia but there is evidence that the maturation time of megakaryocytes is shortened.[5,9] This phenomenon may perhaps be regarded as the operation of a release rate loop containing a brief time-delay of several days.

As an approximation one can regard the cells that are potentially available for release as being in a marrow store or reserve situated at the end of the production line. The size C of this reserve is large for granulopoiesis and small for erythropoiesis. Since it can rapidly vary the release rate of cells, it acts to damp any oscillation produced by the production rate loop. The degree of damping depends on the magnitude of C and is therefore great for granulopoiesis and small for erythropoiesis. Although it also depends on the gain of the release rate loop, virtually nothing is known concerning the functional relationships between the circulating levels of the various blood cells and their corresponding rates of release from storage.

The Intravascular Compartment

The intravascular compartment also acts to damp oscillation. The degree of damping is related to the size of the intravascular pool which in turn is a function of the mean cell lifespan, T_E. For example, for red cells there is a long lifespan and a large intravascular pool; considerable variations in the rate of red cell production and release therefore will have relatively little effect on the red cell level. In actual fact, since any change in the production rate is integrated over a time T_M and is diluted in an intravascular pool which has been integrated over a time T_E, the critical factor in damping is the ratio T_E/T_M.

The Normal Steady State

In the normal steady state all 3 systems seem to be oscillating gently or to be close to the borderline of oscillation. In 8 of 11 normal men the neutrophil count showed oscillation with a period of 14–23 days[10] and in 4 of 11 normal men the platelet count showed oscillation with a period of 21–35 days.[11] In a study of normal male dogs[12] we found that in 6 of 11 the rate of erythropoiesis, as judged from the reticulocyte count, showed oscillation with a period of 14–16 days. Figure 4 shows data from one. In 4 of the dogs the phase of oscillation could be altered by bleeding followed by retransfusion. The fact that the phase of oscillation could be influenced by a pulse stimulus to erythropoiesis is direct evidence that the periodicity was the result of a

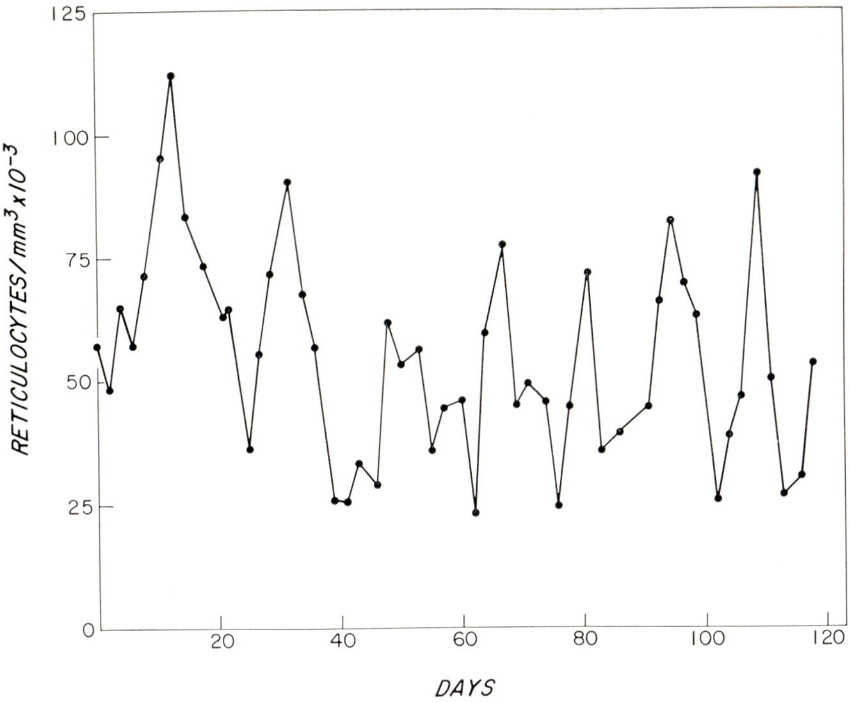

Fig. 4—Serial reticulocyte counts in a normal dog.

negative-feedback loop from red cell to marrow and was not determined by some rhythm external to the marrow.

An analogue computer model of granulopoiesis proposed by us [13] illustrates the genesis of the steady state. Figure 5 shows the functional relationships used for the blood granulocyte level, the granulocyte production rate, and the transit time through the marrow reserve (the total marrow transit time was 8 days plus this value). Figure 6 shows the stable low-amplitude oscillation with a period of 19.2 days that was produced by the model. Gross oscillation resulted when the release-rate loop was switched out of the system, indicating that the steady state was the resultant of the great oscillatory tendency of the high gain production-rate loop and the great damping ability of the marrow reserve loop.

Table I summarizes in a semiquantitative manner the factors favoring and opposing oscillation. For erythropoiesis mild oscillation seems to be the resultant of a high gain production loop, having a large oscillatory tendency, and a large intravascular pool having a high T_E/T_M ratio and thus a large damping ability. For granulopoiesis the oscillatory tendency of a high gain production-rate loop is opposed by the great damping ability of a large marrow reserve. The situation with respect to thrombopoiesis is uncertain. Since the system is probably on the border of oscillation and since the intravascular pool with a T_E/T_M ratio of 1 can only exert moderate damping a release-rate loop

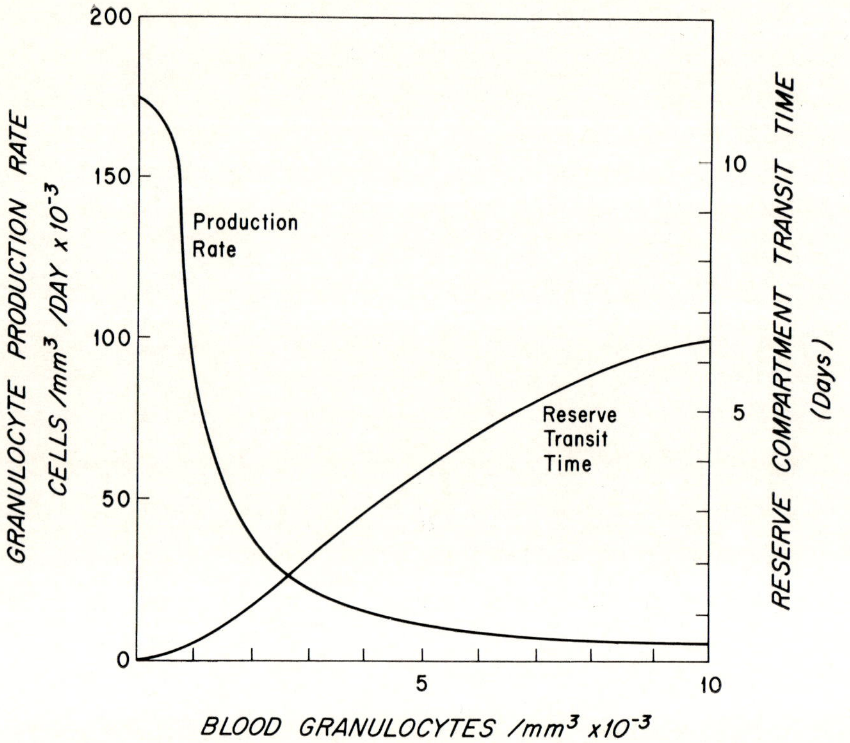

FIG. 5—Functional relationships used in the computer model between the blood granulocyte level, the marrow production rate, and the transit time through the marrow reserve.

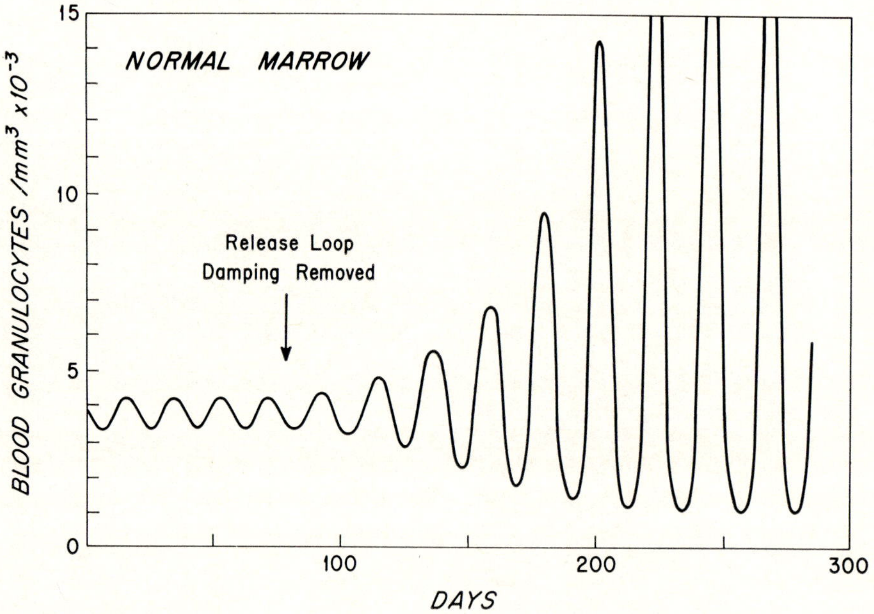

FIG. 6—Results of the simulation of normal granulopoiesis. During the simulation the release-rate loop was switched out of the system.

TABLE 1—*Factors Resulting in Oscillation and Damping* *

	Oscillation		*Damping*				
	Production-Rate Loop Gain		Release-Rate Loop Magnitude of C		Gain	Intravascular Pool T_E/T_M	
Erythropoiesis	High	++++	Very small	±	?	12	++++
Granulopoiesis	High	++++	Very large	++++	?	0.04	±
Thrombopoiesis	?	++	?	++	?	1	++

* The production-rate loop tends to cause oscillation, the release-rate loop and the intravascular pool act to damp the oscillation. The degree of oscillatory tendency or damping ability is indicated by a scale 0 – ++++.

may be of some importance or, alternatively, thrombopoiesis may have a production rate loop of relatively low gain.

PERIODICITY IN DISEASE AND UNDER EXPERIMENTAL CONDITIONS

Cyclical Neutropenia

Cyclical neutropenia is a disease characterized by neutropenia and oscillation of the neutrophil count. The amplitude of oscillation is exaggerated but the period is approximately normal (Fig. 7). Figure 8 shows how computer

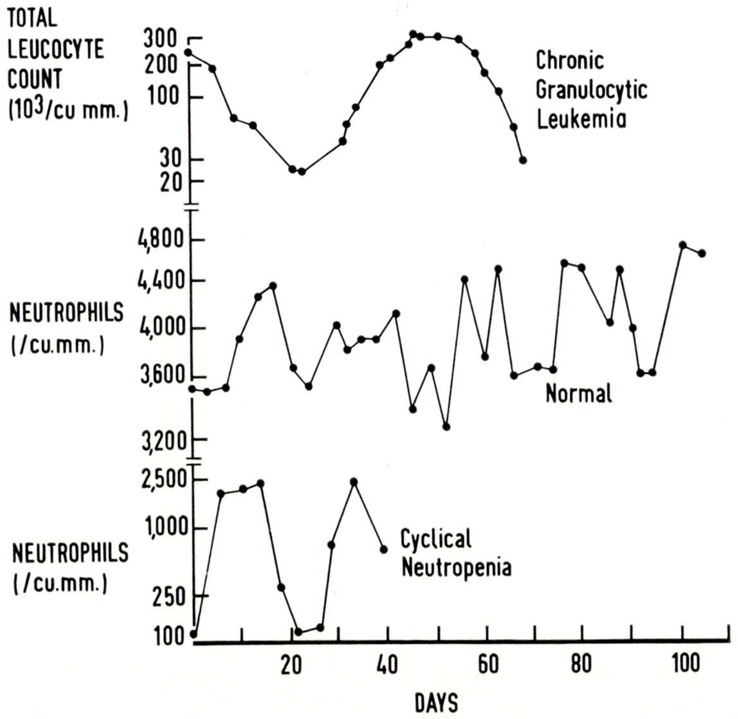

FIG. 7—Oscillation in a normal individual, a patient with cyclical neutropenia, and a patient with chronic granulocytic leukemia.

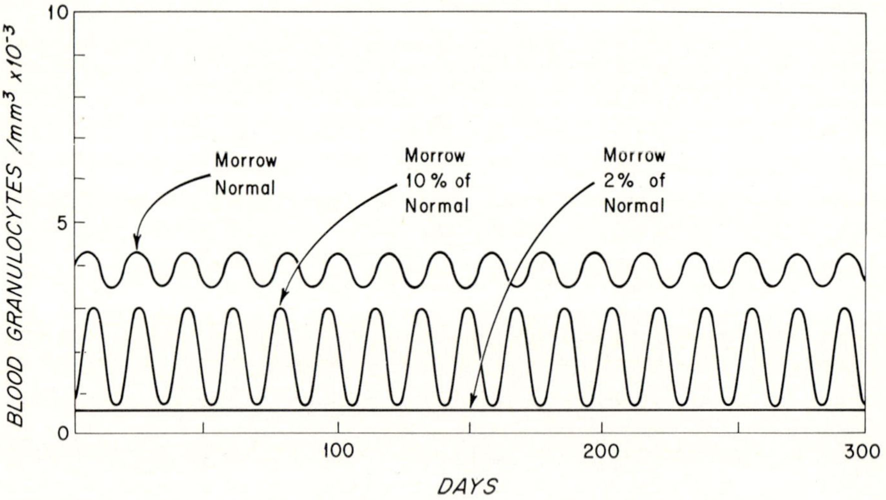

FIG. 8—Computer simulation of mild and severe marrow failure showing how states resembling cyclical neutropenia and severe noncyclical neutropenia result.

simulation of mild marrow failure resulted in cyclical neutropenia whereas simulation of severe failure resulted in chronic noncyclical neutropenia. Mild marrow failure decreased the gain of the production-rate loop but this was largely counteracted by the operating point of the system moving up the curve to a steeper point as the neutrophil count fell. Thus the net result was only a moderate decrease in gain. However, at the same time, the store became so depleted that the release-rate loop lost virtually all damping ability and

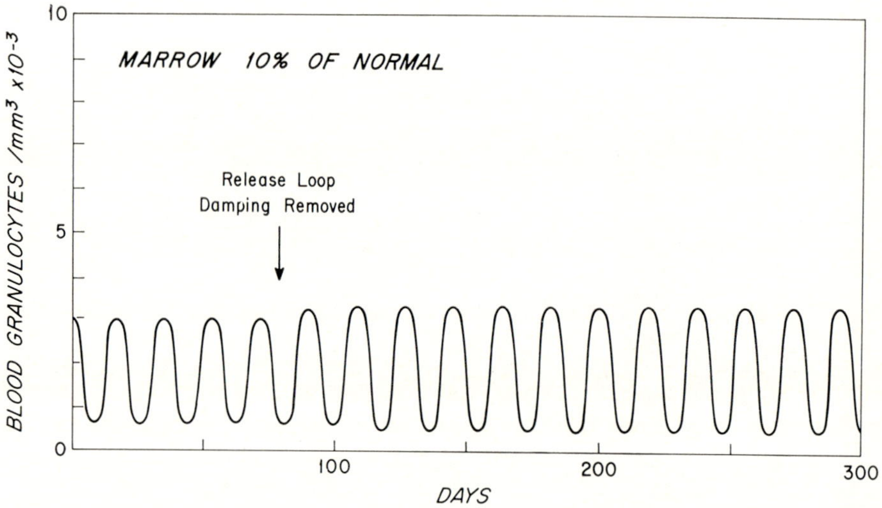

FIG. 9—Effect on the simulation of cyclical neutropenia produced by switching the release-rate loop out of the system.

cyclical neutropenia resulted. Figure 9 shows that at this stage switching the release-rate loop out of the circuit had almost no effect on oscillation, indicating that the loop was almost ineffective. Eventually periodicity disappeared as marrow failure became increasingly severe and this was attributed to the progressively decreasing gain of the production-rate loop.

Experimental support for this hypothesis was obtained by feeding dogs a constant daily dose of cyclophosphamide.[14] Cyclical neutropenia developed in 8 of 13 dogs at a point when the mean neutrophil count had fallen to ½–⅓ of normal. Further suppression in 2 of these 8 dogs changed the neutropenia from cyclical back to noncyclical. Figure 10 shows the data from one of them illustrating how slight marrow suppression failed to produce cyclical neutropenia, further suppression produced cyclical neutropenia and still further suppression produced chronic noncyclical neutropenia.

Cyclical neutropenia therefore does not seem to be a specific entity but may follow an alteration in the balance of factors favoring and opposing oscillation.

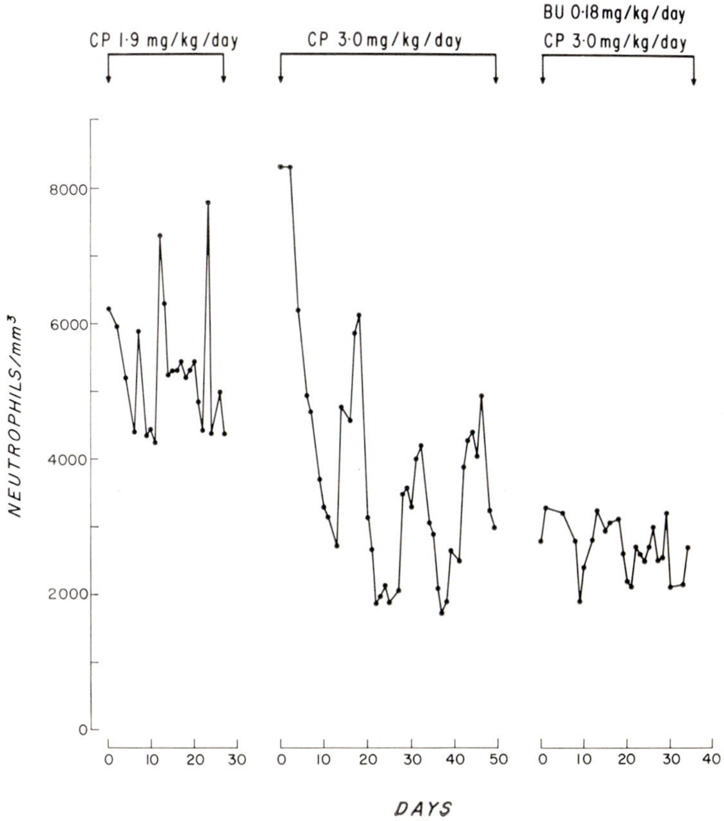

FIG. 10—Data from a normal dog illustrating the effects of slight, moderate, and severe marrow suppression. Suppression was produced by cyclophosphamide (CP) alone or combined with Busulphan (BU).

It may result from:

1. Mild failure at any point around the production-rate loop. Most cases of the disease probably fall into this category, being due to stem-cell deficiency.

2. Primary failure of the release-rate loop. This might for example follow failure of secretion of a releasing hormone. Damping would be lost owing to the very low gain of the release-rate loop.

3. Increased rate of exit of neutrophils from the blood. This situation would be very favorable to oscillation since the production-rate loop would come to operate on a very high gain segment of the curve and the ratio T_E/T_M would be very small.

Cyclical Thrombocytopenia

Cases of this disease probably arise by one of the above 3 mechanisms, perhaps more often by 3.

Chronic Granulocytic Leukemia (CGL) and Polycythemia Vera (P Vera)

As a result of the finding of oscillation in CGL[15] and P Vera,[16] it was suggested that some degree of normal control is retained in both disorders. The period of oscillation is greatly prolonged in CGL (Fig. 7), suggesting that the marrow transit time is increased owing to slowed leukemic cell maturation. On the other hand the period in 1 case of P Vera was reported to be normal which suggests that in this case there was no abnormality of maturation.

Hemolytic Anemia

Oscillation of erythropoiesis has been observed in experimentally produced hemolytic anemia.[17] Several factors would combine to favor oscillation in this situation. A marked fall in the T_E/T_M ratio would decrease damping and because of the fall in hemoglobin level the operating point of the production loop could come to lie on a very steep segment of the curve. It is of interest that in the rat the steepest part of the curve lies over a hemoglobin range 100–75 per cent of normal.[4]

We have speculated that oscillation may account for some of the cases of apparently compensated hemolysis in hereditary spherocytosis. If marked oscillation of erythropoiesis should be present in such cases, the hemoglobin level could at times rise to within the normal range. Blood counts at such times would give a false idea of the mean hemoglobin level and would suggest that the anemia was compensated.

Recovery After Irradiation

The abortive rise of blood elements 6–10 days after irradiation is well recognized.[18] Since there is no abortive rise in the number of erythropoietin responsive stem cells in irradiated polycythemic animals,[19,20] endogenous erythropoietin must be necessary for the abortive erythroid rise to develop. Conceivably the plasma level of erythropoietin may rise shortly after irradiation owing to failure of utilization by the marrow. However, since there seem to be subsequent rises and falls in erythropoiesis before recovery is complete,[21] the abortive rise may in large part be merely a manifestation of the oscillatory nature of hemopoiesis.

Coupling of Oscillators

Erythropoiesis, granulopoiesis, and thrombopoiesis are each actual or potential oscillatory systems. The stem-cell compartment undoubtedly has its own regulatory mechanism and it may therefore also be oscillatory, although the time-characteristics of oscillation would probably be much shorter than those for the differentiated cell lines. Hemopoiesis may therefore be regarded as 3, possibly 4, oscillators which under normal conditions are very loosely coupled at the stem-cell level. Coupling may take the form of interaction between the stem cell and a differentiated cell population,[22] competition for stem cells between differentiated populations and competition between various cell populations for nutrients or space.[23,24] It is, however, a property of coupled autonomous oscillators that as the degree of coupling increases the various oscillators lose their automony and become locked into one another. The rhythmicity of the whole system becomes increased and a variety of new and complex oscillations may appear.[25]

The precise importance of coupling cannot be assessed at present but increased coupling would be expected under conditions of stem-cell depletion. Thus it might be a factor in some cases of cyclical neutropenia, in marrow depression during drug therapy, and during marrow recovery following irradiation. A factor in the oscillation observed in chronic granulocytic leukemia might be competition between the leukemic cell line and the normal stem cells. Perhaps coupling is involved in the reciprocal oscillations between erythropoiesis, granulopoiesis, and thrombopoiesis which we have observed in several dogs with genetically-determined or drug-induced cyclical neutropenia.

Conclusions

It seems unlikely that the control features shown by hemopoiesis are unique. On the contrary, the principles of design that have evolved for hemopoiesis have probably evolved for other systems. The most important principles might be:

1. The steady state borders on oscillation. This setting provides the best compromise between sensitivity and stability.

2. This setting can often be achieved by combining a high gain production loop with a large capacity release loop. The production loop provides sensitivity (although after a delay) and the capability of indefinite response. The release-rate loop provides immediate response and stability.

3. For most biologic systems it is more important to react vigorously in one direction, then another. The production loops of many systems may therefore operate on an inflexion of their input-output curve and a small movement in

the critical direction may move the system into a high gain region with consequent enhancement of rhythmicity.

Finally, if many systems do show the same principles of control as does hemopoiesis, many periodic diseases may arise in a similar fashion to periodic diseases of hemopoiesis.

REFERENCES

1. Milsum, J. H.: Biological Control Systems Analysis. New York, McGraw-Hill, 1966.

2. Stohlman, F., Jr.: Erythropoiesis. New Eng. J. Med. 267:342, 1962.

3. Alpen, E. L., and Cranmore, D.: Observations on the regulation of erythropoiesis and on cellular dynamics by Fe^{59} autoradiography. *In* Stohlman, F., Jr. (Ed.): The Kinetics of Cellular Proliferation. New York, Grune & Stratton, 1959, p. 290.

4. Hodgson, G.: Aplicación de la teoria de control al estudio de la eritropoyesis. Sangre 13:241, 1968.

5. Ebbe, S.: Megakaryocytopoiesis and platelet turnover. Ser. Haemat. 1/2, 65, 1968.

6. Morley, A., and Stohlman, F., Jr.: Studies on the regulation of granulopoiesis I: The response to neutropenia. Blood. 35:312, 1970.

7. Stohlman, F., Jr., Brecher, G., and Moores, R. R.: Humoral regulation of erythropoiesis VIII. The kinetics of red cell production and the effect of erythropoientin. *In* Jacobson, L. O., and Doyle, M. (Eds.): Erythropoiesis. New York, Grune & Stratton, 1962, p. 162.

8. Boggs, D. R., Chervenick, P. A., Marsh, J. C., Cartwright, G. E., and Wintrobe, M. M.: Neutrophil-releasing activity of dogs injected with endotoxin. J. Lab. Clin. Med. 72:177, 1968.

9. Odell, T. T., Jr., Jackson, C. W., Friday, T. J., and Charsha, D. E.: Effect of thrombocytopenia on megakaryocytopoiesis. Brit. J. Haemat. 17:91, 1969.

10. Morley, A. A.: A neutrophil cycle in healthy individuals. Lancet 2:1220, 1966.

11. Morley, A.: A platelet cycle in normal individuals. Aust. Ann. Med. 18:127, 1969.

12. Morley, A., and Stohlman, F., Jr.: Erythropoiesis in the dog: The periodic nature of the steady state. Science 165:1025, 1969.

13. King-Smith, E. A., and Morley, A. A.: A computer model for the mammalian (human) leucocyte control system. Blood. In press.

14. Morley, A., and Stohlman, F., Jr.: Cyclophosphamide-induced cyclical neutropenia: an animal model of a human periodic disease. New Eng. J. Med. 282:643, 1970.

15. Morley, A. A., Baikie, A. G., and Galton, D. A. G.: Cyclical leucocytosis as evidence for retention of normal homeostatic control in chronic granulocytic leukaemia. Lancet 2:1320, 1967.

16. Morley, A.: Blood-cell cycles in polycythaemia vera. Aust. Ann. Med. 18:124, 1969.

17. Orr, J. S., Kirk, J., Gray, K. G., and Anderson, J. R.: A study of the interdependence of red cell and bone marrow stem cell populations. Brit. J. Haemat. 15:23, 1968.

18. Jacobsen, L. O., Marks, E. K., and Lorenz, E.: The hematological effects of ionizing radiations. Radiology 52:37, 1949.

19. Gurney, C. W., Lajtha, L. G., and Oliver, R.: A method for investigation of stem-cell kinetics. Brit. J. Haemat. 8:461, 1962.

20. Weisman, M., Martinson, D., Fried, W., and Gurney, C. W.: A comparison of the rates at which colony forming units and erythropoietin responsive cells recover after 200 r total body irradiation. J. Lab. Clin. Med. 69:438, 1967.

21. Morley, A., and Stohlman, F., Jr.: Periodicity during recovery of erythropoiesis following irradiation. Blood 34:96, 1969.

22. Kirk, J., Orr, T. S., and Hope, C. S.: A mathematical analysis of red blood cell and bone marrow stem cell control mechanisms. Brit. J. Haemat 15:35, 1968.

23. Osgood, E. E.: Ecology of human hemic cells. Perspect. Biol. Med. 6:228, 1963.

24. Okunewick, J. P., and Kretchmar, A. L.: A mathematical model for post-irradiation hematopoietic recovery. Memorandum RM–5272–PR, Rand Corp., 1967.

25. Goodwin, B. G.: Temporal organization in cells. London, Academic Press, 1963.

Cellular Interactions in the Control of Hemopoiesis

By E. A. McCulloch and J. E. Till

Introduction

We are studying the regulation of hemopoiesis because we wish to acquire the capacity to manipulate the system. Particularly, where normal mechanisms of regulation break down in disease, we wish to identify the site of the lesion and to correct it. The first step toward this goal is to identify discrete classes of cells related to each other, as parent is to progeny, but different in significant qualitative or quantitative properties. Our approach to the problem was based on the following concept of the hemopoietic system: "like a bacterial culture, the hemopoietic system consists of large numbers of free living and dividing cells which are not associated together as specific organs. . . . The similarity with microbial populations is sufficient to suggest that the kind of approaches which have been useful in the study of microorganisms may also be helpful in the study of hemopoiesis."[1] Accordingly, our studies were based on the use of clonal analysis and leaned heavily on examination of genetically determined defects. In this way some stages in hemopoiesis have been described, and at least one close parent-progeny pair identified. The approach, however, proved limited when we attempted to understand factors involved in regulation, that is, factors that influence reversible and irreversible transitions between various states of cells. It became apparent that regulation is often a function of cellular interrelations and is obscured by experiments in which clones are allowed to proliferate independently of each other. This has led us to search for models suitable for studies of cellular interactions. While it is hoped that this avenue will lead to practical applications in medicine, it is also obvious that we have acquired a view of the hemopoietic system different from that given above. A more recent quotation is: "Co-operation between various cell classes appears essential for normal function. . . . perhaps the hemopoietic system will play a role in the development of understanding of the government of the cellular community similar to the role played by bacteria and their phages in the growth of molecular biology."[2]

Evidence for Cellular Interaction at Early Stages in Hemopoiesis

A model of early events in granulopoietic and erythropoietic differentiation is given in Figure 1. It is now generally agreed that the mouse hemopoietic system contains a class of pluripotent stem cells (CFU) whose progeny include

From the Departments of Medicine and Medical Biophysics, University of Toronto and Ontario Cancer Institute.)

This work was supported by grants from the National Cancer Institute of Canada, the Medical Research Council of Canada (MT–1420), and the Defence Research Board, Canada (Grant 9350–14 G & C).

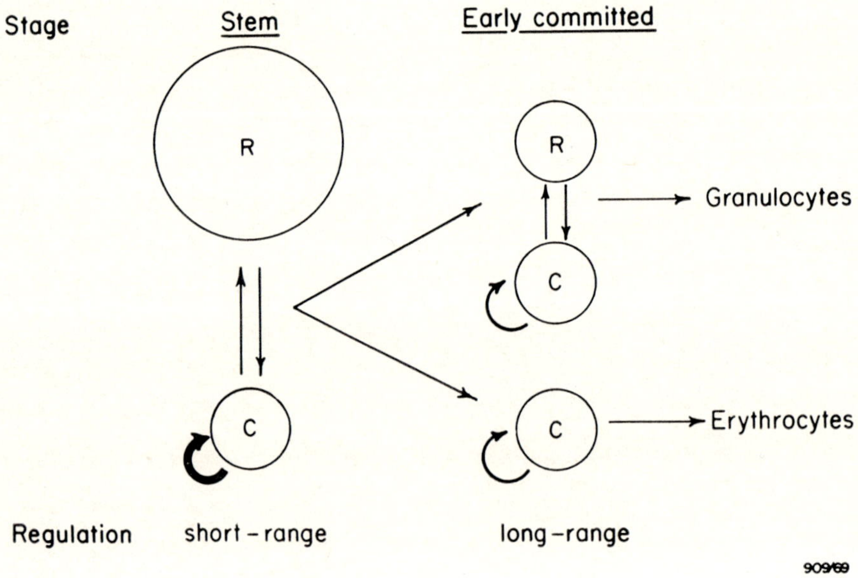

FIG. 1—A diagram representing early events in the myeloid component of the hemopoietic system. Populations of cells at rest (R) or in a state of rapid proliferation (C) are shown as circles. Reversible and irreversible transitions are shown as straight arrows, while circular arrows indicate the process of self-renewal.

at least granulocytes, erythroblasts,[3] thymocytes, and lymph node cells.[4] It is assumed that these stem cells may be either in a state of rest (the G_0 phase of Quastler [5] and Lajtha [6]) or a state of rapid proliferation.[7,8] Under normal conditions the great majority of pluripotent stem cells must be in the resting state since a short exposure of high specific activity tritiated thymidine does not inactivate the colony-forming ability of a detectable fraction of them.[8] There is now also considerable evidence that the granulopoietic descendants of the pluripotent stem cell are derived from an early differentiated progenitor identified by its capacity to give rise to colonies in culture.[9,10,11,12] The proliferative state of these progenitors has been studied using the tritiated thymidine "suicide" method as it was originally applied to the pluripotent stem cells. In contrast to the results obtained for the latter class, when normal marrow was exposed to a pulse of high specific activity tritiated thymidine in vitro and assayed for colony formation in culture, between 35 and 45 per cent of the colony-forming activity was lost. When exponentially growing marrow was tested, however, 80 per cent inactivation was regularly found.[13] These results are consistent with the view that resting marrow contains some granulocytic progenitors at rest and others that are rapidly proliferating. Since cells of this class, like their pluripotent ancestors, are capable of varying rates of proliferation, they are potential targets for control mechanisms.

Figure 1 indicates the way in which cellular properties change during differentiation. The capacity to alter cell cycle parameters is a feature of undifferentiated cells, although it is retained to some degree by the early differentiated progenitors of granulocytes. A different property appears to characterise early differentiated cells; that is, capacity to respond to long-range hormone-like regulators. The most familiar of these is erythropoietin and there is considerable support for the view that this hormone acts on an early differentiated cell committed to erythropoiesis.[14,15,16] Since the erythropoietin-sensitive population does not appear to contain cells at a state of rest, the hormone may act by initiating maturation rather than by regulating cell cycle parameters. No evidence is available for a similar long-range regulation of events that affect the pluripotent stage of hemopoietic differentiation. While a number of nonspecific substances including endotoxin,[17,18] alphaglobulin,[19] and antigen [20] will change the number of pluripotent stem cells in spleen, marrow, or peripheral blood these lack the specificity required for regulation. In contrast to the early differentiated stage of differentiation, therefore, a significant part of regulation of pluripotent stem cells is probably achieved by mechanisms that act by cell-cell interaction at short-range.

Perhaps the most convincing data for the existence of short range regulation mechanisms affecting pluripotent stem cells has been derived from studies of genetically anemic mice of genotype Sl/Sl^d.[21] The defect in these animals affects the capacity of their tissues to support the normal proliferation of pluripotent hemopoietic stem cells.[22] Parabiosis experiments failed to provide evidence that this host factor acts at a distance and it was suggested that the Sl gene might affect a cell-cell interaction.[22] Convincing confirmation of this view has been obtained recently in a series of elegant experiments by Bernstein.[23] He transplanted spleen or splenic fragments from normal coisogenic littermates under the renal capsule or on to the peritoneum of mice of genotype Sl/Sl^d, and found that hemopoiesis occurred in these transplants even though it failed to occur in neighboring host hemopoietic organs. Indeed, the stromal cells of such transplants were found to be sufficiently active in supporting hemopoiesis to provide a significant increase in the hematocrits of genetically anemic Sl/Sl^d recipients. These genetic studies, therefore, provide conclusive evidence of regulation by a mechanism that depends either on cell-cell contact or on an intermediary acting at very short-range. Thus, mice of genotype Sl/Sl^d may be expected to provide an important tool in the uncovering of such mechanisms although at present the transplantation procedure required to manipulate the affects of the genetic defect makes quantitative analysis difficult.

Cell Separation as a Method in the Study of Cellular Interaction

A convincing demonstration of cellular interaction would be obtained if two cell populations, each inactive by itself, could be combined to yield activity. Examples of evidence of this sort are available from studies of the immunologic system. Claman and his collaborators demonstrated marrow-thymus synergism in irradiated mice [24] and these experiments have been extended by Mitchell and his collaborators.[25,26] Mosier has provided evidence that normal spleen can be separated into two cell populations, one adherent to glass and the other nonadherent.[27] Each fails to give rise to antibody producing cells in the

Mishell-Dutton system,[28] but the response is positive when the two populations are recombined. Recently, Goldie and Osoba have shown that the immune response by small numbers of spleen cells in culture could be greatly facilitated if large numbers of heavily irradiated cells were also present; [29] on the basis of data obtained from specificity experiments and experiments at limiting dilution he has suggested that cell-cell contact may be required for this facilitation.

Encouraged by these precedents from the lymphoid component of the hemopoietic system, we searched for similar interactions in the myeloid component. The cell function used as an indicator was the capacity of hemopoietic stem cells to renew themselves. The first experimental approach depended on physical separation of mouse bone marrow into a number of fractions by velocity sedimentation. We found that rapidly sedimenting stem cells were slightly although regularly less efficient in self-renewal than more slowly sedimenting cells.[30] These results can be interpreted in two quite different ways. Either the stem cells are heterogeneous in their intrinsic capacity for self-renewal or else some other cell type with a different sedimentation rate must be present if stem cells are to undergo self-renewal. However, the affect was small and the results of the self-renewal assay were variable; these experimental limitations prevented systematic attempts to distinguish between the two alternatives stated above.

A second approach, somewhat analagous to that of Mosier, was made using suspension cultures of mouse marrow.[31] Two different cultural conditions were used: in one, marrow cells suspended in CMRL 1066 with 10 per cent fetal calf serum were cultivated for two days over an agar layer containing renal tubules; in the other, the renal tubules were omitted from the agar bottom layer. When the marrow stem cells from the two different conditions were compared after two days, it was found that many fewer had survived in the absence of tubules than in their presence; however, these surviving cells were more effective in self-renewal than those cultivated in the presence of tubules. A detailed analysis of the difference, using a growth curve technic, showed that the cells cultivated without tubules continued exponential growth after control cells cultivated with tubules had entered the plateau phase of the growth curve. Again two different interpretations of the results are possible. Either the cultivation of marrow cells in the absence of a feeder layer results in a selective loss of stem cells with a low intrinsic capacity for continued self-renewal, or, alternatively the culture procedure leads to a selective loss of a second cell type that plays some role in regulating stem cell self-renewal. However, the effect was too small, and the assay procedure for detecting self-renewal was too time-consuming and cumbersome to encourage us to attempt to distinguish between these two possibilities. Instead, we chose to search for some other approach to the problem of studying the role of cell-cell interactions in the regulation of stem cell growth.

CFU Repression as a Means for Studying Cellular Interactions

The experiments described above convinced us that some system of amplification was required if regulation of stem cell growth was to be studied experimentally. Accordingly, we compared the growth of cells cultivated without tubules to controls under a number of conditions known to yield abnormalities of growth. The first evidence of a useful difference was found in conditions where marrow derived from C57BL/6 (B6) mice was tested in heavily irradiated F_1 hybrids between C57BL and C3H (B6C3F_1). Under these conditions normal B6 marrow shows greatly reduced colony-forming ability, a phenomenon that we have termed CFU repression.[32] Analogous observations have been made in transplanted tumors [33,34] and in cell culture.[35] The phenomenon is also called hybrid resistance [36] and syngeneic preference.[35] The system is particularly useful since Cudkowicz and Stimpfling determined its genetic basis; [36] reduced growth is exclusively the result of heterozygosity at the H2 locus. In most instances, the donor strain is homozygous $H2^b/H2^b$ while recipients are heterozygotes with the second allele being almost any other H2 type.

We wished to see whether or not the CFU repression phenomenon could be used to amplify the affects on self-renewal observed when marrow fractionated in various ways, was tested in isologous hosts. Accordingly, we compared the colony-forming efficiency of normal B6 marrow and B6 marrow after two days cultivation in the absence of feeders with respect to capacity for colony formation in irradiated B6C3F_1 hosts. In addition, both cell suspensions were also tested in isologous irradiated hosts. The latter procedure provided us with a measurement of the number of CFU in each suspension and permitted us to normalize the results for the loss of colony-forming ability that regularly occurs when marrow is cultivated without feeders. The results of two experiments are presented in Figure 2; the number of CFU injected (obtained from the values observed in isologous hosts) is plotted against the number of colonies in the spleens of B6C3F_1 mice. It may be seen from the figure that the cultivation procedure increased the yield of colonies although this increase did not approach the colony-forming efficiency observed when donor and host are isologous. These experiments, therefore, led us to explore further the use of the CFU repression phenomenon for the study of cellular interaction.

Repression of growth of B6 marrow in F_1 irradiated hosts is a function of an interaction between donor and recipient. Accordingly, we attempted to improve the usefulness of the system as a model by manipulating the recipient. Santos and his collaborators have reported [37,38] that defective growth of parental cells in F_1 recipients is not observed if these recipients are treated with Busulphan and cyclophosphamide rather than with radiation. We, therefore, examined the effect of treating recipients with cyclophosphamide prior to irradiation and marrow injection. The results are presented in Figure 3; in this experiment, groups of B6C3F1 hybrid mice received increasing doses of cyclophosphamide. As soon as possible after injection of the drug, they were irradiated with 900 rads and 10^6 B6 marrow cells were injected intravenously. The number of colonies observed after 9 days is plotted on the vertical axis against dose of cyclophosphamide on the horizontal axis. For amounts of drug in excess of 1 mg. per mouse, colony formation increased linearly as a function of increasing drug dose; mortality in the animals prevented determination of the full extent of cyclophosphamide-increased colony

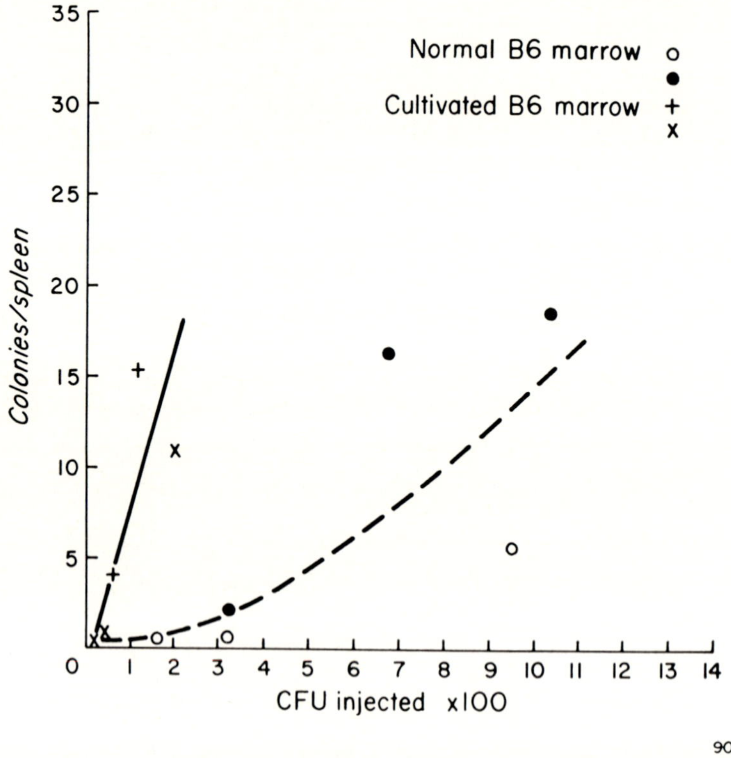

FIG. 2—A comparison of the colony-forming capacity of normal B6 marrow and B6 marrow after 2 days in culture without feeders; both suspensions were transplanted into heavily irradiated B6C3F1 mice. The values are normalized for loss of colony-forming ability in culture by expressing the cell innoculum in terms of CFM. (See text.)

formation. Further, the effect of cultivating the marrow in the absence of tubules on repression was additive with the cyclophosphamide effect. The repression situation, therefore, appeared to have characteristics needed for our studies of cellular interaction; that is, a large quantitative effect was observed in a procedure that occupied only 9 to 10 days.

It was now possible to proceed to the mixing experiments that have been the object of our search for a suitable system. The design of these experiments was guided by the data of Goodman and her collaborators. These workers have shown that the growth of parental marrow cells in F_1 recipients could be improved either by injecting viable parental immunocompetent cells, approximately fourteen days before irradiation [39] or by administering irradiated parental marrow spleen cells or unirradiated parental thymus cells at the time of irradiation.[40,41] In the hands of these workers, treatment with F_1 cells had little or no affect in either regimen. However, since our largest affect was found when F_1 animals were treated with cyclophosphamide, we concentrated on mixing F_1 cells with parental marrow cells before injection. The results of a typical experiment of this design are given in Table I. The addition of intact F_1 thymus cells or heavily irradiated F_1 marrow or spleen cells resulted in an increase in colony formation by B6 cells in

TABLE 1—*Colony Formation by B6 Marrow in Irradiated (900 rad) B6C3F1 Mice*

	Colonies / 10^5 cells	
	Radiation alone	Radiation and Cyclophosphamide*
B6 marrow (2×10^6 cells)	0.16	1.0
B6 marrow + F1 Thymus (2×10^7 cells)		>2.8
B6 marrow + F1 irradiated spleen (2×10^7 cells)		>2.0
B6 marrow + F1 irradiated marrow		> 1.7

*4 mgm/mouse

911/69

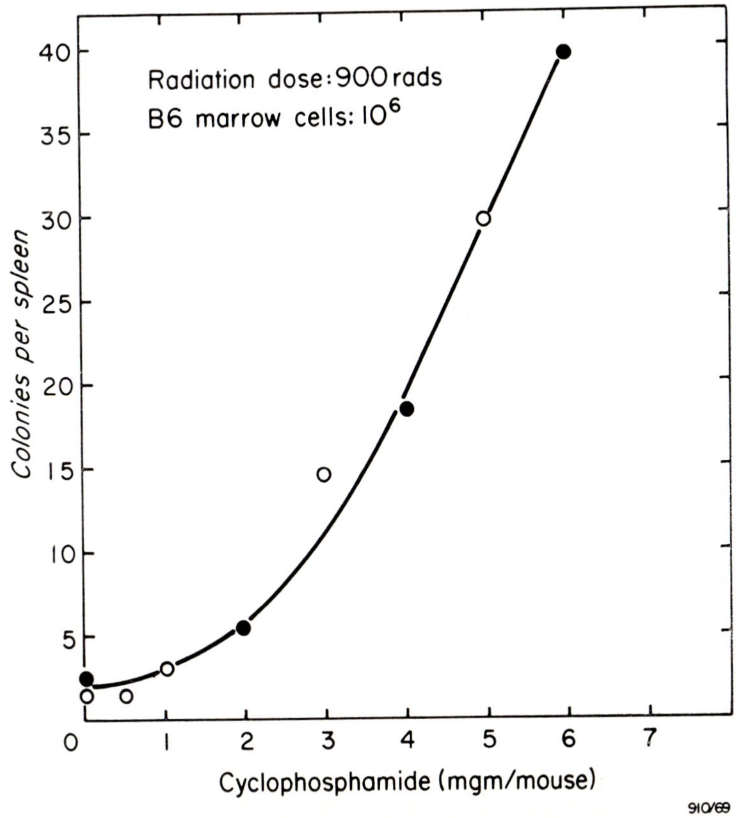

FIG. 3—The affect of cyclophosphamide on the growth of B6 marrow cells in irradiated B6C3F1 mice. Recipient animals received varying doses of cyclophosphamide before irradiation with 900 rads and the injection of 10^6 normal B6 bone marrow cells.

irradiated cyclophosphamide treated B6C3F_1 mice. This observation that addition of F_1 cells, either intact or irradiated, is additive with the cyclophosphamide affect is in contrast to the results of Goodman who failed to observe an affect when F_1 cells were used.

Multiple Aspects of Regulation: A Proposal

The process of differentiation in the hemopoietic system may be considered to occur in a series of discrete steps, characterized by qualitative or quantitative changes in cellular properties. Quantitatively, capacity for cellular proliferation decreases progressively and finally is lost entirely in fully differentiated cells. Qualitatively, potential avenues of differentiation are eliminated as pluripotent stem cells become restricted to single lines of maturation. Normally, these processes are not inevitable but rather are subject to regulation. We now propose that the method of regulation is different at each recognizable stage in hemopoiesis. At the level of mature granulocytes, numbers in peripheral blood are regulated, at least in part, by mechanisms which alter the distribution of these cells in various anatomic sites. At the level of early differentiated progenitors hormonal mechanisms such as erythropoietin are the regulators. At the level of the most undifferentiated identified cells, pluripotent stem cells, regulation appears to be accomplished by mechanisms that depend upon short-range interactions.

The most convincing evidence for this mechanism for control is derived from studies of the S1 gene. This gene specifies a regulator whose role in hemopoiesis appears to be restricted to the level of pluripotent stem cells; that is, the early differentiated descendants of these cells committed to granulopoiesis respond normally to stimulation in mice of genotype $S1/S1^d$ under conditions where pluripotent stem cells are unresponsive.[18] Convincing evidence has been summarized earlier in this paper to indicate that the S1-gene acts by a mechanism that depends on an intimate relationship between stem cells and their environment.

We also propose that a more specific insight into the nature of such short range regulation may emerge from studies of the affect of growth of marrow cells derived from homozygous $H2^b/H2^b$ donors in heterozygous $H2^k/H2^b$ recipients. Evidence has been presented elsewhere that this phenomenon is not readily ascribed to any well-known immunologic reaction such as host-versus-graft or graft-versus-host rejection.[32,39] Rather, the laboratory procedure of transplanting parental cells into F_1 recipients may permit us to observe a mechanism regulating stem cell function. Experimentally, the approach is analogous to the use of genetic methods of demonstrate the existence of normal mechanisms; their roles become obvious when altered by specific mutations.

The genetic basis for defective growth has been demonstrated in the beautiful experiments of Cudkowicz and Stimpfling.[36] These clearly indicate that the phenomenon is under the control of the H2 locus. The gene product of this locus is known to exist at the cell surface.[42] While this surface component has been identified because of its role in graft rejection, we now suggest that H2 "antigen" is in fact a specialized region of the cell surface mediating a cellular interaction required for the normal regulation of pluripotent stem cells. From this point of view, modifications in the growth of parental cells in hybrid hosts, either by drugs or by cell mixtures, may be considered to result from effects of these agents on specific cell surface components. The availability of such technics for modifying growth make the parent-into-F_1-transplantation system an attractive model for studies of regulation by cellular interaction. Experiments in this system will prove or disprove the hypothesis here advanced: that a physiologic role of the structures genetically determined by the H2 genes is in the regulation of hemopoiesis at the level of pluripotent stem cells.

REFERENCES

1. McCulloch, E. A., Till, J. E., and Siminovitch, L.: Genetic factors affecting the control of hemopoiesis. Proc. VI Canad. Cancer Res. Conf., 1965, pp. 336–356.

2. McCulloch, E. A.: Control of hematopoiesis at the cellular level. *In* Albert S. Gordon (Ed.): The Regulation of Hematopoiesis. In press.

3. Wu, A. M., Till, J. E., Siminovitch, L., and McCulloch, E. A.: A cytological study of the capacity for differentiation of normal hemopoietic colony-forming cells. J. Cell. Physiol. 69:177, 1967.

4. Wu, A. M., Till, J. E., Siminovitch, L., and McCulloch, E. A.: Cytological evidence for a relationship between normal hematopoietic colony-forming cells and cells of the lymphoid system. J. Exp. Med. 127:455, 1968.

5. Quastler, H.: The analysis of cell population kinetics. *In* Lamerton, L. F., and Fry, R. J. M. (Eds.): Cell Proliferation, Vol. 18. Oxford, Blackwell, 1963.

6. Lajtha, L. G.: On the concept of the cell cycle. J. Cell. Physiol. 62:143, 1963.

7. McCulloch, E. A., and Till, J. E.: Proliferation of hemopoietic colony-forming cells transplanted into irradiated mice. Radiat. Res. 22:383, 1964.

8. Becker, A. J., McCulloch, E. A., Siminovitch, L., and Till, J. E.: The effect of differing demands for blood cell production on DNA synthesis by hemopoietic colony-forming cells of mice. Blood 26:296, 1965.

9. Pluznik, D. H., and Sachs, L.: The induction of clones of normal mast cells by a substance from condition medium. Exp. Cell. Res. 43:553, 1966.

10. Bradley, T. R., and Metcalf, D.: The growth of mouse bone marrow cells in vitro. Aust. J. Exp. Biol. Med. Sci. 44:287, 1966.

11. Wu, A. M., Siminovitch, L., Till, J. E., and McCulloch, E. A.: Evidence for a relationship between mouse hemopoietic stem cells and cells forming colonies in culture. Proc. Nat. Acad. Sci. U.S.A. 59:1209, 1968.

12. Worton, R. G., McCulloch, E. A., and Till, J. E.: Physical separation of hemopoietic stem cells from cells forming colonies in culture. J. Cell. Physiol. 74:171, 1969.

13. Iscove, N. N., Till, J. E., and McCulloch, E. A.: The proliferative states of mouse granulopoietic progenitor cells. Proc. Soc. Exp. Biol. Med. In press.

14. Bruce, W. R., and McCulloch, E. A.: The effect of erythropoietic stimulation on the hemopoietic colony-forming cells of mice. Blood 23:216, 1964.

15. Lajtha, L. G.: Bone marrow stem cell kinetics. Seminars Hemat. 4:293, 1967.

16. Hodgson, G.: Effect of Vinblastin and 4-amino-N^{10} methyl pteroyl glutamic acid on the erythropoietin response cells. Proc. Soc. Exp. Biol. Med. 125:1206, 1967.

17. Smith, W. W., Brecher, G., Budd, R. A., and Fred, S.: Effects of bacterial endotoxin on the occurrence of spleen colonies in irradiated mice. Radiat. Res. 27:369, 1966.

18. McCulloch, E. A., Thompson, M. W., Siminovitch, L., and Till, J. E.: Effects of bacterial endotoxin on hemopoietic colony-forming cells in the spleens of normal mice and mice of genotype Sl/Sl^d. Cell and Tissue Kinetics. In press.

19. Hannah, M. G., Jr., Nettesheim, P., Fisher, W. D., Peters, L. C., and Francis, M. W.: Serum alpha-globulin fraction: Survival-and-recovery effect in irradiated mice. Science 157:1458, 1967.

20. Barnes, D. W. H., and Loutit, J. F.: Effects of irradiation and antigenic stimula-

tion on circulating hemopoietic stem cells of the mouse. Nature (London) 213:1142, 1967.

21. Russell, E. S., and Bernstein, S. E.: Blood and blood formation. *In* Green, E. L. (Ed.): The Biology of the Laboratory Mouse (ed. 2), Vol. 351. New York, McGraw-Hill, 1966.

22. McCulloch, E. A., Siminovitch, L., Till, J. E., Russell, E. S., and Bernstein, S. E.: The cellular basis of the genetically determined hemopoietic defect in anemic mice of genotype $S1/S1^d$. Blood 26:399, 1965.

23. Bernstein, S. E.: Tissue transplantation as an analytic and therapeutic tool in the treatment of hereditary anemias. Amer. J. Surg. In press.

24. Claman, H. N., Chaperon, E. A., and Triplett, R. F.: Thymus-marrow cell combinations. Synergism in antibody production. Proc. Soc. Exp. Biol. Med. 122:1167, 1966.

25. Mitchell, G. F., and Miller, J. F. A. P.: Immunological activity of thymus and thoracic duct lymphocytes. Proc. Nat. Acad. Sci. U.S.A. 59:296, 1968.

26. Mitchell, G. F., and Miller, J. F. A. P.: Cell to cell interaction in the immune response. J. Exp. Med. 128:821, 1968.

27. Mosier, D. E.: A requirement for two cell types for antibody formation in vitro. Science 158:1573, 1967.

28. Mishell, R. I., and Dutton, R. W.: Immunization of dissociated spleen cell cultures for normal mice. J. Exp. Med. 126:423, 1967.

29. Goldie, J. H., and Osoba, D.: Requirement of a nonproliferating class of cells for generation of immune responses in cell culture. Proc. Soc. Exp. Biol. Med. In press.

30. Worton, R. G., McCulloch, E. A., and Till, J. E.: Physical separation of hemopoietic stem cells differing in their capacity for self-renewal. J. Exp. Med. 130:1, 1969.

31. McCulloch, E. A., and Till, J. E.: Effects of short-term cultures on populations of hemopoietic progenitor cells from mouse marrow: in preparation.

32. McCulloch, E. A., and Till, J. E.: Repression of colony-forming ability of C57BL hematopoietic cells transplanted into non-isologous hosts. J. Cell. Physiol. 61: 301, 1963.

33. Snell, G. D.: Histocompatibility genes of the mouse. II. Production and analysis of isogenic resistant lines. J. Nat. Cancer Inst. 21:843, 1958.

34. Hellström, K. E.: Differential behaviour of transplanted mouse lymphoma lines in genetically compatible homozygous and F1 hybrid mice. Nature (London) 199: 614, 1963.

35. Hellström, K. E., Hellström, I., and Haughton, G.: Demonstration of syngeneic preference in vitro. Nature (London) 204: 661, 1964.

36. Cudkowicz, G., and Stimpfling, J. H.: Deficient growth of C57BL marrow transplanted in F1 hybrid mice. Association with histocompatibility-2 locus. Immunology 7: 291, 1964.

37. Santos, G. W., and Hagshenass, M.: Cloning of syngeneic hematopoietic cells in the spleens of mice and rats pretreated with cytotoxic drugs. Blood 32:629, 1968.

38. Sensenbrenner, L. L., and Santos, G. W.: Cloning syngeneic, allogeneic, parent-to-hybrid, and hybrid-to-parent bone marrow cells in normal spleens. Exp. Hemat. 18:14, 1969.

39. Goodman, J. W., and Wheeler, H. B.: Factors influencing growth of parental marrow grafts in irradiated F1 hybrid mice. Transplantation 6:173, 1968.

40. Goodman, J. W., and Wheeler, H. B.: Factors that modify poor growth of parental marrow cells in F1 hybrid mice. *In* Dansset, J., Hamburger, J., and Mathe, G. (Eds.): Advances in Transplantation. Copenhagen, Munksgaard, 1968.

41. Goodman, J. W., and Shinpock, S. G.: Influence of thymus cells on erythropoiesis of parental marrow in irradiated hybrid mice. Proc. Soc. Exp. Biol. Med. 129:417, 1968.

42. Kahan, B. D., and Reisfeld, R. A.: Transplantation antigens. Science 164:514, 1969.

ACKNOWLEDGMENTS

The authors are grateful to Messrs. Richard Course, Robert Howell, and David Reynolds for excellent technical assistance.

DISCUSSION

JOAN W. GOODMAN, M.D.: In reply to Dr. McCulloch's invitation for me to comment, I would first say that the postulation of a role for cellular interaction involving surface antigens in the control of hemopoiesis is much like what George Klein very effectively set forth in summarizing a symposium on *Isoantigens and Cell Interactions* in 1965 (Philadelphia, Wistar Institute Press, pp. 95–101).

With respect to the contrast Dr. McCulloch suggested between his work and our own, the difference is more apparent than real. The data he has presented shows a 2- to 2½-fold potentiation of parental marrow growth by addition of F_1 hybrid thymocytes or irradiated F_1 marrow cells. We have seen very often and have published (P.S.E.B.M. 129:417–22; and in this Symposium on Cellular Proliferation) a similar 2-fold augmentation of parental marrow growth by F_1 hybrid cells. We have not made much of this small increase but have pointed out that it contrasts with the rather large (6- to 8-fold) effect, within the same experiments, of parental thymocytes and of dead parental spleen or marrow cells.

The Relationship of the Thymus to Erythropoiesis

By JOAN WRIGHT GOODMAN AND CHAREEN G. GRUBBS

A RELATIONSHIP between the thymus and erythropoiesis has been suggested by observations of several kinds over the years. These observations were recently reviewed in brief,[1] and they include association of thymoma with erythroid aplasia in man,[2,3] anemia following neonatal thymectomy in mice,[4] and maturation arrest of erythroid cells in the neonatally thymectomized opossum.[5] Thymus was shown to have a beneficial effect on all elements of spleen growth in vitro by Auerbach [6] and in vivo by Metcalf.[7] Both of these investigators used the mouse, a species in which the spleen is normally very active erythropoietically.

In 1968 two papers were published which showed that thymic cells improved bone marrow growth: Goodman and Shinpock demonstrated improvement of erythropoiesis in irradiated mice treated with thymocytes plus bone marrow,[1] and Metcalf reported improved growth in vitro of colonies which contained only cells of the granulocytic series and monocytes.[8] Metcalf saw extensive reutilization of nuclear material in his experiments and attributed the potentiation of colony growth by lymphocytes, thymocytes, and dead marrow cells to trephocytic effects on the proliferating hemopoietic cells. In the experiments of Goodman and Shinpock,[1] dead (heavily irradiated) thymocytes were ineffective, a finding seemingly inconsistent with a trephocytic effect. However, in another transplantation experiment [9] erythropoiesis was improved by adding dead bone marrow or spleen cells to the viable marrow inoculum. The difference in results suggests two separate mechanisms by which hemopoiesis is potentiated by dead marrow and spleen cells on the one hand and viable thymocytes on the other.

This paper describes results from three kinds of experiments designed to elucidate the potentiating effect of thymic cells on marrow growth. One of these approaches has shown that the augmentation measured in transplantation experiments is not restricted to erythropoiesis but includes all proliferative elements of bone marrow. The presence of a spleen in the recipient was shown to be essential to the potentiating effect of thymocytes, but the lack of a thymus in the host made no difference.

MATERIALS AND METHODS

Inbred mice and their F_1 hybrid offspring used in these studies were C57BL/Cum (hereafter called B), and (B♀X C3H/Anf Cum ♂) F_1 (or $BC3F_1$) females. Donors and recipients were 12–20 weeks old.

The procedure for ^{59}Fe-uptake experiments was as follows: mice were exposed (whole-body) to 900 R of x-rays on the afternoon of day 0; 18–24 hours

From the Biology Division, Oak Ridge National Laboratory, Oak Ridge, Tennessee.
Research sponsored by the U. S. Atomic Energy Commission under contract with the Union Carbide Corporation.

later, cell suspensions were administered intravenously. One week after treatment with cells, ^{59}Fe was given intravenously to each chimera, and 24-hour uptake into circulating erythrocytes and spleens was measured. Details of these procedures have been described.[1]

For histologic study of chimeric spleens and counting of colonies, mice were killed with ether at intervals from 1 to 11 days after marrow, and their spleens were weighed and fixed in Tellyzinsky's fluid. A longitudinal section (5 μ thick), adequately representative of the whole spleen,[10] was cut from the middle portion of the whole organ and stained with hemotoxylin and eosin. Colonies were identified and counted microscopically.

Splenectomy was performed through a lateral incision in the abdominal wall under Metofane (Pittman-Moore Division of Dow Chemical Company) anesthesia.[11] Blood vessels were cauterized with heat, and wound clips closed the skin incision. Sham splenectomy was performed by exteriorizing the spleen briefly and cauterizing a small amount of subsplenic tissue.

Thymectomies were performed under light Diabutal (Diamond Laboratories, Inc.) anesthesia (0.25 ml. of solution containing 6 mg./ml.) through a small longitudinal opening in the thoracic wall adjacent to the sternum. The incision was closed by securing the skin together with 2 or 3 small wound clips. Neither cautery nor suturing was used. At the termination of each experiment mice were examined for thymic tissue, and only those that were completely thymectomized were included in the results presented below. Sham thymectomy consisted of brief exterioration of the thymus in a similar procedure.

RESULTS

Colonies in the Spleen

From Table 1 it can be seen that hybrid recipients of 10^8 parental thymocytes plus 10^6 parental marrow cells had 8 to 10 colonies per spleen on day 4, whereas 3 colonies were found on day 5 in recipients of marrow only. In the first group, colonies had become confluent after day 7 and were uncountable in the first group, but colonies did not reach confluence in marrow-only recipients within the period of observation in these two experiments. The relatively few colonies seen in this group at the later intervals, however, were very large.

Thymocytes from F_1 hybrid mice were also effective in increasing the numbers of colonies from parental marrow, as shown in the second group of experiment A, although the effect was much less pronounced that that of parental thymus. Colonies were confluent on day 10 in the F_1-thymus group, days 8 and 9 having not been looked at, whereas colonies were already confluent on day 7 in recipients of parental thymus.

One group of mice received thymocytes from baby mice of the parental strain ($<$8 to 54 hours old). All of these chimeras were killed on day 6, and their total and differential colony counts did not differ substantially from those of the adult parental-thymus group (Table 1, Experiment B). Because newborns are immunologically immature, these results suggest that augmentation

TABLE 1—*Microscopic Colonies in Spleens of Heavily Irradiated BC3F₁ Females: Effect of Thymus Cells*

Treatment *			Number of Mice Examined	Mean Number of Colonies per Spleen	Differential Count (percentage)				
BM (B)	Thymus	Day			"Stem" §	Eryth.	Myelo.	Meg.	Mixed
				Experiment A					
10^6	B, 10^8	4	9	10.1	52.7	35.1	11.0	1.2	0
		5	9	17.9	37.3	52.8	9.3	0.6	0
		6†	9	29.7	21.3	64.1	7.9	1.5	5.2
10^6	$BC3F_1$, 10^8	4	9	0.7	83.3	16.7	0	0	0
		5	9	10.6	50.6	41.0	7.3	1.1	0
		6	3	8.0	20.9	75.0	4.1	0	0
		7†	9	19.3	28.1	57.5	13.3	0.5	0.5
10^6	None	5‡	9	3.1	35.7	64.3	0	0	0
		6	9	5.1	17.4	82.6	0	0	0
		7	9	8.7	11.6	87.2	1.2	0	0
		10	9	9.8	18.3	65.8	13.6	2.3	0
		11	3	7.7	13.1	78.4	8.5	0	0
None	None	5‡	3	0.3	100	0	0	0	0
		6	3	0.3	0	0	0	0	0
		7	4	0	0	0	0	0	0
		10	3	0.3	0	100	0	0	0
		11	2	0	0	0	0	0	0
None	B, 10^8	6‡	6	0.3	0	0	100	0	0
		7	5	1.2	0	0	100	0	0
		10	3	0	0	0	0	0	0
		11	2	1.0	0	0	100	0	0

Experiment B

10^6	10^8, B, adult	4‡	10	8.1	43.2	39.5	16.1	1.2	0
		6	7	40.6	16.8	58.3	18.2	0.7	6.0
		7†	10	44.4	5.7	68.5	22.0	0.3	3.5
10^6	10^8, B, newborn	6	9	40.2	15.4	52.8	19.9	0	11.9
10^6	None	4‡	10	0.3	33.3	66.7	0	0	0
		6	10	7.8	20.5	75.6	2.6	1.3	0
		7	8	13.1	20.0	75.2	4.8	0	0
		8	10	9.0	14.5	83.3	2.2	0	0
		9	10	15.0	8.0	84.7	6.0	0	1.3
		10	9	8.2	21.7	73.0	4.0	0	1.3
None	None	1–10	80	0					
None	10^8, B, adult	7‡	10	2.0	15.0	0	85.0	0	0
		8	10	3.5	8.6	0	91.4	0	0
		9	7	0.9	0	0	100	0	0
		10	4	0.3	0	0	100	0	0

* Abbreviations: BM, bone marrow; B, C57BL/Cum strain; $BC3F_1$, (B♀ × C3H/Anf Cum ♂)F1 hybrid. "Day" represents days after cellular treatment. Groups sampled 4, 5, 6, 7, 10, and 11 in experiment A; days 1,2, 4, 6, 8, and 10 in Experiment B.

† Comparable numbers of mice examined at later intervals but colonies confluent and not countable.

‡ Comparable numbers of mice examined at earlier intervals, but no colonies found.

§ Primitive, undifferentiated.

of marrow growth does not depend on immunologic activity by the injected thymocytes.

An occasional colony (total of 3, representing three types, from 15 mice) was seen in radiation controls of Experiment A but none in Experiment B. In both experiments a few colonies were found in mice that had received parental thymus but no bone marrow. In all these cases, where differentiation had occurred, colonies were myelopoietic. The meaning, if any, of this observation is unknown.

Effect of Splenectomy

Because the regenerating F_1 hybrid spleen shows remarkable cellular changes while parental thymus is augmenting parental marrow growth, we asked ourselves whether or not the host's spleen is necessary for this effect. Table 2 gives the clear answer that this organ is essential: in the absence of the F_1 hybrid host's spleen, thymocytes were unable to produce the large augmentation of parental marrow growth usually seen. Parental thymocytes augmented marrow growth slightly in splenectomized hybrids, as indicated by erythrocyte ^{59}Fe uptake, but much less than in animals with spleens. In the absence of injected thymocytes, parental marrow grows even less well (RBC value of 0.8) in splenectomized than in sham (RBC value of 1.8) or unoperated (RBC value of 1.8) F_1 hybrid mice. Although ^{59}Fe uptake by red cells was markedly greater in the presence of the spleen, ^{59}Fe uptake by femurs was slightly greater in splenectomized mice in both experiments.

Effect of Thymectomy

The importance of the host's splenic tissue in the potentiation of marrow growth raised the question of whether or not the thymus must also be present. Table 3 presents data showing that injected thymocytes are as effective in improving marrow growth in the absence of a host thymus as they are in its presence.

DISCUSSION

Microscopic study of hemopoietic colonies in the spleens of hybrid recipients showed clearly that the potentiating effect of thymocytes is not limited to red cell formation but extends also to granulopoiesis and megakaryocytopoiesis (Table 1). This finding rules out the possibility that the thymus-induced augmentation of ^{59}Fe uptake reported earlier [1] reflected a diverting of hemopoiesis into increased red cell formation at the expense of other types of differentiation. Although more data are needed to establish the details of the pattern of differentiation, the differential values in Table 1 are very much like counts similarly made after transplantation of isogenic marrow.[12] The present study also showed that the number of colonies produced by a standard inoculum of parental cells (10^6) was increased by administration of thymocytes, whereas thymocytes alone gave rise to essentially no hemopoiesis. Furthermore,

TABLE 2—*Effect of Splenectomy of $BC3F_1$ Host on Marrow-Potentiating Effect of Parental Thymus*

Pretreatment *	Postirradiation Treatment with Parental Cells		No. of Mice	^{59}Fe Uptake (% Injected Dose; Mean ± S.E.)		
	Thymus	Bone marrow		RBC	Spleen	2 femurs
			Experiment A			
Splenectomy	10^8	10^6	24	2.0 ± 0.2		0.7 ± 0.03
Sham splenectomy	10^8	10^6	18	6.2 ± 0.9	3.4 ± 0.2	0.5 ± 0.03
			Experiment B			
Splenectomy	10^8	10^6	26	1.3 ± 0.07		0.6 ± 0.02
Splenectomy	None	10^6	24	0.8 ± 0.06		0.6 ± 0.03
Sham splenectomy	10^8	10^6	23	6.0 ± 0.6	3.9 ± 0.3	0.5 ± 0.02
Sham splenectomy	None	10^6	21	1.8 ± 0.2	1.5 ± 0.2	0.5 ± 0.03
None	10^8	10^6	22	5.8 ± 0.9	3.7 ± 0.4	0.5 ± 0.02
None	None	10^6	28	1.8 ± 0.3	2.1 ± 0.3	0.5 ± 0.02
None	None	None	4	0.4 ± 0.1	0.3 ± 0.03	0.3 ± 0.04

* At 10 weeks of age, 2 weeks before irradiation.

TABLE 3—*Effect of Thymectomy of BC3F$_1$ Host on Marrow-Potentiating Effect of Parental Thymus*

Pretreatment *	Postirradiation Treatment with Parental Cells: Thymus	Bone marrow	No. of Mice	^{59}Fe Uptake (% Injected Dose; Mean ± S.E.): RBC	Spleen
Thymectomy	10^8	10^6	24	6.2 ± 0.5	3.7 ± 0.2
Thymectomy	None	10^6	25	1.1 ± 0.09	0.7 ± 0.06
Sham thymectomy	10^8	10^6	16	5.5 ± 0.6	3.2 ± 0.2
Thymectomy	None	None	7	0.5 ± 0.09	0.3 ± 0.05

* At 10 weeks of age, 10 weeks before irradiation.

these colonies were countable earlier in thymocyte recipients than in marrow-only recipients, and they appeared to grow more rapidly.

There are similarities and differences between our experiments in the whole mouse and those of Metcalf [8] employing colonies grown in vitro. In both cases, thymocytes increased the rate of growth of colonies derived from marrow cells but did not themselves produce colonies. Metcalf found that allogeneic thymocytes were as effective as those of syngeneic origin, whereas our whole-mouse experiments,[1,13] suggest specificity. The immunocompetence of thymocytes makes it difficult to evaluate the in vivo results, however, inasmuch, as graft-versus-host (GVH) and/or graft-versus-graft (GVG) activity may have obliterated a nonspecific augmentation by allogeneic cells. Nevertheless, the one experimental case in which GVH and GVG were genetically ruled out (F$_1$ hybrid thymocytes plus parental marrow given to irradiated F$_1$ hybrid mice) produced only a small, around twofold, increase in growth of marrow cells in the recipient, whereas the increase produced by parental thymus was six- to eightfold.

Metcalf found no potentiation at a ratio of 3 thymocytes to 1 marrow cell in vitro without the addition of stimulating factor, which is found in the serum of leukemic patients or of certain normal mice.[8] However, thymocytes at relatively high doses had proved effective as feeder layer cells in earlier experiments.[14] Our in vivo data were obtained at relatively high thymocyte-to-marrow ratios (between 40:1 and 100:1) which could be considered in the feeder-layer range. There seems to be no analogy in our experiments to Metcalf's colony-stimulating factor: marrow grows poorly in the absence and well in the presence of thymocytes. If this factor exists normally in the serum, then it was presumably present in all of our experiments.

Heavily irradiated marrow and spleen cells were effective in potentiating the growth of parental marrow in vivo,[9] and it seems reasonable to assume that Metcalf's [8] interpretation of trephocytic function in his in vitro experiments is applicable to both situations. The failure of 10^8 heavily irradiated thymocytes to augment marrow growth in vivo is difficult to understand, for one would assume that the localization, degradation, and possible reutilization of components of these cell types are similar. Another difficulty raised by a simple trephocytic explanation of our thymic data [1,9] is the specificity required. In the irradiated F$_1$ hybrid mouse there are large numbers of dead host cells available at many sites, and yet they are apparently unable, as are dead hybrid cells given intravenously to the irradiated mouse, to serve effectively as trephocytes for transplanted parental, allogeneic, or xenogeneic marrow. [1,13] There is no reason to suppose that only one interpretation will be found to fit all the data just mentioned. It may well be that the increased growth in one case results from availability of nuclear material for reutilization; in another, from a stimulatory substance produced by feeder-layer cells; and in still another, from increased plating efficiency on isogeneic (as opposed to nonidentical) cell surfaces.

The failure of thymocytes to potentiate hemopoiesis in the splenectomized mouse suggests that the spleen is the site of action of thymus-derived cells. Splenic ^{59}Fe-uptake data confirm that erythropoiesis was increased in that organ after administration of thymocytes. Viable thymocytes injected intravenously are known to localize principally in splenic white pulp and liver and to some extent in lymph nodes [15]; they are not found in appreciable numbers in the marrow.[4,16] Even when the spleen was removed in the present experiments, thymus-derived cells apparently did not localize in the host's marrow, because erythropoiesis was not increased in the femurs of recipients of thymocytes. If reutilization of nuclear material is in any way involved in the potentiating effect of thymocytes, the material to be reutilized must be released in the immediate vicinity of the proliferating hemopoietic cell. Perhaps reutilization does not occur, but a stimulating substance is produced by living thymocytes. Again, our experiments indicate that the producing cell must be very near the target cell. There remains, of course, the possibility that neither a product of thymic cells nor reutilization of their nuclear material is required for the potentiation we are measuring. Instead, perhaps a cell-cell interaction requiring close contact of the two surfaces is necessary. That thymocytes were effective as a feeder layer in the experiments of Bradley and Metcalf [14] might be an argument against this possibility, because their hemopoietic cells growing in agar, although close to, were not in direct contact with feeder layer cells. It is not known for sure whether or not the effect we are measuring is related to what others have studied and described as syngeneic preference and allogeneic inhibition.[17]

Complete lack of effect of thymectomy on marrow potentiation by thymocytes was not unexpected, as thymocytes do not relocalize in the thymus, once they have left the organ.[18,19] The atrophic histologic appearance of the thymus in the mouse within the first week after irradiation, regardless of cellular treatment or lack of it, attests to the lack of activity of this organ in the early proliferation of transplanted cells.[20]

It is probably too early to generalize about the role of the thymus in hemopoiesis. There are enough data to indicate that a relationship exists, but not enough to define the relationship. The experiments reported here and elsewhere [13] showing that thymocytes augment hemopoiesis apply to transplantation of parental, allogeneic, and xenogeneic marrow in the mouse. They do not apply to syngeneic marrow transplantation, and it remains to be seen whether or not the syngeneic experiments inadequately test the hypothesis that there is *normally* a relationship between thymus and hemopoiesis or whether or not the inferred relationship exists only in abnormal circumstances, such as the transplantation of nonsyngeneic cells, or in disease states.

The fact that marked potentiation is demonstrated in nonisogeneic transplantation suggests that thymocytes may be involved in a process of recognition that is important to the initiation of cell division. Thymus-derived cells have been shown to be intimately involved in the antigen-recognition step of the immune process, [21] so the postulation of a recognition role in control of hemopoietic stem cell growth seems reasonable. Whether or not close contact of cell surfaces is required for this recognition function is not known, although proximity of thymocytes to hemopoietic stem cells appeared to be a necessary condition in our experiments.

SUMMARY

Parental thymocytes given to irradiated F_1 hybrid mice along with parental bone marrow increased the numbers of all kinds of hemopoietic colonies in the spleen. Colonies appeared earlier and grew more rapidly in these chimeras than in recipients of marrow only. When recipient mice were splenectomized two weeks before irradiation, the potentiating effect of thymocytes was not seen. Thymectomy of recipients, on the other hand, did not change the outcome of the experiment.

ACKNOWLEDGMENT

The authors are grateful to Dr. Maxie L. Davis for help in classifying and counting microscopic colonies.

REFERENCES

1. Goodman, J. W., and Shinpock, S. G.: Influence of thymus cells on erythropoiesis of parental marrow in irradiated hybrid mice. Proc. Soc. Exp. Biol. Med. 129:417, 1968.

2. Fisher, E. R.: Pathology of the thymus and its relation to human disease. *In* Good, R. A., and Gabrielsen, A. E. (Eds.): The Thymus in Immunobiology. New York, Hoeber-Harper, 1964, pp. 676–726.

3. Roland, A. S.: The syndrome of benign thymoma and primary aregenerative anemia. Amer. J. Med. Sci. 247:719, 1964.

4. Metcalf, D.: The Thymus, Its Role in Immune Responses, Leukaemia, Development and Carcinogenesis. Recent Results in Cancer Research, Vol. 5. New York, Springer-Verlag, 1966.

5. Miller, J. F. A. P., Block, M., Rowlands, D. T., Jr., and Kind, P.: Effect of thymectomy on hematopoietic organs of the opossum "embryo." Proc. Soc. Exp. Biol. Med. 118:916, 1965.

6. Auerbach, R.: Developmental studies of mouse thymus and spleen. Nat. Cancer Inst. Monogr. 11:23, 1963.

7. Metcalf, D: Functional interactions between the thymus and other organs *In* Defendi, V, and Metcalf, D. (Eds.): The Thymus, The Wistar Institute Symposium Monograph Number 2. Philadelphia, The Wistar Institute Press, 1964.

8. Metcalf, D.: Potentiation of bone marrow colony growth *in vitro* by the addition of lymphoid or bone marrow cells. J. Cell. Physiol. 72:9, 1968.

9. Goodman, J. W., and Wheeler, H. B.: Factors that modify poor growth of parental marrow cells in F_1 hybrid mice. *In* Dausset, J., Hamburger, J., and Mathé, G. (Eds.): Advance in Transplantation. Copenhagen, Munksgaard, 1968, pp. 427–430.

10. Davis, M. L., Upton, A. C., Cosgrove, G. E., and Satterfield, L. C.: Effect of age of recipient mice on growth and differentiation of transplanted bone marrow cells. Proc. Soc. Exp. Biol. Med. 128:1149, 1968.

11. Simmons, M. L. and Smith, L. H.: An anesthetic unit for small laboratory animals. J. Appl. Physiol. 25:324, 1968.

12. Lewis, J. P., Passovoy, M., Freeman, M., and Trobaugh, F. E., Jr.: The repopulating potential and differentiation capacity of hematopoietic stem cells from the blood and bone marrow of normal mice. J. Cell. Physiol. 71:121, 1968.

13. Goodman, J. W., and Shinpock, S. G.: Further studies on the relationship of the thymus to hemopoiesis. Manuscript in preparation, 1970.

14. Bradley, T. R., and Metcalf, D.: The growth of mouse bone marrow cells *in vitro*. Aust. J. Exp. Biol. Med. Sci. 44:287, 1966.

15. Fichtelius, K. E.: On the destination of thymus lymphocytes. *In* Wolstenholme, G. E. W., and O'Connor, M. (Eds.): CIBA Foundation Symposium on Haemopoiesis: Cell Production and its Regulation. London, Churchill, 1960, pp. 204–220.

16. Cronkite, E. P.: Personal communication, 1969.

17. Hellström, K. E., and Hellström, I.: Syngeneic preference and allogeneic inhibition. *In* Palm, J. (Ed.): Isoantigens and Cell Interactions. Philadelphia, The Wistar Institute Press, 1965, pp. 79–91.

18. Micklem, H. S., and Ford, C. E.: Proliferation of injected lymph node and thymus cells in lethally irradiated mice. Transplantation 26:436, 1960.

19. Iorio, R. J., Chanana, A. D. Cronkite, E. P., and Joel, D. D.: Studies on lymphocytes. XVI. Distribution of bovine thymic lymphocytes in the spleen and lymph node. Cell and Tissue Kinetics, submitted for publication, 1969.

20. Goodman, J. W., Martin, F. B., and Congdon, C. C: Histology of poor growth of parental marrow cells in F_1 hybrid mice: Absence of cellular immune response. Arch. Path. (Chicago), 89:226, 1970.

21. Miller, J. F. A. P., and Mitchell, G. F.: Cell to cell interaction in the immune response. I. Hemolysin-forming cells in neonatally thymectomized mice reconstituted with thymus or thoracic duct lymphocytes. J. Exp. Med. 128:801–820, 1968.

Hematopoietic Stem Cell Compartment: Patterns of Differentiation Following Radiation or Cyclophosphamide

By Samuel Hellman, Helen E. Grate, John T. Chaffey, and Richard Carmel

Introduction

In order to assay the ability of progenitor cells * in the bone marrow to give rise to specific differentiated forms, two methods can be employed. Both measure the ability of transplanted stem cells to produce specific mature progeny in irradiated recipient animals. In the system described by Hodgson,[1] irradiated recipient animals are given syngeneic bone marrow following a dose of radiation sufficiently large to reduce endogenous stem cells to low levels. Time is allowed for proliferation and differentiation; then incorporation of ^{59}Fe into newly formed erythrocytes is determined. This incorporation is proportional to the number of bone marrow cells injected earlier. Using such a system, one can study the effect of cytotoxic agents such as radiation on the stem cell pool. We have described a similar technic for measuring the granulocytic progeny of injected marrow.[2] The system is similar to the Hodgson technic except that the granulocytic progeny of the injected marrow are measured by the peripheral white blood cell response to bacterial endotoxin. Such a response has been shown to be a measure of the marrow granulocyte reserve.[3] This response is proportional to the bone marrow cells injected earlier and can be used to assess the progenitor compartment of normal or treated marrow.[2] The average response to endotoxin 9 days after bone marrow injection is approximately 5×10^6 granulocytes for 1×10^6 bone marrow cells injected. Since the bone marrow contains a large number of recognizable differentiated forms destined to mature long before 9 days, a large number of which are in the erythroid compartment, this granulocyte response must indicate significant proliferation of the injected progenitor cells.

In this report, we will first review our evidence for an interrelation between the two assay systems as evidence for the existence of a common stem cell pool that is at least a part of what is measured by both systems. Further, the effects of x-ray on the stem cell pool and some preliminary data with cyclophosphamide will be presented. All of these will be used to present a model of the stem cell compartment and to indicate what is measured by these and some other assay systems.

From the Department of Radiation Therapy, Harvard Medical School, Boston, Massachusetts.

Supported by USPHS Grants CA–06519 and CA–10941.

* "Stem cell" in this paper is used synonymously with "progenitor cell" to mean a cell which is capable of extensive self-replication as well as progressive differentiation to one or more mature cell types.

Evidence for Competition for a Common Stem Cell

There is increasing evidence that the cell responsive to erythropoietin is not a true stem cell but rather committed only to erythropoiesis and is the progeny of a more primitive, multipotent stem cell.[4,5,6,7] The pool measured by the erythrocytic repopulation technic of Hodgson [1] is not the same as that measured by the response to erythropoietin. This has been demonstrated by Blackett using the response to continuous irradiation [8] and by Porteous and Lajtha measuring the response to fractionated acute radiation.[9] It may be that the repopulation technic measures in part this erythropoietin sensitive pool but it does not do so exclusively. As evidence for both the repopulation technics measuring a common stem cell are experiments in which irradiated recipient animals were bled or injected with erythropoietin.[10] Under such conditions, transplanted bone marrow gave rise to fewer granulocytic progeny than when the same marrow was injected into irradiated recipients not so treated. Such experiments can be interpreted as indicating that, under conditions of maximum demand on a limited stem cell pool, increased demand for erythropoiesis results in decreased ability to produce granulocytes. This would argue for a common stem cell on which can be made competing proliferative demands. Further evidence for the recipient affecting the path of differentiation of bone marrow progenitor cells is the response of the "pre-irradiated" mouse. Mice recovering from 150–160 R 6 to 7 days before 700 R and bone marrow transplantation can be shown to preferentially support erythropoietic differentiation [11] as compared to recipients not so "pre-irradiated." Concomitant with this, such recipients have a diminished ability to support granulopoiesis.[12] Both these phenomena suggest that the assays at least in part measure a common stem cell pool on which can be made competing proliferative demands. The response of the stem cell compartment, as assayed by these procedures, to radiation and cyclophosphamide will be presented as this gives some insight into the fine structure of the progenitor compartment.

METHODS

Animals used in all experiments were male C3H/HEJ mice 8–12 weeks old.* X-irradiation factors were 250 kv., 15 ma. HVL 1.5 mm. of copper, FSD of 50 cm. The dose rate for whole animal irradiation was approximately 114 R per minute. Dose measurements were made in a mouse "phantom" using a Baldwin-Farmer dosimeter whose calibration was confirmed by the National Bureau of Standards. Before and after irradiation, animals were placed in cages of 8 animals and supplied with food and water ad libitum.

Cyclophosphamide was prepared by dissolving 100 mg. of the drug † and 45 mg. of sodium chloride in 5 cc. of distilled water. The appropriate dose of drug was injected intravenously when marrow was to be removed three hours

* Supplied by Jackson Laboratories, Bar Harbor, Maine.

† Cytoxan by Mead Johnson, Inc.

later and intraperitoneally when the marrow was to be removed 24 hours later. All injections were made with freshly prepared drug.

Bone marrow cell suspensions were prepared by flushing the medullary cavity of the tibia and femur with cold sterile Tyrode solution. Cell suspensions were kept on ice until used, and were injected intravenously into the lateral tail vein not longer than 90 minutes following removal. A sufficient volume of solution was used so that the injected bolus varied between 0.05 and 0.4 cc. Cell counts were made using a hemocytometer. Donor marrow was taken immediately or at scheduled times following irradiation. All donor marrow was pooled from both hind limbs of at least two donors.

Assay of Erythrocytic Repopulating Ability of the Transplanted Bone Marrow

Recipient animals were exposed to 700 R whole body irradiation and within 2 hours injected with graded doses of syngeneic normal bone marrow suspended in Tyrode solution. Similar littermate animals were injected with treated bone marrow. This marrow was taken either from donor mice immediately or at indicated times after irradiation or cyclophosphamide. Seven days were allowed for the cells to proliferate. This period of time appears sufficient to assay the continued proliferative ability of progenitor cells.[13] The animals then received 0.5 microcuries of ^{59}Fe as ferrous citrate. Two days later, a cardiac puncture was made and 0.5 ml. of blood removed and counted in a well scintillation counter. The per cent of ^{59}Fe incorporation was then calculated assuming a blood volume of 0.66 cc. per 10 grams of body weight. The per cent of ^{59}Fe incorporation was then plotted against the number of bone marrow cells injected. This can be used to estimate the number of normal cells necessary to produced the same ^{59}Fe uptake as that produced by a larger number of treated cells. From this, a survival fraction can be determined. Using this analysis as described by Hodgson,[1,14] a survival fraction was determined. Each determination used a test group of 6 to 8 mice injected with treated marrow and 3 groups of 6 mice, each group receiving a different dose of normal bone marrow. An additional group of 6 mice served as an uninjected control.

Assay of the Granulocytic Repopulating Ability of the Transplanted Bone Marrow

This system has been described in detail previously.[2,13] Recipient animals were exposed to 700 R whole body irradiation and within 2 hours injected intravenously with graded doses of unirradiated syngeneic bone marrow. Similar littermate animals were injected with treated marrow. This marrow was taken from donor mice either immediately or at indicated times after irradiation or cyclophosphamide. Nine days were allowed for the bone marrow cells to proliferate and then the animals were injected intravenously with either 10^{-2} gamma of Pyrexal, a bacterial endotoxin isolated from *Salmonella* abortus-equi * or in later experiments, 10 gamma of a similar preparation of lipopoly-

* Originally prepared by the Dorsey Laboratories, Lincoln, Nebraska.

saccharide isolated from this bacteria.* More consistent results were obtained with this latter preparation and it is currently the only one used. Nine days appears sufficient to assay the continued proliferative ability of progenitor cells.[13] The white blood cell count of each animal was measured immediately before and at 2, 4, and 6 hours following the injection using a Coulter Electronic Counter model B or Fisher Autocytometer. The maximum white blood cell responses in the animals were then grouped and plotted against the number of cells injected earlier. Compared to this was the white blood cell response in that group of animals injected with treated bone marrow. From this, one can estimate the number of normal cells necessary to produce the same white blood cell count that a larger number of treated cells produced, and a survival fraction can be determined. White blood cell counts were used since the response to endotoxin in this method has been shown to be a granulocytic response.[2] This obviates the need for differential counts. Each determination of survival fraction used a test group of 6 to 8 mice injected with treated marrow and 3 groups of 6 mice, each group receiving a different dose of normal bone marrow. An additional group of 6 mice served as an uninjected control.

Spleen Colony Method

Recipient mice were exposed to 865 R whole body irradiation and within 2 hours injected with normal or treated bone marrow. In later experiments, recipient animals received 650 R, then, 6 hours later, 490 R and within 2 hours following this second dose they received the injected marrow. Delivering the radiation in this divided fashion allowed more surviving animals with the background colony number still less than 0.1 per animal. Nine days later the surviving animals were sacrificed, the spleens removed and placed in Bouin's solution. Twenty-four hours later, macroscopically-visible colonies were counted. The number of colony-forming units was calculated and a survival fraction determined. Each determination was based on counting at least 10 spleens in each normal and treated bone marrow group. An uninjected control group was scored with each experiment and the colony number subtracted from the experimental groups.

Results and Discussion

The repopulating ability of irradiated marrow can be compared to unirradiated marrow and from this a survival fraction determined.[13,14] These can be displayed as a radiation survival curve. If animals are irradiated, the marrow removed immediately and injected into irradiated recipients and compared with normal marrow injected into similar recipients, the survival curves displayed in Figure 1 are observed. Similar results obtain when the marrow is irradiated in vitro as well. These data are presented in greater detail elsewhere.[13] Since the previously described evidence favors both assay procedures measuring a common pool, it is surprising that the pool as measured by erythrocytic repopulating ability appears more sensitive to radiation than when

* Supplied by the Difco Co., Detroit, Michigan.

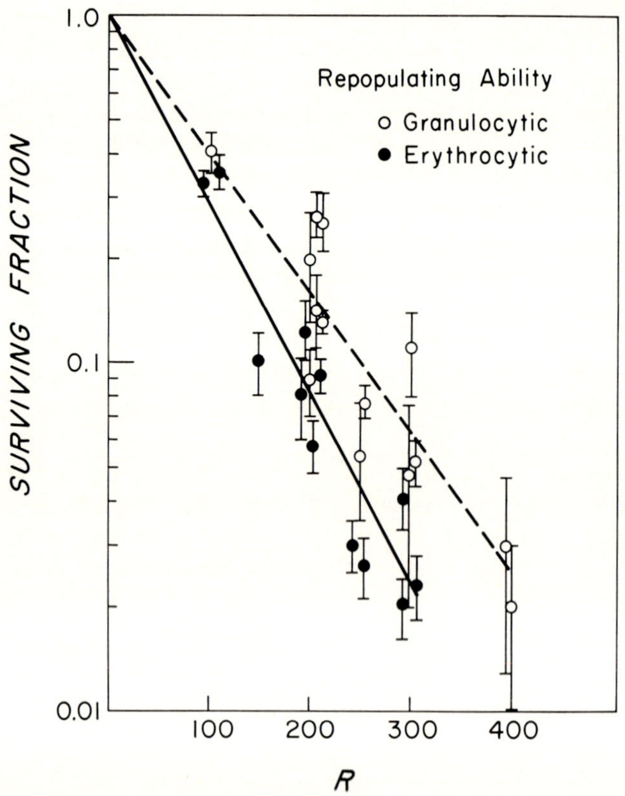

FIG. 1—Survival curve of bone marrow irradiated in vivo as assayed by either the erythrocytic or granulocytic repopulating ability.

Lines fitted by regression analysis using the method of least squares. All points are indicated with their standard errors. The granulocytic repopulating ability resulted in a D_o of 100(± 14)R and the erythrocytic repopulating ability resulted in a D_o of 77(± 9)R. In both, the extrapolation number is 1.

assayed by granulocytic repopulating ability. If one gives 200 R and at various times thereafter measures the survival fraction, the recovery curve is quite different for the 2 assays (Fig. 2). These curves can be compared and plotted as the survival ratio of granulocytic repopulating ability as compared to erythrocytic repopulating ability. This is shown in Figure 3. This demonstrates initial favoring of the granulocytic pathway. This then reverses and slowly returns to the pre-irradiated state. These differences can be interpreted as consistent with the notion of a common stem cell by suggesting that, initially following irradiation, granulopoiesis is preserved at the expense of erythropoiesis. However, after two or three days, the trend reverses in an attempt to compensate for this initial reduction in erythropoiesis. It is in this latter period of preferential differentiation toward erythropoiesis that the "pre-irradiated" recipients described earlier are to be found. This could explain the preferential effect of 6-day pre-irradiated recipients on erythropoiesis. After further recovery, the ratio approaches unity. These responses to radiation will be discussed later with a model of the stem-cell pool. Operationally, however, regardless of mechanism, it appears that not only does radiation affect the number of surviving cells but also the pattern of differentiated progeny seen.

If cyclophosphamide, instead of x-ray, is used as the cytotoxic agent, different results are seen. The drug is given and 3 hours later bone marrow removed and a survival fraction determined as described for x-ray. The data collected is seen in Table 1. The erythropoietic repopulating ability is 5 times more sensitive than the granulocytic to 4 mg. of drug. For this same dose, when marrow is removed 24 hours following Cytoxan, this ratio increases to 9.4. This is similar to the results with x-ray; however, the ratios are greater with cyclophosphamide. A more complete curve was done with marrow removed 24 hours after graded doses of cyclophosphamide and the results are shown in Figures 4 and 5. These clearly indicate marked increase in sensitivity as measured by the erythrocytic as compared to the granulocytic repopulating ability. The spleen colony assay yields a survival similar to the granulocytic assay with both x-ray or cyclophosphamide. Since the repopulation ratio is dependent on dose, we have chosen, when comparing x-ray to cyclophosphamide, doses which give similar survival fractions per 10^6 cells injected as measured by the erythrocytic repopulating assay. These comparisons are per

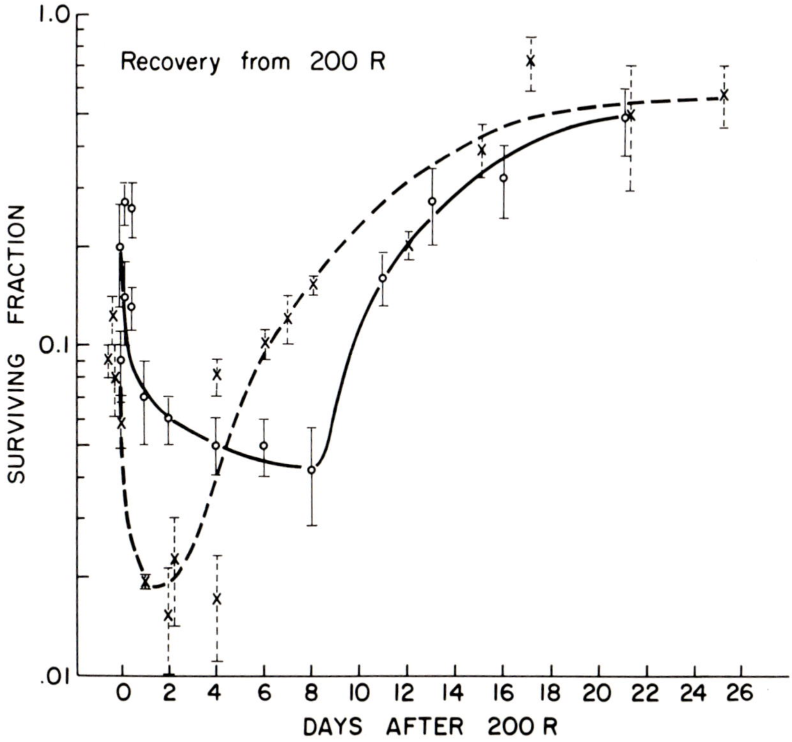

FIG. 2—Fraction of cells surviving 200 R as measured by the erythrocytic (x ---- x) and granulocytic (o —— o) repopulating ability at various times after radiation. All points are indicated with their standard errors.

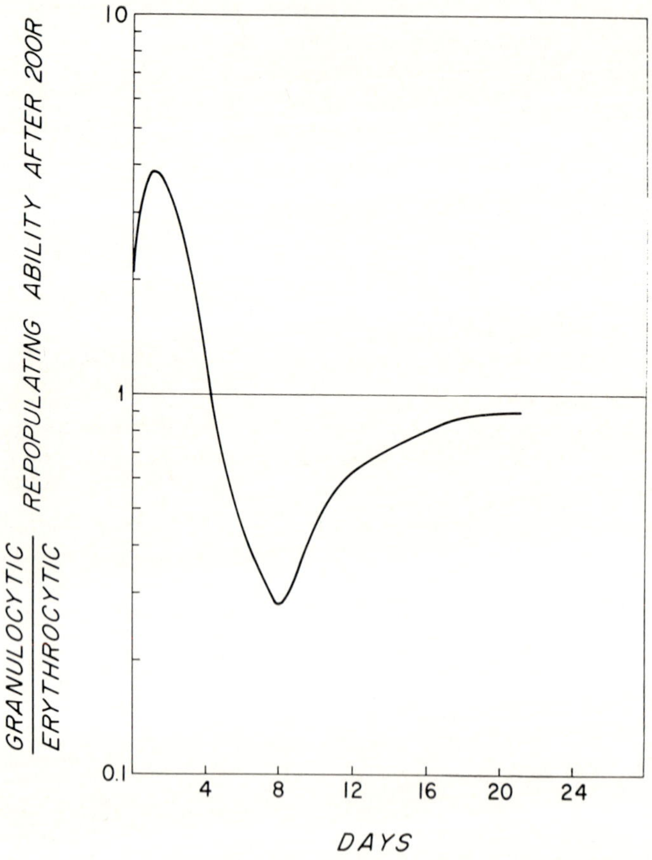

FIG. 3—Ratio of survival fractions measured by the granulocytic and erythrocytic repopulating ability following 200 R. Calculated from the curves shown in Figure 2.

nucleated cell injected, not per animal. The results would need scaling down, as in Figure 5, due to the marrow cell depletion following treatment. However, for any dose of drug or x-ray the cell depletion will be constant when comparing the assay technics; thus, the ratio is identical.

Qualitatively, the initial response to cyclophosphamide is similar to that with x-ray; however, there are significant quantitative differences to be explained. Bruce et al.[15] have shown that lymphoma cells are more sensitive to this drug than hematopoietic colony-forming cells and suggest that this difference is due to increased sensitivity of rapidly-proliferating versus slowly-proliferating cells to cyclophosphamide. If this is true, then it suggests that the pool measured by the erythrocytic repopulating ability has within it not only the same pool of progenitor cells measured by the granulocytic repopulation assay, but, in addition, another compartment of rapidly-proliferating cells committed to erythropoiesis. From this and the above studies, we present the following model of the progenitor compartments with some idea of what cells are measured by each assay technic.

TABLE 1

Agent	Time After	Survival Fraction/10^6 Cells Injected: Repopulating Ability: Erythrocytic	Survival Fraction/10^6 Cells Injected: Repopulating Ability: Granulocytic	Survival Fraction/10^6 Cells Injected: CFU	Ratio of Repopulating Ability G/E
200 R	Immediately	0.12(± .02), 0.08(± .02) 0.09(± .02), 0.06(± .01) *mean 0.087*	0.20(± .07), 0.27(± .04) 0.09(± .02), 0.26(± .05) 0.13(± .01), 0.14(± .04) *mean 0.18*	0.26(± .04) 0.15(± .06) 0.14(± .06) *mean 0.18*	2.1
	1 day	0.019(± .001)	0.07(± .02)		3.9
	6 days	0.10(± .01)	0.05(± .01)		.50
Cyclophosphamide 4 mg.	3 hours	0.14(± .03) 0.12(± .02)	0.71(± .21) 0.59(± .07)	0.72(± .04)	5
	1 day	0.16(± .01) 0.13(± .03) 0.11(± .03)	1.22(± .36)		9.4

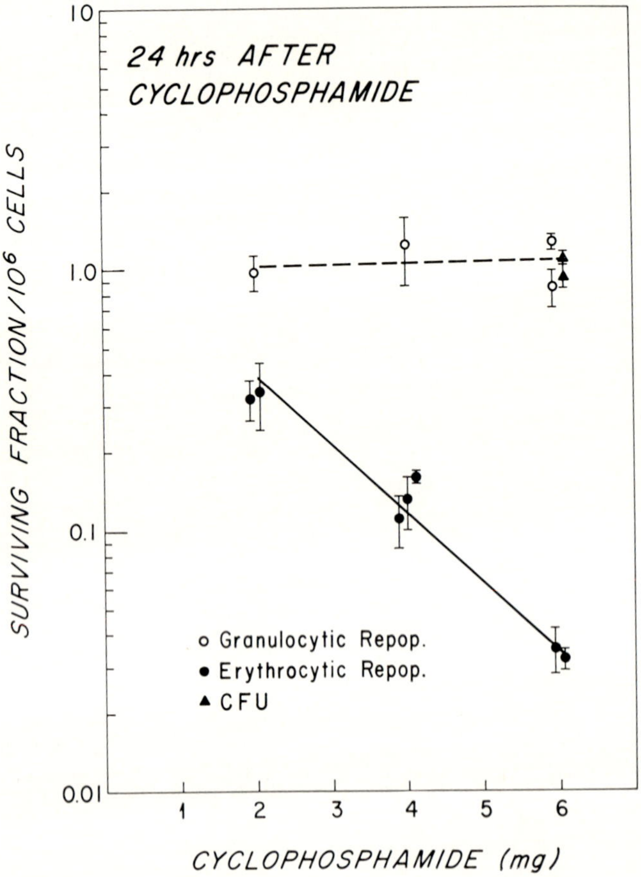

FIG. 4—Survival fraction measured 24 hours after cyclophosphamide measured as a function of the number of cells injected. Lines are drawn by eye. Each point is indicated with its standard error.

Model

Figure 6 gives a graphic outline. The stem-cell compartment is normally slowly proliferating, requiring little, if any, increase in proliferative rate to meet end-cell depletion; but it responds with a marked increase in proliferative rate when the pool or the adjacent compartment is diminished by cytotoxic drugs or x-ray. It is at least bipotent and probably can give rise to platelets and lymphocytes as well. From this compartment, cells entering the erythroid pathway become committed cells, rapidly proliferating, capable of extensive self renewal, but eventually destined toward erythroid differentiation. It is this latter compartment that is sensitive to erythropoietin, and it is what is measured by the incorporation of ^{59}Fe into RBC in polycythemic mice given erythropoietin.[16] Following this, cells enter the recognizable red blood cell precursor compartment. This compartment is well known and not measured in these assay procedures. Similarly, in granulopoiesis, there is some evidence of a committed compartment of cells capable of some cell renewal but committed to granulopoiesis. Presumably, such cells are measured by the soft agar technic.[17,18] Increased demand for erythropoiesis under certain circumstances appears to reduce size of this compartment, presumably by competition for the stem cell.[19] This compartment may also give rise to monocytes since soft agar colonies contain these cells in increasing numbers with time after plating. Their significance is unclear at present. We suggest the assays measure the compartments as outlined in Table 2.

The spleen colony assay seems to measure only the number of "S" cells. Soft agar colony cells do not give rise to such colonies. Erythropoietin or polycythemia may affect the growth, but not the number of foci of colonies in the spleen.[20] Both the erythrocytic and granulocytic repopulating ability appear to measure in part a common stem-cell compartment. Evidence for this is the reciprocal affect of bleeding, erythropoietin, and the pre-irradiated state[10,12] on these pools. These assays do, however, demonstrate differences in initial

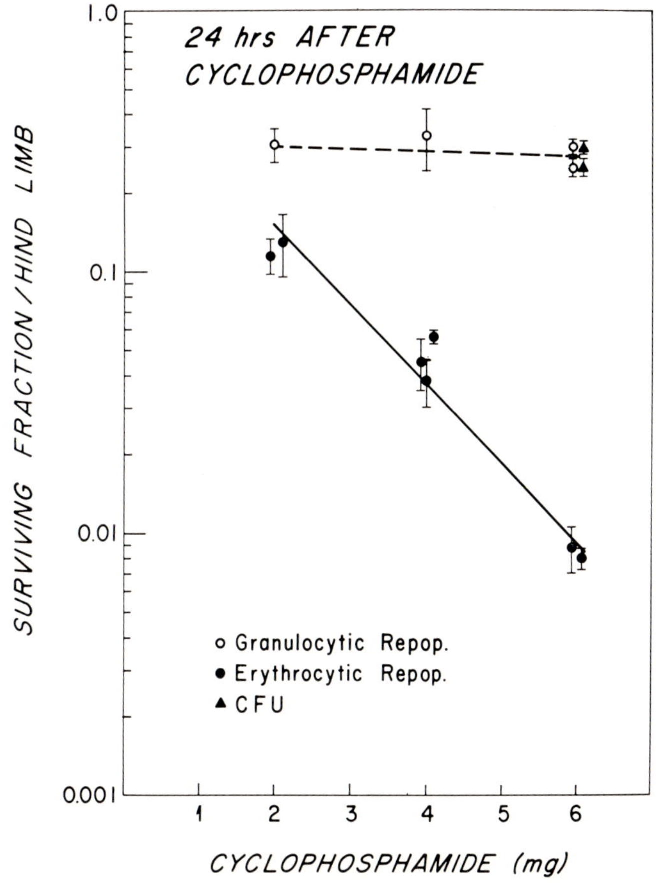

FIG. 5—Survival fraction measured 24 hours after cyclophosphamide measured as a function of the repopulating ability of the hind limb. Lines are drawn. Each point is indicated with its standard error.

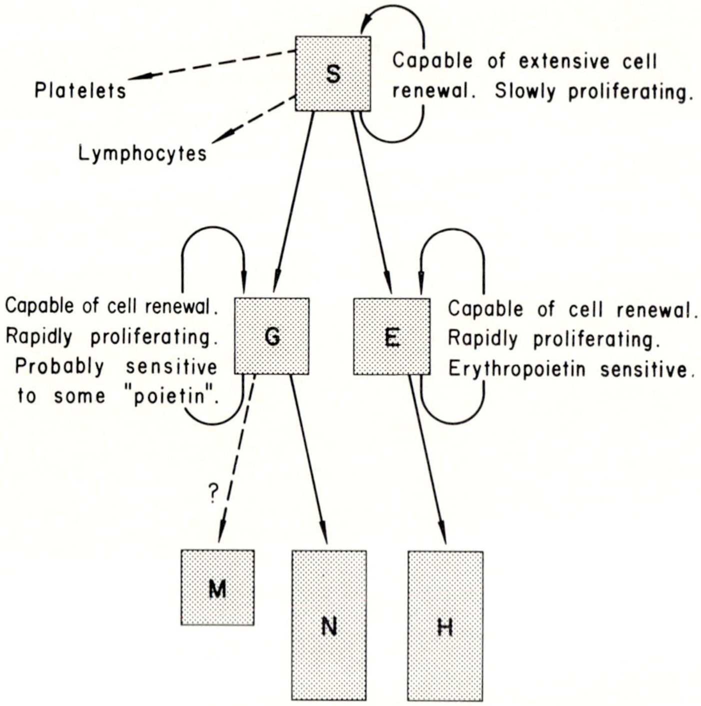

FIG. 6—A model of the early stages of hemopoiesis: S—stem cell; E—committed early erythropoietic cell; H—committed recognizable erythrocytic precursors; G—committed early granulopoietic cell; N—committed recognizable granulocytic precursors; M—monocytes.

response to radiation. This response to radiation must indicate either altered differentiation of stem cells following irradiation or selectively greater sensitivity of cells in compartment "E" as compared to "S." This latter, while possible, does seem unlikely. If this were the case, with time following irradiation, one would expect slower recovery of the erythrocytic repopulating ability due to such increased destruction of this pathway. This is not the case (Fig. 2). Thus, the most likely reason for the differences in response to radiation is sequential changes in differentiation patterns. The mechanism by which this is accomplished is unknown but likely due to changes in production or response to appropriate "poietins." These initial changes are also noted with cyclophos-

TABLE 2

Assay	Compartments Measured
Spleen colony	S
Erythrocyte repopulating	S + E
Granulocyte repopulating	S (± G)
Erythropoietin response of plethoric mice	E
Soft agar	G

phamide but to them must be added the increased sensitivity of proliferating cells to this agent. Thus, compartment "E" is far more sensitive to this drug than "S"; hence, the greater ratio of granulocytic to erythrocytic repopulating ability than seen with x-ray.

The granulocytic repopulating assay probably does not measure "G" since Rickard et al. (these proceedings) and McCulloch et al. (these proceedings) present evidence to indicate significant proliferative activity in this compartment. If the increased sensitivity to cyclophosphamide of the erythrocytic repopulating ability is due to "E" being rapidly proliferating, similar sensitivity of the granulocytic repopulating ability would be expected if "G" were measured by this technic.

This model is simple and perhaps only a small modification of those presented by others.[5,6] It does, however, suggest what is measured by the different assay methods and thus allows experimental testing of the characteristics and further fine structure of these compartments. The results are also pertinent as they begin to explain differences in the effect of cytotoxic agents on the early hematopoietic cells. Such differences in these experiments can be explained by an increase in sensitivity of proliferating cells to cyclophosphamide and by the favoring of granulopoietic differentiation as an immediate response to stem-cell depletion by any agent.

Summary

The granulocytic and erythrocytic repopulating ability of bone marrow were compared and contrasted following in vivo treatment of this marrow with either x-ray or cyclophosphamide. X-ray resulted in a lower survival fraction for the erythrocytic repopulating ability but, following a marked postirradiation dip, there appeared more prompt recovery. Such data appeared best explained by assuming that the radiation not only resulted in cell kill but also affected the pattern of differentiation of the reduced population of the stem-cell compartment. Cyclophosphamide showed further increased sensitivity of erythrocytic repopulating ability. This was explained by the apparent increased sensitivity of rapidly-proliferating cells to this drug. The results of these agents were used to construct a model of the early compartments of the hematopoietic system and to indicate what compartments appear to be measured by some of the "stem cell" assay procedures.

REFERENCES

1. Hodgson, G. S.: Erythrocyte Fe^{59} uptake as a function of bone marrow dose injected in lethally irradiated mice. Blood 19:460, 1963.

2. Hellman, S., and Grate, H. E.: Production of granulocytic progeny by transplanted bone marrow in irradiated mice. Blood 30:103, 1967.

3. Craddock, G. S., Jr., Perry, S., Ventske, L. E.: and Lawrence, J. S.: Evaluation of marrow reserves in normal and disease states. Blood 15:840, 1960.

4. Bruce, W. R., and McCulloch, E. A.: The effect of erythropoietic stimulation on the hematopoietic colony-forming cells of mice. Blood 23:216, 1964.

5. Morse, B. S., and Stohlman, F., Jr.: Regulation of erythropoiesis XVIII. The effect of vincristine and erythropoietin on bone marrow. J. Clin. Invest. 45:1241, 1966.

6. Stohlman, F., Jr., Ebbe, S., Morse, B., Howard, D., and Donovan, J.: Regulation of erythropoiesis XX. Kinetics of red cell pro-

duction. Ann. N. Y. Acad. Sci. 149;156, 1968.

7. Schooley, J. C., Hayes, J. M., Cantor, L. N., and Havens, V. W.: Studies on the behavior of erythropoietin-sensitive cells in the mouse during recovery from 200 roentgens of whole body irradiation. Radiat. Res. 32:875, 1967.

8. Blackett, N. M.: Erythropoiesis in the rat under continuous gamma irradiation at 45 rads/day. Brit. J. Haemat. 13:915, 1967.

9. Porteous, L. L., and Lajtha, L. G.: On stem cell recovery after irradiation. Brit. J. Haemat. 12:177, 1966.

10. Hellman, S., and Grate, H. E.: Hematopoietic stem cells: Evidence for competing proliferative demands. Nature (London). 216:65, 1967.

11. Blackett, N. M., and Hellman, S.: Increased proliferation of transplanted mouse bone marrow cells by preirradiation of the recipient. Nature (London). 210:1284, 1966.

12. Hellman, S., and Grate, H. E.: Enhanced erythropoiesis with concomitant diminished granulopoiesis in pre-irradiated recipient mice; Evidence for a common stem cell. J. Exp. Med. 127:605, 1968.

13. Hellman, S., Grate, H. E., and Chaffey, J. T.: Effects of radiation on the capacity of the stem cell compartment to differentiate into granulocytic and erythrocytic progeny. Blood 34:141, 1969.

14. Hodgson, G. S.: Radiosensitivity of marrow cells responsible for reestablishing erythropoiesis in lethally irradiated mice. Acta Physiol. Lat. Amer. 12:365, 1962.

15. Bruce, W. R., Meeker, B. E., and Valeroite, F. A.: Comparison of the sensitivity of normal hematopoietic and transplanted lymphoma colony-forming cells to chemotherapeutic agents administered in vivo. J. Nat. Cancer Inst. 37:233, 1966.

16. Gurney, G. W., Lajtha, L. G., and Oliver, R.: A method for investigation of stem-cell kinetics. Brit. J. Haemat. 8:461, 1962.

17. Pluznik, D. H., and Sachs, L.: The cloning of normal "mast" cells in tissue culture. J. Cell. Physiol. 66:317, 1967.

18. Bradley, T. R., and Metcalf, D.: The growth of mouse bone marrow cells in vitro. Aust. J. Exp. Biol. Med. 44:287, 1966.

19. Metcalf, D.: The effect of bleeding on the number of in vitro colony forming cells in the bone marrow. Brit. J. Haemat. 16: 397, 1969.

20. Curry, J. L., Trentin, J. J., and Wolf, N.: Hemopoietic spleen colony studies. II Erythropoiesis. J. Exp. Med. 125:703. 1967.

Persistence of Radiation Damage: Hemopoietic Effects in Irradiated Mice Treated with Actinomycin D

By WILLIE W. SMITH AND SHIRLEY M. WILSON

TWO SUBLETHAL DOSES of radiation separated by several hours result in a higher survival fraction than the same total dose given within a few minutes. This early recovery, first described by Elkind and Sutton [1] for mammalian cells in culture, has been observed with hemopoietic colony forming cells [2] and intestinal crypt cells [3–4] as well as 4–5-day and 30-day survival in mice.[5–7] Evidence of this repair is not seen, however, when actinomycin D (AD) replaces radiation as the second treatment.[8] When sublethally irradiated mice are given AD 6 to 24 hours later, the survival fraction remains low. In contrast, when mice injected with AD are irradiated 6 to 24 hours later the survival fraction increases with increasing time between treatments. The low survival in irradiated mice challenged with AD must, therefore, be the result of persisting radiation effects.

The nature of the persisting radiation damage has not yet been identified. We are presently studying the extent of hemopoietic and gastrointestinal involvement, the experiments reported here being primarily concerned with joint radiation-AD effects on the hemopoietic system. In evaluating the influence of hemopoietic damage on lethality, effects of bone marrow injection were studied in these mice. Other studies include effects on leukocyte and endogenous colony counts in mice given graded doses of AD followed by sublethal radiation, and effects of the interval between 2 irradiations or irradiation and AD injection. An additional experiment deals with femoral CFU in mice given radiation, AD or both.

MATERIALS AND METHODS

Mice were (BALB/c×DBA/2)F_1 females 14–18 wks old, caged singly from the beginning of an experiment. ^{60}Co radiation was delivered at a rate of approximately 120 rads/min. to groups of 30 mice or less, which were then distributed equally among the appropriate treatment groups. AD was injected I.P. and bone marrow cells I.V. Differential leukocyte counts, based on cover slip preparations stained with Giemsa, are expressed as geometric means. For endogenous colony counts, spleens were collected 10 days after irradiation and fixed in Bouin's and median values estimated as previously described.[9] CFU were determined by the method of Till and McCulloch[10] on cell suspensions prepared as usual.[11] Error terms are given as ± 1 standard error (SE).

RESULTS

It was previously noted that for a given survival fraction, disproportionate numbers of deaths occurred 5.5–7.5 * days after treatment in the mice given

From the Laboratory of Physiology, National Cancer Institute, Bethesda, Maryland.

* The survival time of mice found dead at the first morning check of the sixth day, for example, was taken to be 5.5 days.

Table 1—*Per Cent Survival in Mice Given a First Treatment of 700 Rads and a Second Treatment of 300 Rads or 500 μg./Kg. Actinomycin D a Few Minutes or 6 Hours Later*

Time after 700 Rads	Second Treatment	
	300 Rads	500 μg./Kg. AD
30-day per cent survival *		
Minutes	0	10
6 hours	60	0
7.5-day per cent survival		
Minutes	90	60
6 hours	100	60

* 10 mice per group.

both radiation and AD as compared to those given radiation alone.[8] Table 1 shows results for mice given 700 rads followed in a few minutes or 6 hours by 300 rads or an injection of 500 μg./Kg. AD. None of the mice survived when given the total dose of 1000 rads within a few minutes: one died in 7.5 days, the remainder in 10–12 days. Six mice survived when the exposures were separated by 6 hours, the four deaths occurring in the third week. None of the mice given AD 6 hours after irradiation survived. Of those given both treatments either together or separated by 6 hours, 8 of 19 deaths occurred in 5.5–7.5 days, the remainder in 10–12 days.

The earlier times of death are unusually short for lethal hemopoietic and long for lethal gastrointestinal damage as judged by mice given radiation alone. If hemopoietic damage is responsible, the number of deaths should be reduced by bone marrow injection. On this assumption, experiments were done in which mice given radiation and AD also received an injection of bone marrow, with results as shown in Table 2. Both 7.5-day and 30-day survival were higher in

Table 2—*Survival and Granulocyte Counts in Mice Treated with Radiation and Actinomycin D Separated by 5–6 Hours, Without and With a Bone Marrow Injection*

Treatments *			Percent Surviving †		Granulocytes per Cu. Mm., Day 7
1st	2nd	Marrow	30 Days	7.5 Days	
750 rads	AD	0	0	20	12 ± 6
		+	33	44‡	910 ± 170
700 rads	AD	0	0	60	
		+	89	89	410 ± 130
AD	640 rads	0	10	40	40 ± 12
		+	70	80	1760 ± 190
AD	720 rads	0	0	30	28 ± 6
		+	20	50	1500 ± 280
AD	800 rads	0	0	20	
		+	30	40	

* AD dose 500 μg./Kg.; time between irradiation and AD 5 hours for the first pair, marrow immediately after irradiation (750 rads); time between treatments in remaining groups 6 hours, marrow immediately after the second treatment; marrow dose half the cells collected from 1 femur, injected i.v.

† First two groups given marrow, 9 mice each; remaining groups, 10 mice each.

‡ Two of these mice died in 3.5–4.5 days.

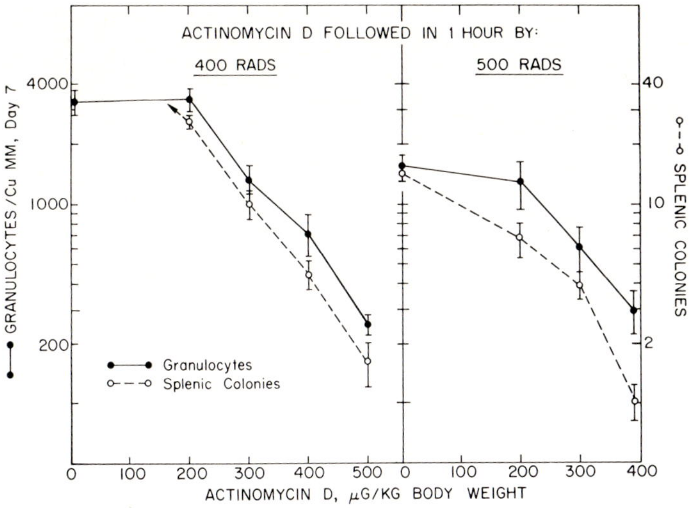

FIG. 1—Granulocyte and splenic colony counts in mice given increasing doses of actinomycin D followed in 1 hour by 400 or 500 rads.

the mice given marrow. Pooling the results, 7.5-day survival increased from 36 to 60 per cent, 30-day survival from 4 to 48 per cent. The granulocyte levels in these mice would lead to a high expectation of survival if they had been given radiation alone. The fact that 5.5–7.5-day mortality was reduced but not abolished by the injected bone marrow suggests that hemopoietic damage was the determining factor in some but not all of the deaths.

Effect of AD and Radiation Dose on Leukocyte and Splenic Colony Counts

Compared to its lethal effect,[8] actinomycin D given alone had only a small effect on leukocyte counts and CFU. Although 4 per cent of the mice given 500 μg./Kg. AD failed to survive, their granulocyte counts a week later (3400 ± 640) were in the normal range, while the reduction in lymphocyte counts (to 5000 ± 1100) could be expected from a 160 rad dose of radiation. A dose of 700 μg./Kg. AD (LD_{50}) reduced the femoral CFU to 30 per cent, although 400 rads (200 rads below the approximate minimal lethal dose) reduced the CFU to 1.3 per cent.* (See Table 5.)

Effects of radiation in mice that had been given graded injections of AD are shown in Figure 1. A dose of 200 μg./Kg. AD prior to 400 rads did not perceptibly reduce either granulocyte or lymphocyte counts. With increasing AD dose the counts dropped steadily, until with 500 μg./Kg. 7-day counts for granulocytes were 250 and for lymphocytes 730/cu. mm. The reduction in splenic colonies paralleled the reduction in granulocytes, with counts falling from 26 to 1.6 as the AD dose was increased from 200 to 500 μg./Kg.** When the radiation dose was increased to 500 rads granulocyte counts were approxi-

* Estimations made without regard to transplantation fractions,[12] whose values are not known for these conditions.

** Colonies were too numerous and coalesced to count in mice given 400 rads alone, and must therefore have been reduced by the 200 μg./Kg. AD dose.

mately halved. With 400 μg./Kg. AD and 500 rads granulocyte, lymphocyte and colony counts were nearly the same as with 500 μg./Kg. AD and 400 rads.

Effects of Time Between Treatments on Leukocyte and Spleen Colony Counts and Femoral CFU

A 5-hour separation between 500 and 300 rad exposures resulted in higher leukocyte counts than were observed when the total dose was given within a few minutes. As seen in Table 3, the increase in 7-day count appears greater for granulocytes than for lymphocytes, but in terms of equivalent rads the effects were essentially the same—a reduction to about 650 rads. On the other hand, with 400 μg./Kg. AD as the second treatment the counts were lower with a 5-hour separation; for granulocytes approximately equivalent to an additional 140 rads and for lymphocytes an additional 70 rads. The granulocyte count had dropped to 50/cu. mm. 10 days after treatment, but on day 14 had risen to 460 ± 130/cu. mm. One of the mice failed to survive. Lymphocyte counts were lower on days 7, 10, and 14 by an average of 42 per cent.

Athough not a major part of this study, we were interested in seeing the effect of a 5-hour interval between a first treatment with 400 μg./Kg. AD and a second treatment with 500 rads. The direction of the effect was, in fact, like that seen with first and second irradiations, although differences were not as great. Granulocyte counts for minute and 5-hour intervals, respectively, were 112 ± 26 and 164 ± 34 on day 7, 75 ± 38 and 350 ± 160 on day 10, and 770 ± 210 and 1570 ± 550 on day 14. Lymphocyte counts were about 30 per cent higher for the 5 hour than for the minutes interval.

Changes in endogenous splenic colony count in mice given 400 rads initially followed in 1 to 48 hours by a second irradiation (300 or 350 rads) or an injection of 333 μg./Kg. AD are shown in Table 4. With radiation as the second treatment, counts were highest for the 5-hour or the 24- or 48-hour interval.

TABLE 3—*Leukocyte Counts in Mice Given 500 Rads Followed in a Few Minutes or 5 Hours by 300 Rads or 400 μg./Kg. Actinomycin D*

Day of * Counts	Time After 500 Rads	Second Treatment	
		300 Rads	400 μg./Kg. AD
Granulocytes/Cu. Mm.			
7	Minutes	87 ± 20	164 ± 32
	5 hours	593 ± 106	64 ± 14
10	Minutes	73 ± 28	173 ± 52
	5 hours	738 ± 72	50 ± 20
Lymphocytes/Cu. Mm.			
7	Minutes	210 ± 39	365 ± 47
	5 hours	581 ± 64	242 ± 31
10	Minutes	204 ± 42	417 ± 58
	5 hours	588 ± 82	220 ± 40

* 10 mice per group.

TABLE 4—*Endogenous Splenic Colonies in Mice Given 400 Rads Followed in a Few Minutes to 48 Hours by 300 or 350 Rads or 333 μg./Kg. Actinomycin D*

Time After 400 Rads	Second Treatment *		
	300 Rads	350 Rads	333 μg./Kg. AD
Minutes	3.1 ± 0.8	0.3 ± 0.2	9.4 ± 1.0
1 hour	2.3 ± 0.2	1.8 ± 0.8	8.3 ± 1.3
5 hours	6.4 ± 0.8	4.3 ± 1.4	4.1 ± 1.0
10 hours	1.7 ± 0.6	1.1 ± 1.4	4.4 ± 0.8
24 hours	6.7 ± 1.8	0.9 ± 0.4	3.8 ± 0.9
48 hours		4.6 ± 0.9	7.7 ± 3.7

* Radiation, 10 mice; AD, 20 mice per treatment group.

TABLE 5—*Femoral CFU and Nucleated Cells 3 Days after Treatment in Mice Given Radiation, Actinomycin D, or Both*

Rads	AD μg./Kg.	CFU/femur	Cells/femur ÷ 10^6
0	0	7260 ± 200	13.0 ± 0.5
0	300	6500 ± 270	11.1
0	700	2160 ± 290	6.2
400	0	93 ± 6	3.8 ± 1.4
400	300, 1 hr. pre-irrad.	23 ± 3	3.4 ± 2.5
400	300, 6 hrs. post-irrad.	15 ± 2	2.1

With AD as the second treatment counts were lower for 5, 10, and 24 hours than for minutes, 1 hour or 48 hours between treatments.

Effects on femoral CFU and nucleated cell counts are illustrated in Table 5. Values determined 3 days after the injection of 300 μg./Kg. AD * showed only a 15 per cent reduction. As mentioned above, the 400 rad dose of radiation had a much greater effect on CFU and nucleated cells. The CFU suffered an additional 75 per cent reduction when 300 μg./Kg. AD was given an hour before irradiation, or an 84 per cent additional reduction, to 0.2 per cent of the control, when AD was given 6 hours after irradiation. Again no evidence of the repair of radiation damage was apparent.

DISCUSSION

The small effect of AD alone on granulocyte count, and the relatively early deaths (5.5–7.5 days) frequently observed in mice given both radiation and AD, led us to anticipate that gastrointestinal rather than hemopoietic damage would be found mainly responsible for the pattern of deaths and possibly for other departures from effects of radiation alone. The departure that seemed most interesting was the failure of sublethally irradiated mice to show the same evidence of repair when given AD as when given radiation as the second treatment.[8] We wished to locate the responsible tissue, attempting first to

* Note again reference 12.

exclude the hemopoietic system. Measurements of hemopoietic damage followed the same pattern as mortality, however, with no evidence of the early recovery seen in the mice tested with a second irradiation. Indeed, granulocyte counts in mice given 500 rads and 400 μg./Kg. AD 5 hours later fell to a range normally associated with radiation lethality. Bone marrow injections resulted in a sufficient increase in 7.5-day survival to reinforce the impression that hemopoietic damage is responsible for at least some of the relatively early deaths. Results of these experiments did not eliminate gastrointestinal damage as an additional cause of death. Keeping in mind that the early repair has already been accomplished at the time of an AD injection 6 hours after irradiation, it cannot be said that the AD prevented the repair. It seems possible, on the other hand, that AD injected into irradiated mice persists longer in the tissues than when injected in normal mice.[13] Elimination may be slower, or AD binding to DNA in the cells may be increased in irradiated animals. Shortly after the postirradiation mitotic arrest, processes leading to cell division begin. In many cases, greater than normal mitotic activity continues for a number of days.

In considering the possibility that changes in AD sensitivity in the surviving partially synchronized cell population [14] may be responsible for the observed lethality, it should be noted that, with the exception of the 1-hour separation between radiation and AD, survival remained low, with no significant variations observed between groups given AD a few minutes, 3, 6, 9, 12 and 24 hours after irradiation.[8] In mice treated the same but given a second irradiation instead of AD, the survival pattern showed the structure expected from partially synchronized cells progressing through the division cycle. Since the cells had progressed equally at the time of AD administration, it seems likely that persisting high concentrations of AD in the cells may have counterbalanced potential decreases in AD sensitivity.

Interpretation of spleen colony data is subject to many pitfalls. This is especially true in the case of superimposed effects of sequential treatments. If cells have divided between treatments multiplicity is a problem; if cells have migrated this introduces other difficulties. In studies of the effect of time between conditioning and second irradiation, Frindel et al.[15] call attention to cyclic variations in dose response characteristics of spleen nodules, but offer no simple explanation of their observations.

Investigation of the reasons for changes in toxicity with varying conditions of treatment is particularly needed in combined radiation therapy-chemotherapy. In the present experiments the correspondence between the patterns of hemopoietic damage and lethality, and the severity of hemopoietic as distinct from gastrointestinal damage, indicates that hemopoietic damage may be a determining factor in the deaths which result from AD administration in sublethally irradiated animals.

Summary

For a given lethality, a substantial number of irradiated mice given actinomycin D die in 5.5–7.5 days. Some of these deaths were prevented by an injection of bone marrow, thus implicating the hemopoietic system as a determining factor in survival. Although granulocyte counts were not greatly affected by AD alone, in mice given graded doses of AD and 400 rads an hour later, 7-day counts fell from 3300 with 200 μg./Kg. and to 250 with 500 μg./Kg.

Although a separation of 5–6 hours between two sublethal doses of radiation led to higher leukocyte and splenic colony counts and survival than a separation of only a few minutes, with AD as the second treatment no evidence of recovery from the radiation damage was seen. In fact, these measurements of hemopoietic activity, including femoral CFU, showed evidence of greater injury when AD was given 5–6 hours after irradiation than when only a few minutes separated the two treatments.

REFERENCES

1. Elkind, M. M., and Sutton, H.: Radiation response of mammalian cells grown in culture. I. Repair of X-ray damage in surviving Chinese hamster cells. Radiat. Res. 13:559, 1960.

2. Till, J. E., and McCulloch, E. A.: Early repair processes in marrow cells irradiated and proliferating in vivo. Radiat. Res. 18:96, 1963.

3. Winters, H. R., and Elkind, M. M.: Dose-survival characteristics of mouse intestinal mucosa. Radiology 91:998, 1968.

4. Winters, H. R., and Elkind, M. M.: Radiosensitivity and fractionation response of crypt cells of mouse jejunum. Radiat. Res. 38:598, 1969.

5. Hornsey, S.: Some characteristics of the survival curve of crypt cells of the small intestine of the mouse deduced after whole-body X-irradiation. Brit. J. Radiol. 36:795, 1963.

6. Kallman, R. F., and Silini, G.: Recuperation from lethal injury by whole-body irradiation. I. Kinetic aspects and the relationship with conditioning dose in C_{57}B1 mice. Radiat. Res. 22:622, 1964.

7. Mole, R. H.: The relevance of intracellular damage to effects in the complex organism. Proc. Roy. Soc. Edinburgh 70:152, 1968.

8. Smith, W. W., Carter, S. K., Wilson, S. M., Newman, J. W., and Cornfield, J.: Joint lethal effects of actinomycin D and radiation in mice. Cancer Res. In press.

9. Smith, W. W., Budd, R. A., and Cornfield, J.: Estimation of radiation dose-reduction factor for β-mercaptoethylamine by endogenous spleen colony counts. Radiat. Res. 27:363, 1966.

10. Till, J. E., and McCulloch, E. A.: A direct measurement of the radiation sensitivity of normal bone marrow cells. Radiat. Res. 14:213, 1961.

11. Smith, W. W., Brecher, G., Fred. S., and Budd, R. A.: Effect of endotoxin on the kinetics of hemopoietic colony-forming cells in irradiated mice. Radiat. Res. 27:710, 1966.

12. Fred, S. S., and Smith, W. W.: Induced changes in transplantability of hemopoietic colony forming cells. Proc. Soc. Exp. Biol. Med. 128:364, 1968.

13. Schwartz, H. S., Sodergren, J. E., and Ambaye, R. V.: Actinomycin D: Drug concentration and actions in mouse tissue and tumors. Cancer Res. 28:192, 1968.

14. Elkind, M. M., Kano, E., and Sutton-Gilbert, H.: Cell killing by actinomycin D in relation to the growth cycle of Chinese hamster cells. J. Cell. Biol. 42:366, 1969.

15. Frindel, E., Charruyer, F., Tubiana, M., Kaplan, H. S., and Alpen, E. L.: Radiation effects on DNA synthesis and cell division in the bone-marrow of the mouse. Int. J. Radiat. Biol. 11:435, 1966.

Reconstitution of Bone Marrow in a Depleted Medullary Cavity

By Harvey M. Patt and Mary A. Maloney

In most studies of bone marrow regeneration, the principal focus is on repopulation of hemic cells rather than on replacement of marrow as an organized tissue. For example, regeneration of marrow after moderate irradiation depends merely upon the survival or seeding of a certain number of hematopoietic stem cells within an existing matrix. The problem is vastly different, however, when marrow is missing or its structure is disrupted. Here, regeneration involves the interplay of a number of reparative processes which, in many respects, parallel those seen in the healing of a fracture. This was clearly shown in the classic study by Röhlich [1] and confirmed by Steinberg.[2]

Several years ago we noted as an incidental observation that the medullary cavity in the femur diaphysis of the rabbit could be emptied simply by aspiration and simultaneous perfusion with dextran.[3] This has provided a convenient technic for study of the phenomenon of bone marrow reconstitution. The regenerative process in the evacuated femur, like that in fracture repair, tends to recapitulate ontogenic mechanisms by which the original marrow developed.[1,2] Bone formation and resorption, and reconstruction of the marrow reticulum and vasculature are a prelude to hemic cell repopulation. We believe that the entire regenerative process is basically a local phenomenon, which depends upon an induction system involving mesenchymal cells derived from bone.[4] Bone, unlike muscle and skin, retains a primordial capacity to regenerate injured or missing substance,[5] and, in a way, marrow is part of the substance of bone. In this paper, we shall review our studies with the evacuated femur system and related work in an attempt to define essential steps and problems in the program of bone marrow reconstitution.

General Features of the Regenerative Response

Reconstitution of a depleted medullary cavity involves an initial period of reorganization that blends with a more prolonged period of hemic cell repopulation. The entire regenerative process is completed within a matter of several weeks when the marrow is indistinguishable from normal in its organization and cellular composition. The pattern is similar whether marrow is removed surgically,[1] by perfusion,[3] or by a combination of both methods.[2] Immediately after femur evacuation, the cavity is filled with a polymorph-infiltrated blood coagulum. Macrophages and undifferentiated cells soon appear and lead to the formation of a "granulation tissue." This initial nonspecific phase is associated with activation of connective tissue in enlarging haversian canals. Mesenchymal elements within the canals label avidly with tritiated thymidine and apparently

From the Laboratory of Radiobiology, University of California Medical Center, San Francisco, California.

Work performed under the auspices of the U.S. Atomic Energy Commission.

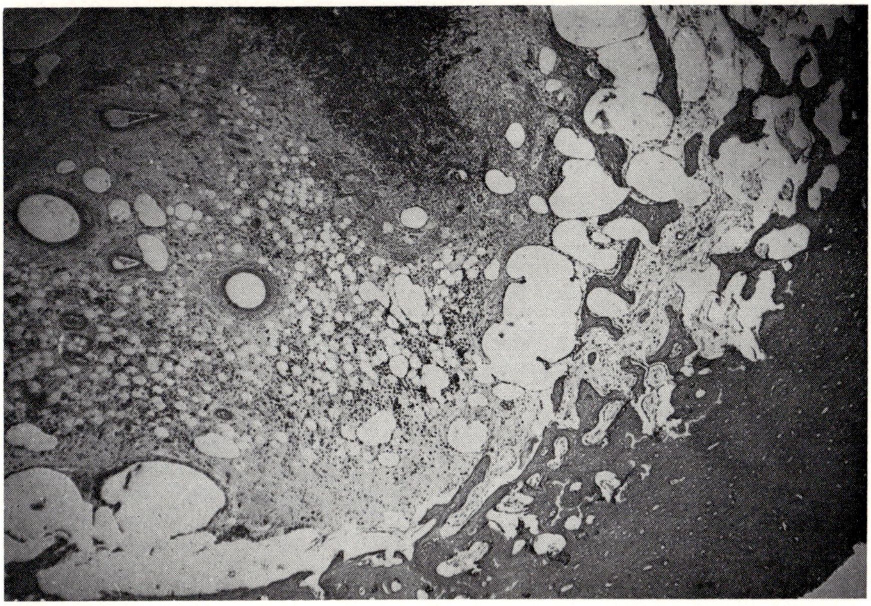

FIG. 1—Development of stroma, vasculature and trabecular bone 8 days after marrow removal (× 50).

play an important role in the formation of what may be thought of as a regenerative blastema. At the end of the first week the picture is that of an emerging connective tissue. A well formed connective tissue matrix with blood vessels and numerous bone trabeculae are clearly evident soon thereafter (Fig. 1). Trabecular bone encroaches only temporarily on the evacuated cavity, and the onset of obvious marrow restoration toward the end of the second week coincides with the beginning of bone resorption.

RESTORATION OF BLOOD FLOW

An immediate consequence of marrow removal is disruption of the vascular supply to the endosteal region of the diaphyseal bone. This occurs because of breakage of the nutrient vessels, which provide the principal blood supply not only to the marrow but also to the surrounding bone.[6] Studies in our laboratory with radioactive plastic microspheres[7] (in collaboration with R. Forsyth) reveal that there is no blood flow to the medullary cavity immediately after marrow extrusion, or after ligation of vessels that enter the nutrient foramen. However, a collateral circulation, presumably through the periosteal blood supply and the haversian system, is soon established in each instance. Blood flow, as judged by the medullary uptake of labeled beads of varying size in the experimental relative to the contralateral femur, returns more rapidly after ligation of the nutrient vessels than after extrusion of marrow. In the former, the control level is achieved within 4 days. In the latter, the uptake of the small beads (15 μ) is also within the normal range by 4 days, but that of the larger beads (50 μ) is still incomplete by 8 days. The apparent difference in return of flow must be due to the requirement for blood vessel regeneration in the evacuated femur; when the nutrient vessels are ligated, it is only necessary for collateral flow to be established in an existing vasculature. Since the uptake of labeled microspheres is expressed per unit weight of medullary tissue and regeneration occurs sooner in some areas than in others, it follows that the blood flow to active sites in the evacuated femur may even be greater than normal by the end of the first week.

Our work with labeled beads indicates that there is considerably less blood flow to bone than to marrow on a unit weight basis in a normal femur shaft. Similar studies of bone blood flow have not yet been made in the evacuated femur, but changes in uptake of ^{18}F (kindly determined for us by D. Van Dyke) are suggestive of an increased rate of bone perfusion a few days after marrow removal. This is to be expected since the periosteal blood supply is thought to be greatly increased when the medullary arterial supply is damaged. Increased periosteal blood flow may account for the proliferation of periosteal bone which is a concomitant of marrow regeneration. However, bone resorption during enlargement of haversian canals may also be a factor in the stimulation of periosteal bone growth.[1]

Cytokinetic Parameters

Repopulation of the medullary cavity portrays a typical growth curve in that the rate of increase of cellularity decreases as the cell population enlarges.[3] The pattern of regeneration as revealed by tritiated thymidine uptake into DNA and the number of hemic cells is depicted in Figure 2. (The term hemic

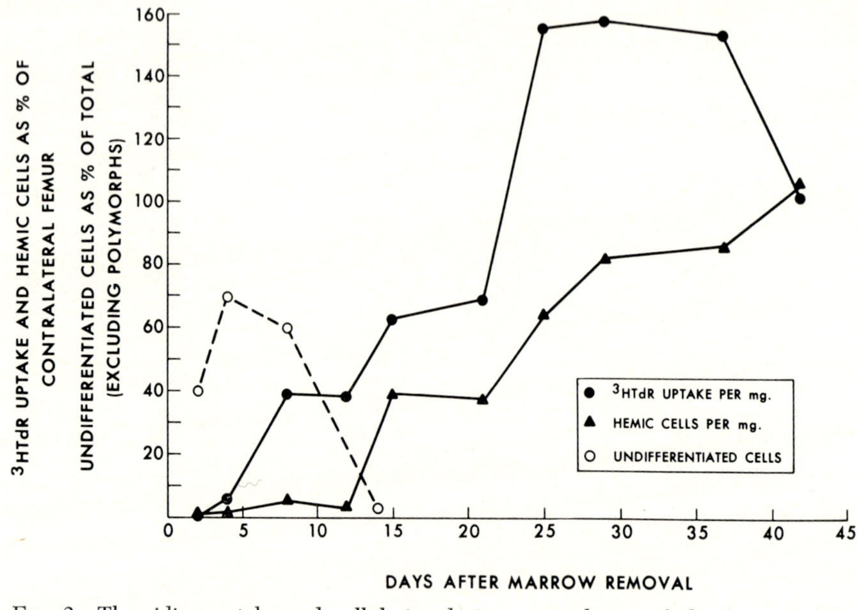

FIG. 2—Thymidine uptake and cellularity during repopulation of the femur diaphysis. (Each point represents the mean of determinations on at least 3 rabbits, the total number of animals studied being 54).

cells will be used in this paper to signify nucleated erythroid cells and myeloblasts through metamyelocytes.)

The character of the cell population changes rapidly during the first 2 weeks after marrow ablation, from undifferentiated cells, fibroblasts and osteoblasts to immature and maturing hematopoietic elements. Many presumptive proliferating cells are seen during this period, but by the third week the ratio of proliferating to nonproliferating cell types is similar in the regenerating and contralateral marrows. Maturation of hemic cells seems to occur almost simultaneously with the evolution of recognizable immature forms. The more mature hemic cells increase at about the same rate as the immature cells with a separation of only a few days even though the level of the latter is still greatly reduced. Thus, postponement of maturation to the nonproliferative state is not an important factor in marrow repopulation. Myeloid elements recover somewhat more rapidly than erythroid, which apparently denotes a greater capacity of the myeloid system for augmentation of cell production.[3]

The uptake of tritiated thymidine (^{3}HTdR) per cell reaches normal levels by 6 to 7 days after marrow removal and then is approximately twice normal for several weeks.[3] This overshoot is due to an increase in labeling intensity and labeling index; the former is increased by about 65 per cent and the latter by 45 per cent on days 18 to 22 after femur evacuation. The change in DNA synthesis rate as reflected in the grain count corresponds to a 40 per cent reduction in the duration of S. If all cells were in a proliferative cycle, an increase in flash labeling index in the presence of an increased rate of DNA synthesis would indicate a proportionately greater decrease in cycle duration than in S period. Since it is likely that all cells are not in a proliferative cycle, the elevated labeling index could also be due, in part at least, to a change in growth fraction.

Origin and Nature of the Regenerative Cells

Although it is generally held that hematopoietic stem cells are present in circulating blood and can move from one marrow region to another, the migration of such cells does not seem to be a critical factor in the regeneration of a small volume of depleted marrow as described here. We have observed that there is no delay in repopulation of a shielded femur shaft when the remainder of the body is exposed to 800 R. In contrast, a similar dose only to the femur leads to a definite impairment of regeneration.[4,8] Further evidence for the local origin of the regenerative cells has been provided in recent experiments with marrow transplants. In these studies (in collaboration with P. Fong), rat marrow, the myeloid elements of which can be identified by alkaline phosphatase positivity, was given to mice X-irradiated over the entire body with exception of a single femur shaft. The shielded shaft was subsequently depopulated and only mouse cells were seen in the regenerating marrow, even though rat cells were evident in hematopoietic tissue elsewhere including the epiphyseal marrow adjacent to the evacuated area. It is of interest to recall that repair of a bone fracture is also basically a local phenomenon; an unirradiated bone transplanted to an irradiated site and subsequently fractured heals normally, whereas an irradiated bone placed in an unirradiated site fails to heal.[9]

While it is conceivable that regeneration is accomplished by a few marrow cells that may have been left behind in the femur diaphysis, microscopic examination of the endosteal surface immediately after dextran perfusion has failed to disclose nests of adherent marrow cells. In any event, since stem-type cells must represent a small part of the total marrow population, the probability that any retained cells are hematopoietic stem cells should be correspondingly low. Autoradiographic analysis of marrow imprints after ^{3}HTdR labeling reveals that regeneration is not initiated by a growing edge from the epiphyseal marrow; labeled cells, when first seen, are distributed randomly throughout the shaft.[3] It seems, therefore, that the regenerative process is initiated by cells resident in bone. Such cells could serve as progenitors of the hemic colony-forming cells normally present in the marrow proper. Yet, the fact that donor marrow cells can more or less permanently recolonize part of the recipients bone marrow [10] tends to suggest that any influx of cells from bone is ordinarily very small, or limited to certain conditions of marrow regeneration. Further work is necessary, however, to determine the normal interplay, if any, between cells in bone and marrow.

Regeneration may emanate from one or more cells associated with osseous tissue. There are two obvious alternatives: (1) one cell type that can be transformed directly or through a branching differentiation into cells responsible for matrix formation and hematopoiesis; (2) two or more cell types already committed to a given stem line. We believe that regeneration of the marrow tissue stems from mesenchymal cells within the haversian canals which are implanted in the organizing blood clot. In point of fact, this was suggested by Röhlich [1] nearly 30 years ago. In this connection, it is of interest to recall that primitive marrow is intimately associated with formation of the haversian system during embryogenesis.

Development of the marrow structure apparently depends upon interactions between mesenchymal cells and their derivatives in a manner not unlike that after bone fracture. We have observed that many of the cells in the haversian canals are labeled with ^{3}HTdR during the first few days after femur evacuation (Fig. 3); only an occasional flash-labeled cell is seen in the canals of the contralateral femur. Labeled cells are also evident at this time in the developing granulation tissue. When ^{3}HTdR is injected at 2 days, 10 per cent of undifferentiated cells in imprints label immediately and over 40 per cent of these cells are found to be labeled 24 hours later. This large increase probably reflects the entry of labeled cells from the haversian system. The influx of undifferentiated cells continues at 4 days and subsides at 8 days after marrow extrusion. These changes contribute to the rise and fall of undifferentiated cells as seen in imprints of medullary tissue (Fig. 2).

The nature of the stimulus that provokes connective tissue cells in bone to reconstitute marrow, or for that matter to repair a fracture, is unknown. The tissue that is removed represents only a small part of the total marrow and does not lead to any significant perturbation of peripheral blood cells.[3]

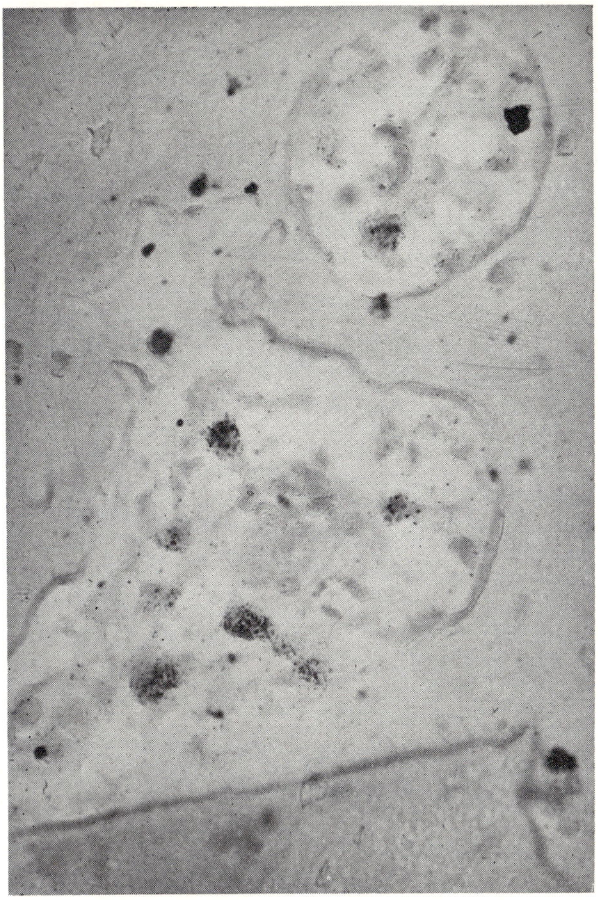

FIG. 3—^{3}HTdR-labeled cells in haversian canals 4 days after marrow removal (× 1000) (0.5 μCi/Gm. of a 360 μCi/m mole solution was given IV and the animal sacrificed 1 hour later).

We may conclude, therefore, that its restoration is a local phenomenon in regard to both the stimulus and the origin of the responsible cells.

ROLE OF BONE FORMATION

A fundamental question pertains to the significance of trabecular bone formation and resorption during marrow regeneration. Invasion of the medullary cavity by trabecular bone could represent an isolated event of osteogenic stimulation. On the other hand, the formation and resorption of bone might be a necessary condition for reconstitution of the marrow structure. This question was considered by Röhlich [1] who concluded that bone resorption provided the stimulus which induced the granulation tissue to form blood cells. It was his belief that factors derived from bone were also essential for normal hematopoiesis. In more recent work, Amsel et al.[11] suggested that formation of trabecular bone was a critical step in reconstruction of marrow sinusoids.

Disruption of a swath of marrow without its removal led to an osteogenic response similar to that after marrow removal. Marrow sinusoids were formed in the interstices of the cancellous bone produced by osteoblasts already present or derived from mesenchymal cells in the immediate environment. It would seem that bone formation in this instance is a response to injury along the path of disrupted marrow.

In some circumstances bone formation and resorption is not a necessary event for reconstruction of a sinusoidal marrow. For example, in the work of Knospe et al.,[12] regeneration apparently occurred without bone formation when autologous marrow cells were injected into a radiation-induced aplastic medullary cavity. Preliminary studies in our laboratory by Brecher[13] also point to the absence of bone formation in an area of femur which was merely evacuated by dextran perfusion and injected with autologous marrow. These findings are in sharp contrast to the results of transplantation to extramedullary sites. In this case, hemic cells disappear and the reticular cells that remain interact with mesodermal elements of the host to form bone and a microcirculation before hematopoietic repopulation of the implant.[14–16]

The absence of bone formation in medullary sites containing transplanted marrow may be due to the more favorable environment for survival of the constituent cells. It could represent a mechanical effect since it is known that endosteal proliferation can be prevented by packing the medullary cavity with bone wax[17] or by insertion of a steel rod for fixation of a fracture.[5] Another explanation of the failure of bone formation after marrow injection might revolve around the interplay of cells in the transplant and cells of the endosteum and haversian canals. Apropos of this, we believe that cells in the haversian canals are mainly responsible for initiation of bone formation, since trabeculization occurs regardless of the procedure for femur evacuation, which may or may not leave an endosteal layer.

Although marrow can be reconstituted without the formation of cancellous bone when stromal and hematopoietic cells are present in their natural environment, bone formation seems to be an important feature of the regenerative process when such cells are removed, injured or placed in a foreign setting. The short-lived trabecular bone could contribute inducing or stimulating factors. By analogy with the callus at a fracture site, it could provide a scaffolding for reconstruction of sinusoidal marrow. Whether or not trabeculization creates an environment or perhaps provides a source of progenitor elements, or both, it is abundantly clear that a sinusoidal marrow begins to emerge as bone resorption begins. Little is known about the stimulus for bone formation; even less is known about the stimulus for bone resorption.

Radiosensitivity of the Regenerative Process

In general, healing or regenerative processes are delayed by irradiation, and it is instructive to consider the radiation response of the transient phenomenon elicited by femur evacuation. When the rabbit femur is X-irradiated just prior to dextran perfusion, a dose of about 250 R can delay regeneration as judged by ^{3}HTdR uptake and the number of hemic cells.[8] The ED_{50} is

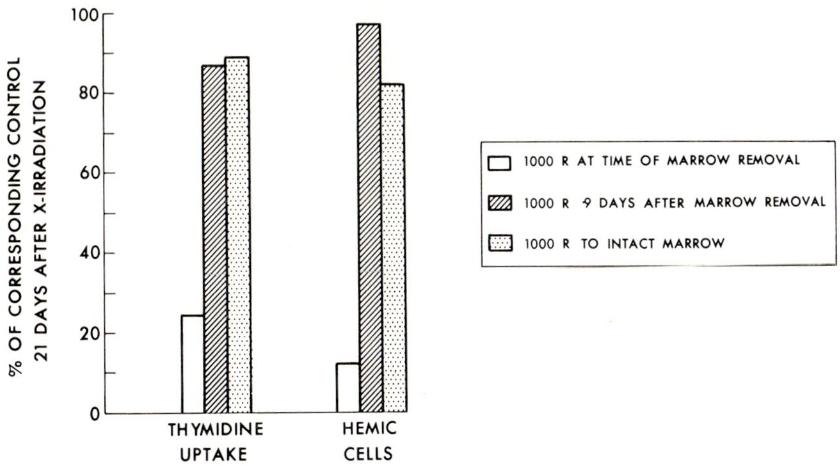

FIG 4—Effect of local X-irradiation before and after initiation of femoral marrow regeneration.

about 400 R. A given dose, i.e., 1000 R. has a greater effect on the evacuated than on the normal femur (Fig. 4).[18] Significantly, when irradiation takes place 9 days after femur evacuation, rather than immediately before or after, the response to 1000 R closely resembles that of intact marrow. The histologic picture at 9 days reveals an emerging connective tissue matrix and vasculature but very few recognizable hemic cells. If recovery of marrow in the normal femur irradiated with 1000 R is essentially a local phenomenon, as in the case of the evacuated femur, we might speculate that the 9-day regenerating marrow already contains a normal complement of hemogenic cells. However, other explanations are possible. The basis for the greater effect of 1000 R at the time of marrow removal than at 9 days after the ablation is also a matter of conjecture. It need not signify a differential radiosensitivity, for example, in respect to matrix formation and hematopoiesis. Rather, it may merely reflect differences in cell population dynamics and in progenitor cell pools during the course of marrow reconstitution.

Evaluation of the dose required for any radiation effect depends upon the period selected for analysis. It is expected that a "threshold" dose for a regenerating system would tend to decrease when the analysis is confined to the first days after irradiation and to increase as the period is extended. This may account for the conclusion that 1000 R does not affect the repair process after mechanical disruption of marrow in the rat tibia;[11] there was a gap in sampling from 12 to 45 days which could preclude histologic demonstration of an impairment of marrow regeneration by this dose. In our studies with prior irradiation of the depopulated rabbit femur, the effect observed over the range of 250 to 1000 R was manifest as a delay and not as a permanent impairment of marrow restoration. The latter can occur in the rat tibia with 4000 R as shown by Amsel.[11]

It is known that doses in excess of 2000 R may lead to a persistent aplasia of locally irradiated normal marrow.[19] The aplasia is thought to be associated with disappearance of the sinusoidal matrix over a period of several months. Some hematopoiesis recurs in the interim, even after 10,000 R, which suggests an influx of stem-type cells from unirradiated areas. While this no doubt occurs, surviving and even damaged cells in situ must also contribute to the abortive attempt at recovery. If this were not so, one would anticipate a rather equivalent transient increase of marrow cellularity at all dose levels, whereas in point of fact cellularity 2 weeks after 10,000 R was only about half that after 2000 R.[19] It is plausible to assume that the relative contribution to recovery by cells migrating from remote areas will depend upon the magnitude of local radiation damage. Inferences about stem cell migration drawn from differences in speed of marrow recovery after local and total irradiation may be misleading, owing to the possible influence of a variety of feedback mechanisms with exposure of a large part of the body.

Knospe and associates [19] have proposed that the eventual development of an aplastic marrow after heavy local X-irradition is a consequence of the failure of replacement of slowly-turning over stromal cells leading to sinusoidal breakdown. Presumably, the immediate progenitors of stromal cells, unlike hematopoietic stem cells, cannot migrate from distant sites. These primitive reticular elements are present in the marrow matrix, endosteum, and haversian canals. They are usually considered to be radioresistant, but this could be due to the much slower rate at which such cells enter a degenerative phase after irradiation, perhaps because of a very slow rate of division.[20] That mesenchymal cells and derivatives can be quite sensitive is shown by the delay of marrow restoration in an empty femur shaft with exposures of 1000 R and less.[8] In this instance, damage to connective tissue cells in the haversian canals is apparently unmasked by stimulation of proliferative activity. Moreover, in recent work we have noted that local irradiation of a normal femur even with 1000 R can result in a secondary wave of decreased proliferative activity and hypocellularity within a few months after the exposure. Delayed vascular effects in this dose range have been observed in other tissues.[21]

Fibrosis is almost invariably associated with late effects of irradiation on vasculature. Can we conclude from this that myelofibrosis unassociated with irradiation has a similar origin? If this disease state arose primarily because of a faulty matrix, removal of the fibrous marrow might lead to reconstitution of normal tissue in the medullary cavity. A few attempts have been made along these lines with rather disappointing results.[22,23] Since normal marrow can be reconstituted through the interaction of perivascular connective tissue cells in adjacent bone, this finding suggests that a defect in such cells or in their induction could be responsible for the evolution of idiopathic myelofibrosis. It is noteworthy that a fibrotic marrow develops on occasion when normal marrow is removed from the rabbit femur diaphysis. While this is a rare occurrence, it lends credence to our suggestion that the etiology of myelofibrosis involves mesenchymal elements in bone.

Summary

The femur diaphysis can be evacuated in vivo simply by aspiration and simultaneous perfusion with dextran. This has provided a useful experimental model for study of the phenomenon of bone marrow reconstitution. Marrow restoration in the evacuated femur has been examined by cytokinetic and histologic technics in conjunction with radiation studies. The results of these investigations, along with the work of others, indicate that:

1. The restoration of marrow in a mechanically depleted site is a local phenomenon in regard to both the stimulus for regeneration and the origin of the responsible cells.
2. The regenerative process stems from mesenchymal cells in the connective tissue of the haversian canals. Development of the marrow structure depends upon subsequent interactions of these cells and their derivatives in a manner not unlike that after bone fracture.
3. Trabeculization of the medullary cavity seems to be an important feature of the regenerative process, perhaps because it creates an environment or provides a source of progenitor cells or both.
4. Mesenchymal cells can be quite radiosensitive as shown by an ED_{50} of about 400 R for delay of marrow restoration in an empty femur shaft.
5. X-irradiation with 1000 R has a greater effect on the just evacuated femur than on the normal femur. However, when irradiation takes place 9 days after femur evacuation, even though there are few recognizable hemic cells, the response to 1000 R closely resembles that of intact marrow.
6. Development of a fibrotic marrow after removal of normal marrow is a rare occurrence. Yet, the fact that it can occur suggests that an aberration of mesenchymal elements in bone may play an important role in the genesis of idiopathic myelofibrosis.

ACKNOWLEDGMENT

We thank Professors J. M. Yoffey and G. Brecher for their interest and advice, particularly in connection with our histologic studies.

REFERENCES

1. Röhlich, D.: On the relationship between the bone substance and hematopoiesis in the bone marrow. Z. Mikr. Anat. Forsch. 49:425, 1941.

2. Steinberg, B., and Hufford, V.: Development of bone marrow in adult animals. Arch. Path. (Chicago), 43:117, 1947.

3. Maloney, M. A., and Patt, H. M.: Bone marrow restoration after localized depletion. Cell Tissue Kinet. 2:29, 1969.

4. Maloney, M. A., and Patt, H. M.: Origin of repopulating cells after localized bone marrow depletion. Science 165:71, 1969.

5. McLean, F. C., and Urist, M. R.: Bone—Fundamentals of the Physiology of Skeletal Tissue (ed. 3). Chicago, The University of Chicago Press, 1968.

6. Branemark, P. I.: Vital microscopy of bone marrow in rabbit. Scand. J. Clin. Lab. Invest. 11 (Suppl. 38):1, 1959.

7. Rudolph, A. M., and Heymann, M. A.: The circulation of the fetus in utero: Methods for studying distribution of blood flow, cardiac output, and organ flow. Circ. Res. 21:163, 1967.

8. Patt, H. M., and Maloney, M. A.: Radiosensitivity of the process initiating bone marrow regeneration. Radiat. Res. In press.

9. Cooley, L. M., and Goss, R. J.: The effects of transplantation and X-irradiation on the repair of fractured bones. Amer. J. Anat., 102:167, 1958.

10. Ford, C. E., Micklem, H. S., Evans, E. P., Gray, J. G., and Ogden, D. A.: The inflow of bone marrow cells to the thymus: Studies with part-body irradiated mice injected with chromosome-marked bone marrow and subjected to antigenic stimulation. Ann. N.Y. Acad. Sci. 129:283, 1966.

11. Amsel, S., Maniatis, A., Tavassoli, M., and Crosby, W. H.: The significance of intramedullary cancellous bone formation in the repair of bone marrow tissue. Anat. Rec. 164:101, 1969.

12. Knospe, W. H., Blom, J., and Crosby, W. H.: Regeneration of locally irradiated bone marrow. II. Induction of regeneration in permanently aplastic medullary cavities. Blood 31:400, 1968.

13. Brecher, G.: Unpublished data.

14. Jacob, S. W., Moloney, W. C., Barsamian, E. M., Owen, O. E., and Dunphy, J. E.: Histologic sequences following implantation of the autologous canine rib into the spleen. Surg. Gynec. Obstet. 109:697, 1959.

15. Tavassoli, M., and Crosby, W. H.: Transplantation of marrow to extramedullary sites. Science 161:54, 1968.

16. Friedenstein, A. J., Petrakova, K. V., Kurolesova, A. I., and Frolova, G. P.: Heterotopic transplants of bone marrow: Analysis of precursor cells for osteogenic and hematopoietic tissues. Transplantation 6:230, 1968.

17. Steinberg, B., and Martin, R.A.: Permanent ablation of bone marrow in long bones of living rabbits. Proc. Soc. Exp. Biol. Med. 63:390, 1946.

18. Patt, H. M., and Maloney, M. A.: Influence of time of X-irradiation on the regenerative process after localized bone marrow depletion. Unpublished data.

19. Knospe, W. H., Blom, J., and Crosby, W. H.: Regeneration of locally irradiated bone marrow. I. Dose dependent long-term changes in the rat, with particular emphasis upon vascular and stromal reaction. Blood 28:398, 1966.

20. Patt, H. M., and Quastler, H.: Radiation effects on cell renewal and related systems. Physiol. Rev. 43:347, 1963.

21. Maier, J. G., and Casarett, G. W.: Cellular growth and tissue radiosensitivity: Tissue studies *in vivo* and the concept of radiation nephritis. Trans. N.Y. Acad. Sci. 26:599, 1964.

22. Wimer, B. M.: Bone marrow curettage in the treatment of myelofibrosis, Exp. Hemat. 15:3, 1968.

23. Van Dyke, D., and Harris, N.: Bone marrow reactions to trauma: Stimulation of erythropoietic marrow by mechanical disruption, fracture or endosteal curettage. Blood 34:257, 1969.

Morphologic and Cytokinetic Aspects of Bone Marrow Stroma

By T. M. Fliedner, W Calvo, R. Haas, J. Forteza and F. Bohne

Introduction

Effective blood cell formation is bound to be a delicate ecologic relationship between hematopoietic precursor cells and a viable stroma, composed of cellular and noncellular elements. In the adult human being this relationship seems to be granted exclusively to the bone marrow organ, distributed throughout all bones of the body, although only some parts are active hematopoietically throughout life. In other mammals, such as mice, an additional favorable ecology for blood cell production is given to the spleen. However, the effectiveness of splenic versus bone marrow cell production is not known. But in all instances, a self-maintaining and efficient hematopoiesis is bound to particular organ structures. Their significance is also borne out by the failure of all attempts to reproduce a self-maintaining hemopoietic cell proliferation and differentiation for any length of time under in vitro conditions,[29] despite the fact that short-term bone marrow cultures have been used successfully to study some aspects of stem-cells [3,4,5,26] and of cell cycle characteristics.[23,25] Under pathologic conditions, such as leukemia, hemopoietic cell proliferation may occur extramedullary. However, the cells produced are grossly abnormal as far as their reproductive and functional potentialities are concerned. This phenomenon leads to the question of whether or not the leukemic process or other diseases associated with extramedullary hematopoiesis (such as agnogenic myeloid metaplasia) may also be related to a profound disturbance of the ecology of the hematopoietic parenchyma and its supporting stroma.

While a vast amount of information has become available about cytokinetic and functional aspects of normal and abnormal hematopoiesis in the last few years through the availability of radioactive tracer methods,[2,10,28] relatively little is known about the structure and cytokinetic aspects of the bone marrow stroma as related to blood cell formation.

It is the purpose of this paper to review light and electronmicroscopic findings on the bone marrow stroma and its structure, to report about cytokinetic studies on the renewal of the cellular components of the stroma by means of the "complete ^{3}H-thymidine labeling method," [14,15,18,19] and to provide evidence for the hematopoietic potentiality of those bone marrow "lymphocytes" (or small round cells with a dense nucleus) that do not, or rarely, divide in the adult steady-state equilibrium of blood cell formation unless specifically challenged.

From the Biology Division, European Atomic Energy Community, Bruxelles, and the Institute of Radiation Biology, Nuclear Research Center, Karlsruhe.

Research supported by Euratom contracts Nos. 031–64–1 BIAD, 072–68–1 BIOD, and the Bundesministerium für Wissenschaftliche Forschung.

On the Architecture of the Bone Marrow

In Figure 1, the various components of the bone marrow architecture are given as they have become recognized during the last few years.[6,11,12] The marrow organ is contained within a bone which provides a closed rigid capsule, protecting the marrow from external pressure and granting a constant volume, thus forming a peculiar "ecological niche" (Baserga). The marrow vascular system (Fig. 2) is supplied with blood through a nutrient artery which enters a bone, such as the femur proximally and the tibia distally.[12] This artery divides into several branches shortly after entering the marrow cavity and several branches proceed tortuously around the central sinusoid while giving rise to numerous arterioles which are relatively straight and traverse the marrow longitudinally. The sinusoidal system receives blood through these arterioles, as

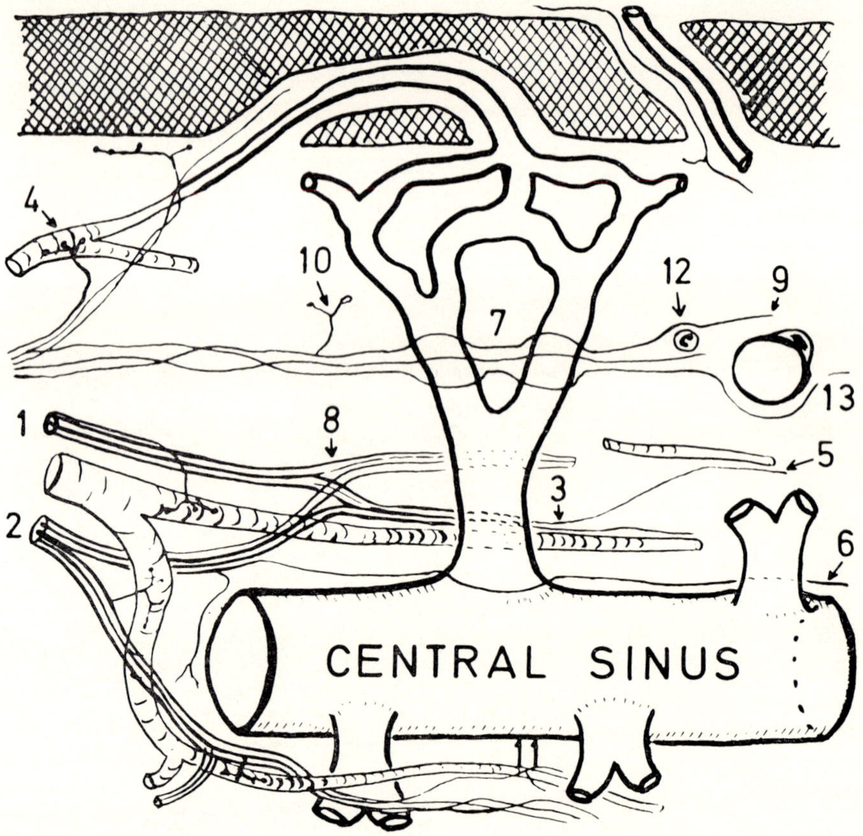

Fig. 1—Schematic representation of angio- and neuroarchitecture of the bone marrow. The central sinus collects the blood from the sinusoids. Arterial vessels follow the length of the marrow cylinder ending in capillaries and the sinusoidal tree. The innervation of the nutrient vessels is shown in numbers 1 to 5. Nerves of the parenchyma correspond to numbers 6 to 13. These nerve bundles are also intimately related to the sinusoidal tree. A few nerve fibers end in the endosteum and some penetrate the largest Haversian canals. (From Calvo, W.[6])

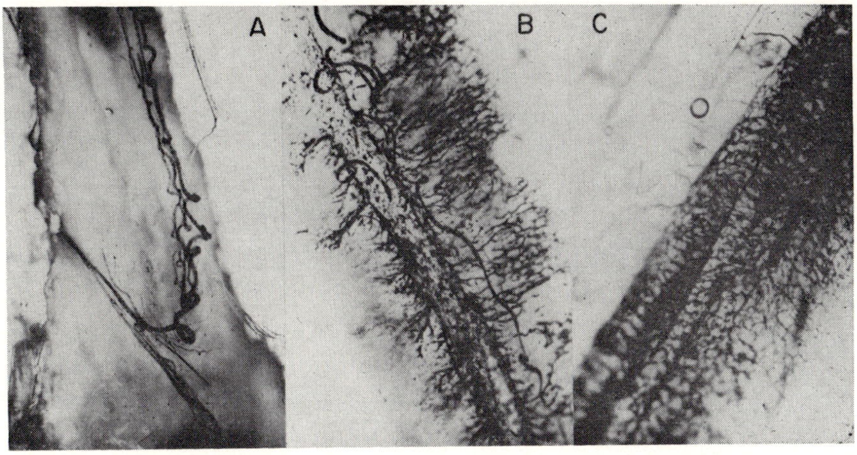

FIG. 2—India ink injection of the bone marrow vessels in the femur of a rat. (A) shows the nutrient artery dividing into hellicoid branches that follow the length of the marrow cylinder. (B) and (C) show the sinusoidal tree with its radially distributed vessels.

does the inner third of the bony cortex to a high extent. The sinusoids of the parenchyma are directed centripetally, as shown by studying longitudinal and cross sections of the marrow as well as India ink injection preparations. These studies have also provided evidence that numerous "contortal sinusoids" communicate and join as the center of the marrow is approached to form larger and larger "rectal sinusoids," the last branches of which enter the central sinusoid at a right angle. By this injection technic it was demonstrated that the basic functional and anatomic unit of the marrow is a "sinusoid segment," which can be compared with a tree, the branches of which are directed toward the marrow periphery and the trunk of which rests on the central sinusoid. The central sinusoid collects the blood from the sinusoidal segments and is drained by a vein which leaves the femur distally and the tibia close to the knee joint.

Recently, a series of studies has been performed by Calvo [6,7,8] to describe the neuroanatomy of the marrow. The main nerve trunks follow the nutrient arteries in their general pattern, remaining close to the arterial walls. From these nerves, very thin nerve fibers detach and reach the blood vessels innervating their walls (Fig. 3A). These nerves are visible especially at the level of the branching of the arterioles and have a possible role in the regulation of blood flow through the organ by stimulating the smooth muscle cells of the arterial walls to contract, with a subsequent reduction of the arterial lumen. Other nerve fibers appear to be independent of the blood vessels, and can be followed in serial sections of the marrow over long distances (Fig. 3B). Occasionally they touch the wall of a capillary but most of their fibers remain in close contact with the blood cell precursors for their total length. Also, if a sinusoidal wall is in their path, a contact, although not a merging, appears to be established. The nerve fibers enter the bone cavity very early in the development of the bone marrow. They can be found shortly after the mesenchymal

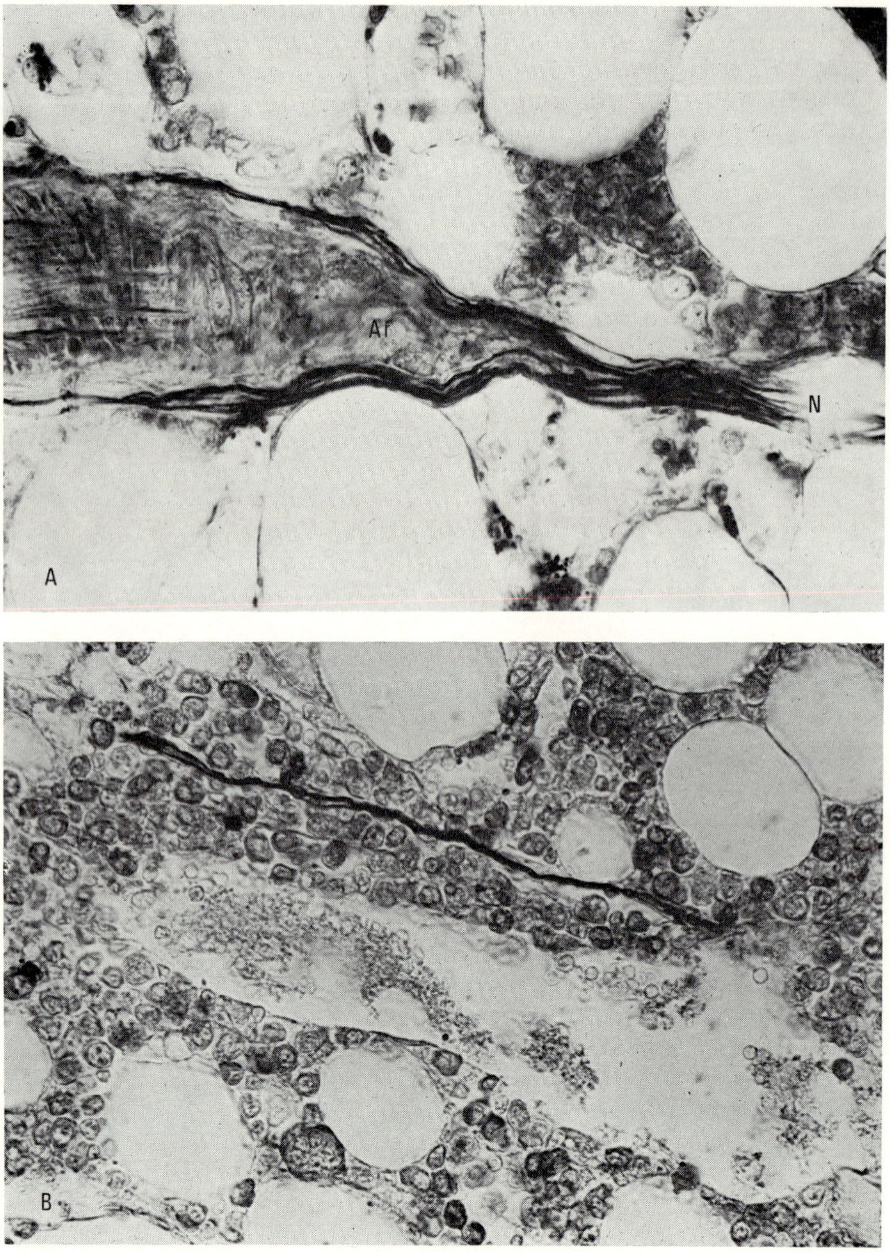

FIG. 3—Femoral marrow of an adult rabbit. Nerve fibers demonstrated with the reduced silver nitrate of CAJAL. (A) A nerve (N) gives off nerve fibers to the wall of the artery (Ar) × 1000. (B) Nerve fibers of the parenchyma. Such fibers can be followed a long distance completely surrounded by blood cells. Occasionally they contact fat cells and sinusoids (× 650).

cells and the first blood vessel penetrate through the windows opened in the calcifying cartilage and before erythropoiesis begins in the marrow.[7] Electron-microscopic studies of the neuro- and of the angioarchitecture have confirmed the intimate relationship between cellular and noncellular components of the marrow stroma and the hematopoietic cells [16,30] and can be summarized as follows:

In many sections of the marrow, the endothelial lining of the sinusoids, arterioles with their characteristic structure, nerve fibers, and hemopoietic cells are found in a close relationship to each other (Fig. 1); note an endothelial cell lining a sinusoid (Fig. 4), which is usually rich in an ergastoplasmic reticulum. The endothelial cells of the rectal and contortal sinusoids are not welded together. This results in a "slit" or even a "window" between them through which cellular traffic from the parenchyma into the sinusoidal lumen and vice versa is possible under appropriate conditions (growth pressure from the parenchyma?). In contrast, the endothelial cells of the central sinusoid are welded together and surrounded by a thick layer of collagen fibers.[16] These findings concerning the fine architecture of the vascular outline of the marrow sinusoids may explain the pattern of marrow hemorrhage within the first 3 days after whole body irradiation [11,13] which was found to be most prominent in the regions of contortal and rectal sinusoids, but not around the central sinusoid.

Figure 5 also indicates the typical structure of a marrow arteriole; note the typical structure as formed by endothelial cells protruding into the lumen. Around these cells there is a thick elastic membrane, a layer of smooth muscle cells and the adventitia. The endothelial cells usually contain multiple pinocytosis vesicles, some mitochondria and a small amount of ergastoplasmic reticulum.

Furthermore, a marrow nerve (Fig. 5) is located adjacent to the arteriole and between two hematopoietic cells. A Schwann cell with its nucleus can be recognized and several unmyelinated axons. In the marrow of adult rats the nerve branches contain both myelinated and unmyelinated fibers in a proportion of approximately 1:6 (Fig. 6). Myelinization of marrow nerve fibers commences about at the end of the second week after birth and is completed about 4 weeks after birth.[8] It is of interest to realize that, in the rat, it is during this time that the marrow cell production comes under the adult-type controlling mechanisms as shown by Lucarelli et al.[9,17,24] Electronmicroscopic studies confirm the close relationship between sinusoidal endothelial cells and nerve fibers as seen in Fig. 4. This picture shows an endothelial cell at the border of a sinusoid attached to a nonmyelinated nerve fiber with several axons.

These studies lead to the hypothesis that the function of the marrow, as an organ fitted to produce cells that are released into the circulation, is related to its particular location within a rigid capsule and to its delicate angio- and neuroarchitecture. One may speculate that one of the functions of the nerve fibers is to receive signals produced by pressure changes due to the growth pressure within the parenchyma (continuing cell division) on the one hand,

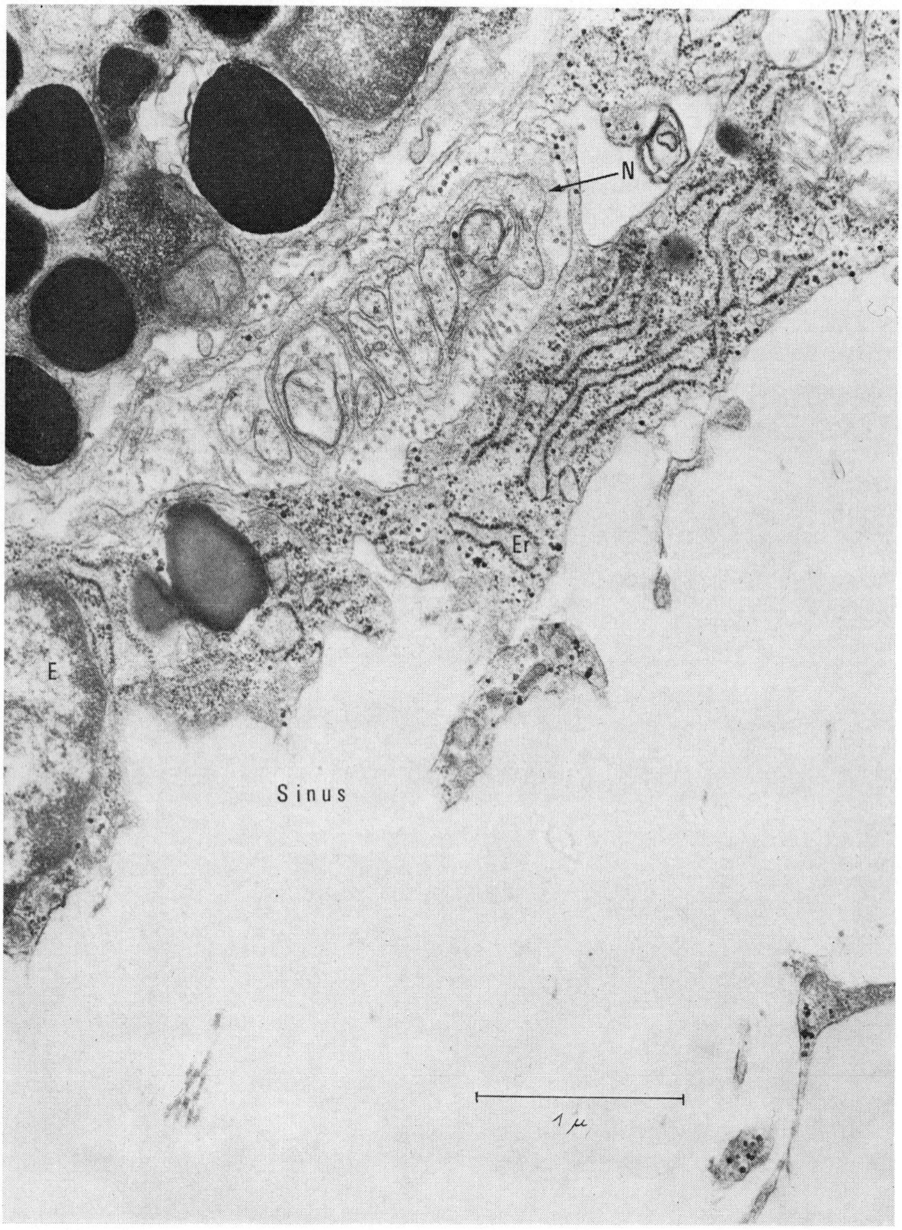

FIG. 4—Femoral marrow of an adult rat. Fine structure of the wall of a sinusoid. Endothelial cell (E) with abundant ergastoplasmic reticulum (Er.) A nerve bundle (N) with 9 nonmyelinated axons is shown in contact with the sinusoidal wall. On the upper left corner a leukocyte of the parenchyma.

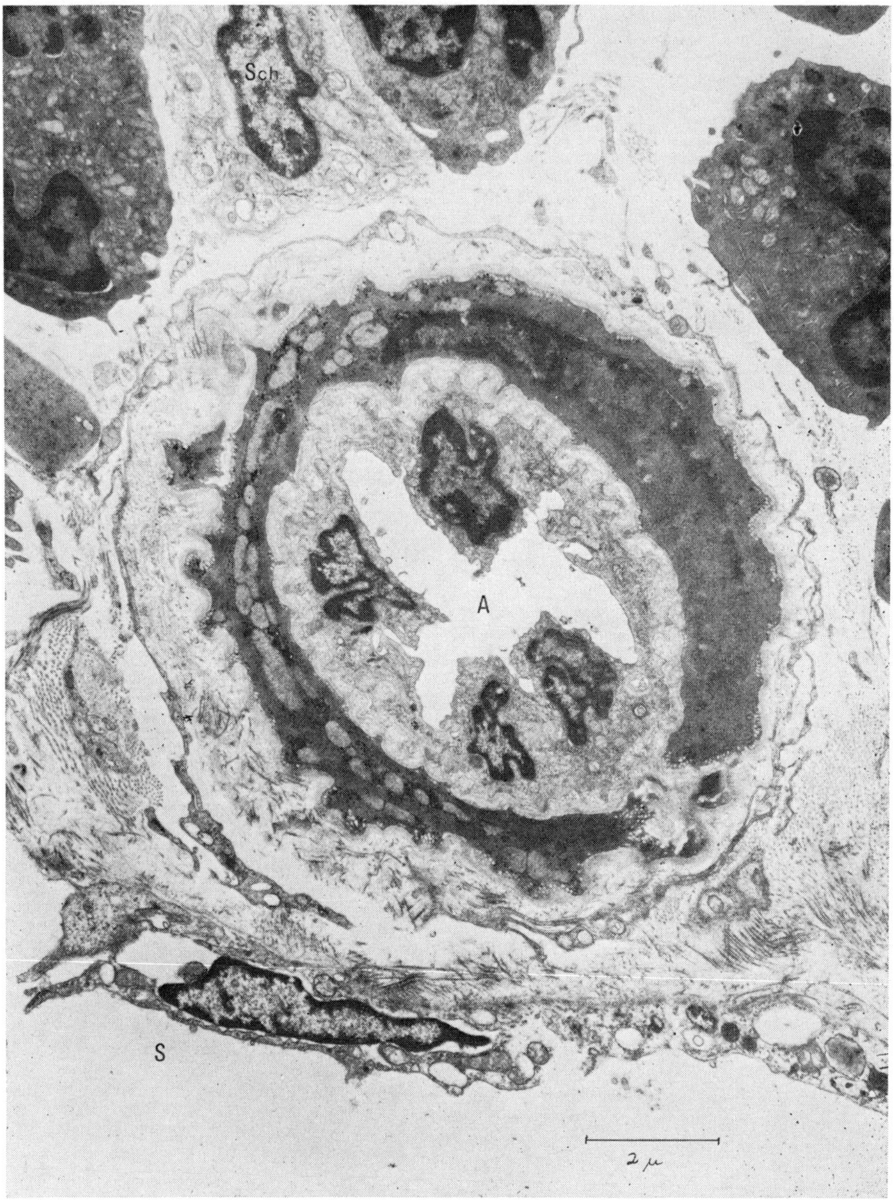

FIG. 5—Femoral marrow of an adult rat. Fine structure of an area showing the cross section of a thin arteriole (ca. 12 microns in diameter). In the upper left a Schwann cell (Sch) with severeal nonmyelinated axons in close relationship with 2 leukocytes. Below the arteriole, the wall of a sinusoid (S).

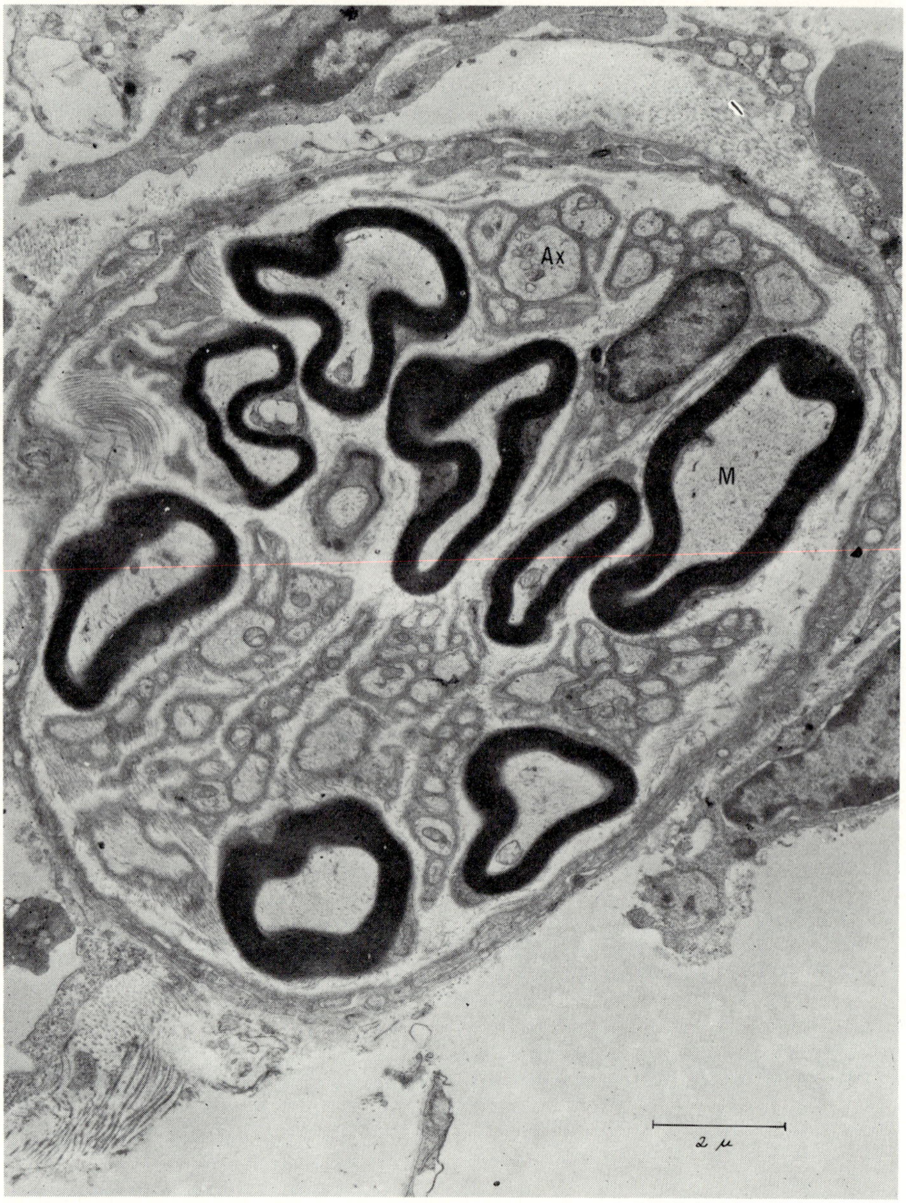

FIG. 6—Femoral marrow of an adult rat. Large nerve containing myelinated (M) and nonmyelinated axons (Ax).

and the pressure within the vascular tree due to the blood flow on the other. These signals may lead to an increase or decrease of blood flow in the particular marrow region, thus altering the blood flow and thereby its pressure and oxygen supply. The rigid bone capsule would serve as a shield against external pressure on the marrow, decrease the environmental "noise," and thus increase the sensitivity to marrow internal pressure changes.

Renewal of the Cellular Stroma of the Bone Marrow

Complete Labeling before Birth

In order to investigate the cytokinetics of the cellular components of the marrow stroma, the "complete ^{3}H-thymidine labeling method" of the bone marrow has been devised.[14,15,18,19] In principle, pregnant rats receive a continuous infusion of ^{3}H-thymidine (^{3}HTdR) via a polyethylene catheter inserted into one tail vein from day 9 to day 22 of pregnancy. The amount of ^{3}HTdR administered is about 288 μCi per day. This procedure results in the ^{3}H labeling of every single cell nucleus in the offspring. If no further ^{3}HTdR is being administered after birth, the labeling intensity, and as a consequence, the fraction of labeled cells decreases in all tissues and organs as a function of the renewal characteristics. In the bone marrow it was shown in these studies that the ^{3}H label disappears from all differentiated myeloid and erythrocytic cells, as well as from "blasts" and "lymphoid cells" within about 10 days. In contrast, endothelial cells, reticular cells, and a small but significant fraction of marrow lymphocytes remain labeled (Figs. 7A and B). The labeling intensity of endothelial and reticular cells decreases within the first 12 hours after birth in the femur to about one-half of the value at birth, indicating, on the average, one division per cell during this time. Thereafter, the labeling intensity decreases very slowly. The average grain count decreases to about one-half during the first 10 weeks of life for the endothelial and reticular cells and the fraction of labeled cells goes down from 100 per cent to about 50–60 per cent. This is the time phase of the most rapid growth of a newborn rat. Beyond the tenth week, the decrease in the labeling intensity and fraction continues even more slowly and by 20 weeks between 40 and 50 per cent of all marrow endothelial and reticular cells are still labeled with an average grain count of 8–11 grains per cell.

Among the small round cells of the marrow that are characterized by a dense chromatin of the nucleus and a small rim of basophilic cytoplasm (in this paper called "bone marrow lymphocytes" since in smears they are indistinguishable morphologically from lymphocytes) is a small fraction of 2–5 per cent that does not lose the ^{3}H label after birth. The mean grain count of this labeled fraction is in the order of 7 to 8 grains per cell with the autoradiographic exposure time used. These labeled cells are clearly present for at least 21 weeks after birth with only a very slow decline between the second and twentieth week after birth.

Thus, this method allows the quantitative characterization of renewal of the cellular elements of the marrow stroma.

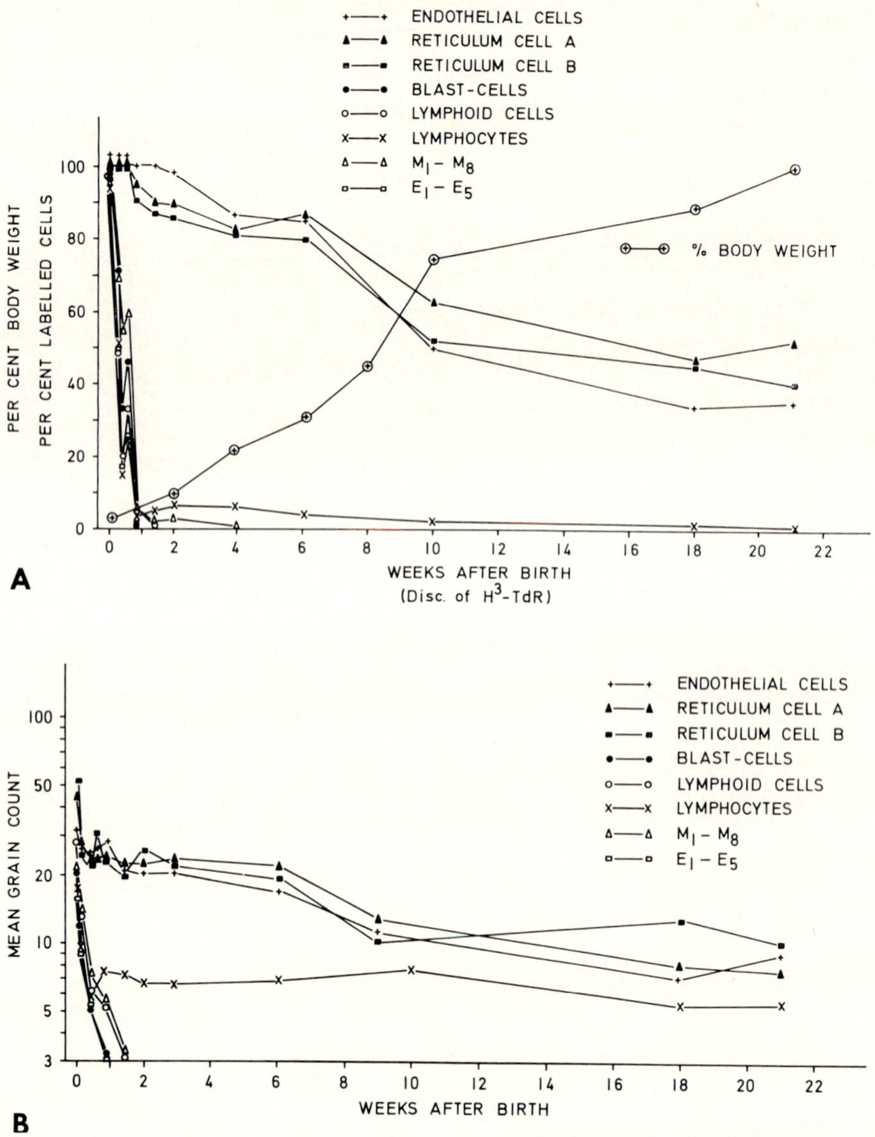

FIG. 7—Labeling pattern of bone marrow cells in rats "completely H^3-TdR labeled" at birth. (A) Fraction of labeled cells and body weight expressed as per cent of weight at 21 weeks after birth as a function of time. (B) Labeling intensity changes, expressed as mean grain count per cell for each cell type. (From Haas et al.[21])

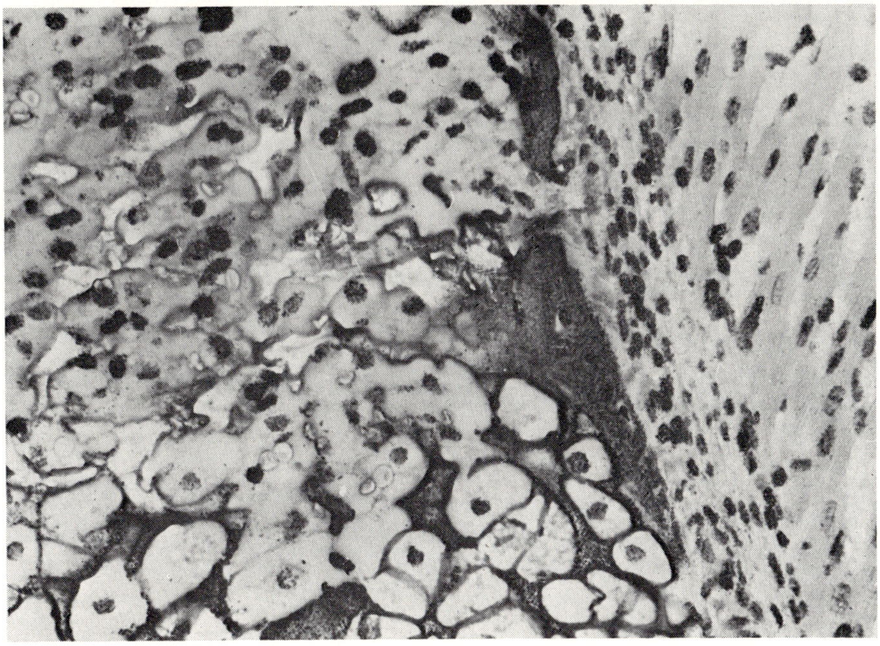

FIG. 8—Autoradiograph of a histological section of a lumbar vertebra 12 hours after birth. Through cortical "windows" 100% labeled mesenchymal cells enter into the primitive bone marrow cavity to form the primary bone marrow. Ilford L 4 liquid emulsion exposed 45 days. Haematoxylin and Eosin (× 705).

It is of interest to note in these studies that the colonization of each bone cavity with marrow cells during fetal and neonatal life commences from a population of completely and highly labeled cells originating, apparently, in the perichondreal zone and migrating into the cavity of the bone formed by the disintegration of cartilage cells through "windows" (Fig. 8). This process repeats itself in every bone, regardless of the time after birth. Thus, at the time of birth several bones of the sternum start to be colonized by cells from the perichondreal region. Heavily-labeled other bones, such as the metacarpal or metatarsal bones, are still in their cartilage status; they start to form a cavity 2–3 days after birth and its colonization commences again from a population of perichondreal cells that are still as heavy and completely labeled as the perichondreal cells were in the sternal bones at the time of birth. This leads to the assumption that some perichondreal cells "rest" during the first days after birth until called upon to colonize the bone cavities to establish the marrow stroma. Whether hemopoiesis itself originates from the descendents of these perichondreal cells or from cells that migrated into the marrow parenchyma out of the blood circulation remains to be determined.

Complete Labeling Before and Several Weeks After Birth

In order to have a suitable model to investigate the proper stimulus for the renewal of bone marrow stroma cells, a second series of experiments were performed to have a complete labeling of all stromal cells in an animal whose marrow would be entirely under the adult regulatory mechanisms.[21] For this purpose ^{3}H-thymidine was given to rats from day 9 of pregnancy until term (288 μc. per day). After birth, the newborn rats were given 0.5 μc. ^{3}HTdR per Gm. body weight, subcutaneously, every 12 hours for 4 weeks. At the point of discontinuation of ^{3}HTdR injections, the rats were completely labeled in every single nucleus of all bone marrow cells and in all other tissues studied so far. Thereafter, rats were sacrificed at regular intervals for 30 weeks after the last ^{3}H-thymidine injection. The bone marrow sections and smears indicate a very rapid disappearance of label from all myelopoietic and erythropoietic cells as well as from all blasts and "lymphoid" cells within about 10 days. Only the endothelial and reticular cells and about 12 per cent of marrow lymphocytes were found to be labeled at two weeks after discontinuation of ^{3}HTdR administration. During the subsequent 28 weeks, the fraction of labeled endothelial and reticular cells diminished gradually from 100 per cent to about 70–80 per cent due to a grain count diminution from about 20 to 15 per cell (about 25 per cent in 28 weeks), which would indicate a life cycle of these cells well in excess of 14–15 months. The fraction of labeled marrow lymphocytes (about 12 per cent 2 weeks after the last ^{3}HTdR injection) diminished until the thirtieth week to about 4–5 per cent, the mean grain count dropping from about 10 to 7 per cell during this time. Thus, these studies indicate that endothelial and reticular cells, as well as a small fraction of marrow lymphocytes of the "adult" bone marrow, of the rat had a very slow turnover once the structure was established, with a division occurring in these cells on the average as rarely as about once a year.

Evidence for Resting Bone Marrow Cells Having Hematopoietic Potentialities

After having established a labeling method by means of ^{3}HTdR all those cells of the marrow that do not or rarely undergo cellular division in the adult rat, the question was raised as to which of these "resting" cells would respond to a variety of stimuli. In previous reports, the effect of ionizing radiation on "resting" cells of the bone marrow has been demonstrated.[22] It was found that after irradiation of one hind limb with 750 R with the rest of the body shielded, there was a progressive fall in the lymphocyte labeling in the irradiated area, but not in the contralateral shielded marrow. These findings were taken as suggestive evidence that the regeneration of bone marrow after local or whole body irradiation was not associated with a proliferation of endothelial or reticular cells. They do not seem to be affected cytokinetically with the radiation doses used. However, regeneratory activity of the marrow is associated with a proliferative burst of marrow lymphocytes, since a diminution in grain count over marrow lymphocytes can be explained for all practical purposes only by their proliferation. But if the shielded marrow areas were large compared to the irradiated site, the stimulus is too small to affect the labeled lymphocyte population grossly.

However, if the shielded area was small (one hind limb as compared to the irradiation of the rest of the body), the lymphocytes in this marrow area started to proliferate and there is actual evidence of the immigration of labeled cells into irradiated marrow sites from other parts of the body as evidenced by the initial decrease and subsequent increase in label over the lymphocytes in the irradiated marrow sites.

Similar evidence for a proliferative activity of marrow lymphocytes in contrast to endothelial and reticular cells was also obtained when 0.9 mg./kg. of nitrogen mustard was administered to rats in which endothelial and retricular cells as well as lymphocytes of the marrow were labeled by means of the "complete ^{3}HTdR labeling methods."[20] There was no change in the labeling intensity of endothelial or reticular cells but the labeling intensity of the fraction of labeled marrow lymphocytes decreased to very low levels within 36 to 48 hours after nitrogen mustard, during a time when a wave of hemopoietic regeneration started to commence in this bone marrow.

In a further experiment, hydroxyurea was given to rats labeled in all endothelial and reticular cells of the marrow as well as in a fraction of marrow lymphocytes. This compound was first introduced into the study of cell kinetics by Morse, Rencricca, and Stohlman[27] since its unique feature seems to be its effect only on DNA synthesizing cells. Thus, the question was raised whether or not the wave of regeneration to be expected in the marrow would actually commence from one of the marrow cells types that are never or very rarely in DNA synthesis normally. Pregnant rats were given 288 μc. of ^{3}HTdR per day from day 9 of pregnancy until term. After delivery the new born rats were given subcutaneously 1 μc. per Gm. body weight of ^{3}HTdR every 8 hours for 6 weeks. At that time ^{3}HTdR administration was stopped. From experience it was known that all label would now disappear from the actively turning over cells and only the slowly turning over or resting cells would retain their label. This was confirmed 6 weeks later when control animals were sacrificed. They showed only labeled endothelial and reticular marrow cells with a mean grain count of about 23 per cell (autoradiographic exposure times of 29 days). The labeled fraction of small marrow lymphocytes showed a mean grain count of about 14 grains per cell (Fig. 9). One group of rats was given four injections of 500 mg. Hydoxyurea at 6-hour intervals. Three animals were sacrificed at each of 6, 12, 24, 36, 52, and 60 hours, thereafter at 3, 4, 5, 6, 7, 8, 12, and 16 days. Smears were made from one femur and one tibia. One tibia was processed for histology and one femur was used for the determination of marrow cells per mg. marrow. In autoradiograms (29 days exposure) the fraction of labeled cells of various categories and their labeling intensity were determined and, in sections, the changes in the absolute number of labeled cells per unit area determined.

The results indicate a profound fall in the total number of marrow cells after 4 × 500 mg. of hydroxyurea. A rapid recovery can be seen commencing about 2–3 days after the last dose of hydroxyurea with normal values being present about 2 weeks later (Fig. 10).

In the autoradiograms of marrow smears (Fig. 11) it was found that the la-

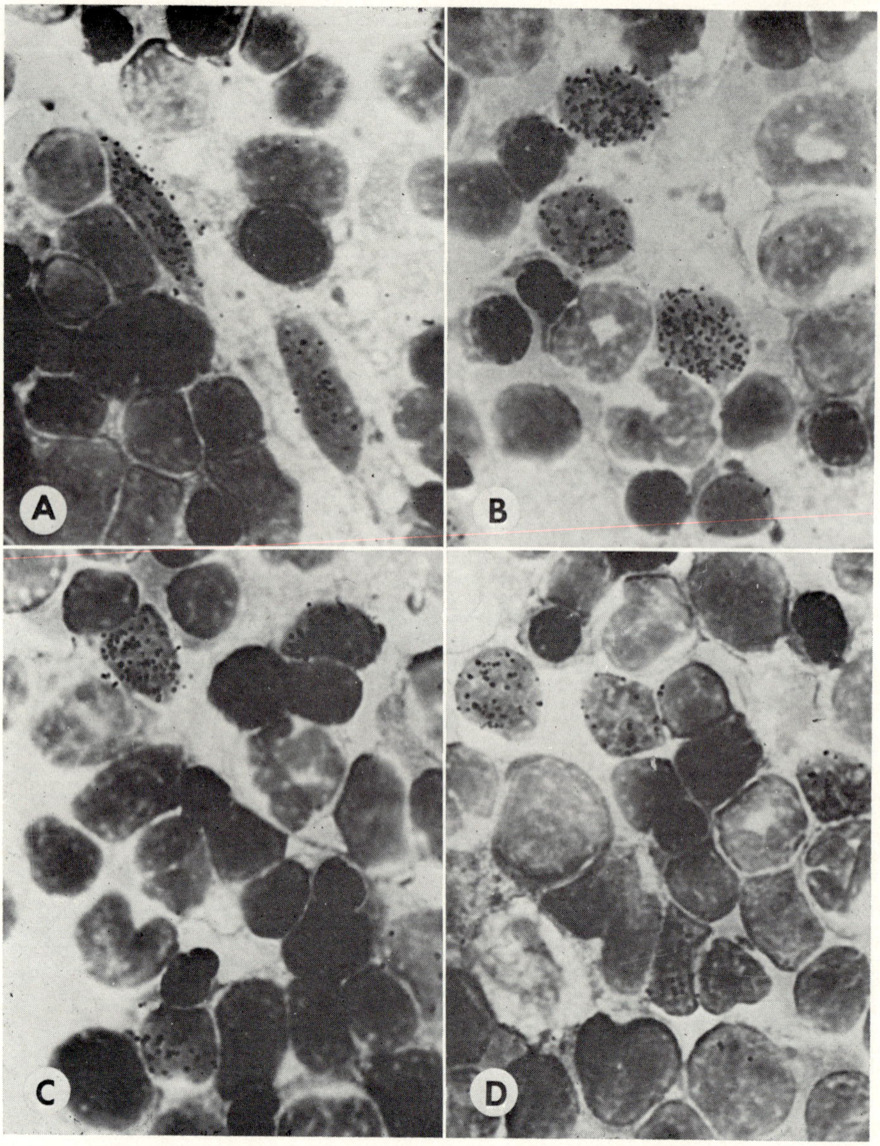

FIG. 9—Examples of bone marrow cells heavily labeled 6 weeks after the last H^3-TdR injection, 12 weeks after birth (AR 10 stripping film, 29 days exposure, × 705). (A) Endothelial cells. (B) Reticular cells type B. (C) "Lymphocyte" lower left corner. (D) Reticular cell type A upper left corner.

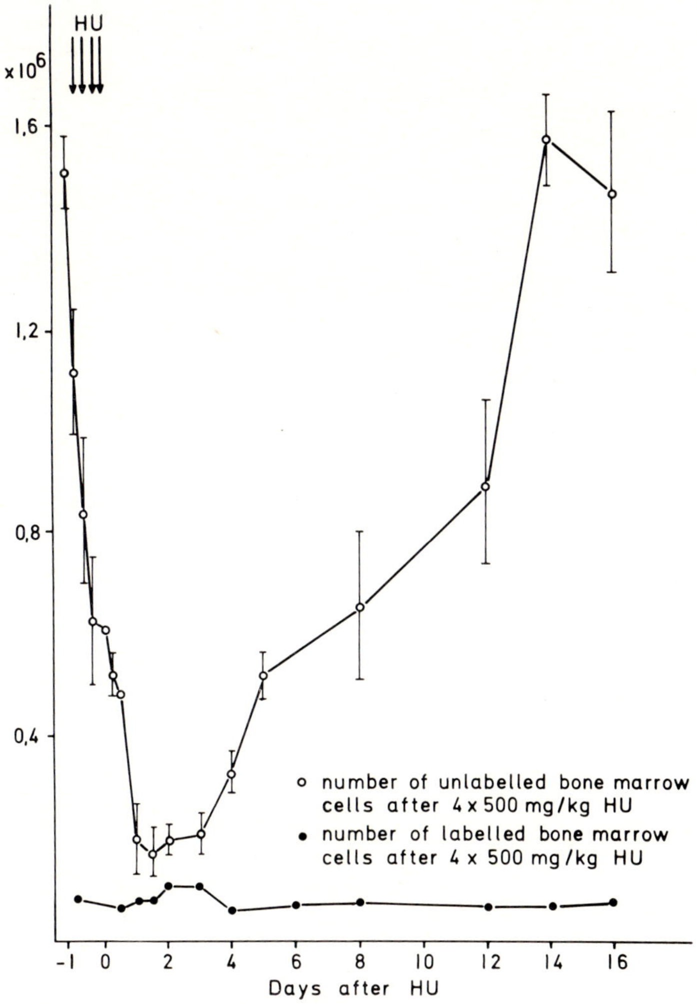

FIG. 10—Millions of labeled and unlabeled nucleated cells per mg. bone marrow (I standard error) at different times after 4 × 500 mg./kg. hydroxyurea. The control group consisted of 10 animals and all other groups of 3 animals.

beling intensity of endothelial and reticular cells did not change appreciably. However, the labeled marrow lymphocytes decreased in their labeling intensity from about 14 to about 3 grains per cell which would indicate, on the average, 2 cell divisions. Of great importance was the finding that concomitant with the dilution of label over lymphocytes a wave of labeled blast cells and, subsequently, of labeled early hemopoietic precursor cells was seen; (Fig. 12). There seems to be no doubt on morphologic grounds that early red cell precursors, white cell precursors, and megakaryoblasts were found to be labeled (Fig. 13). At day 4, of 13×10^3 blast cells mg. bone marrow, 11.5×10^3 were labeled (close to 45 per cent). It may well be that all blasts are labeled if a longer exposure time is used.

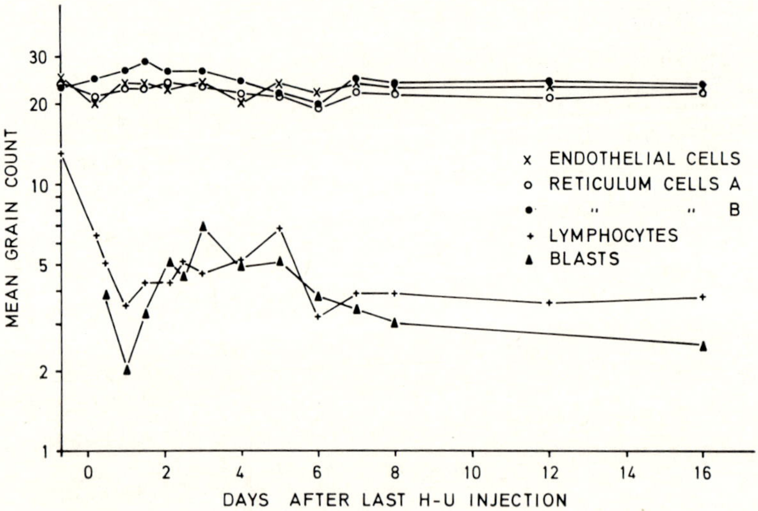

FIG. 11—Labeling intensity of resting bone marrow cells and blast cells as a function of time after 4 × 500 mg./kg. hydroxyurea.

These findings give support to the hypothesis that there is a "resting" stem-cell compartment in the marrow, composed of small round cells ("marrow lymphocytes"), that has hemopoietic potentialities when called upon—as after the eradication of the majority of bone marrow cells "in cycle" by the administration of 500 mg. hydroxyurea four times. This stem-cell pool can apparently feed cells into erythropoiesis, mylopoiesis, and megakaryocytopoiesis.

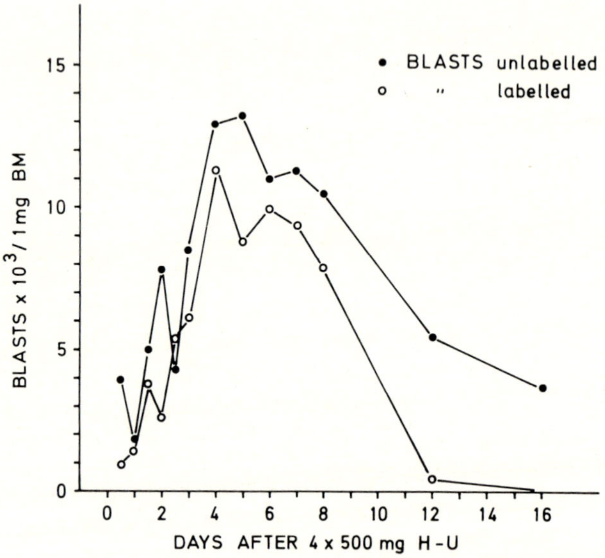

FIG. 12—Absolute number of labeled and unlabeled undifferentiated blast cells per mg. bone marrow at different times after 4 × 500 mg./kg. hydroxyurea. Each symbol represents values obtained from 3 animals.

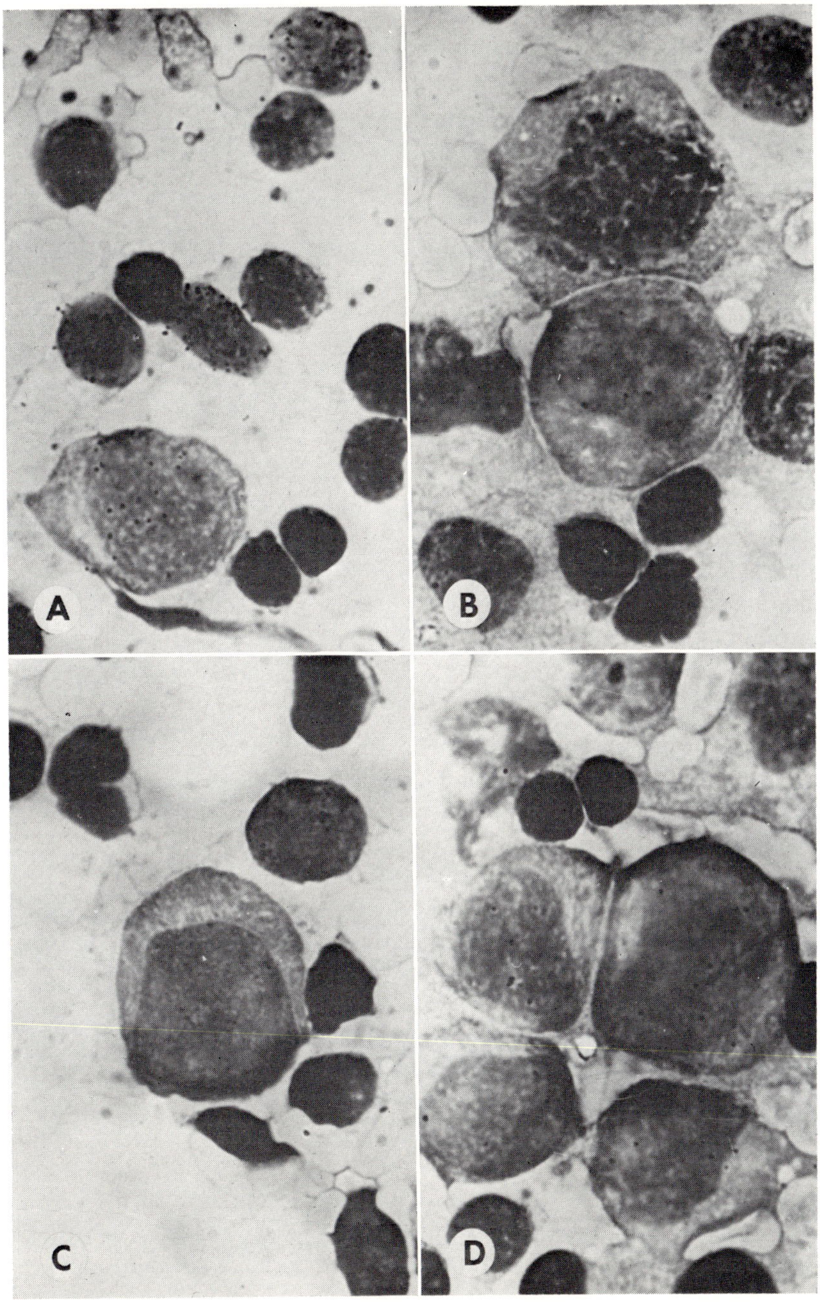

FIG. 13—Examples of labeled blast and early myelocytic precursor cells 3 and 4 days after 4 × 500 mg./kg. hydroxyurea. Autoradiograph of a femoral bone marrow smear. AR 10 stripping film exposed 29 days (× 705). (A) Blast cell (proerythroblast?) 18 grains plus heavily labeled reticular and endothelial cells. (B) Blast cell (early promyelocyte?) 6 grains, labeled mitosis 12 grains. (C) Blast cell 14 grains. (D) 4 labeled early myelocytic precursor cells 4–12 grains.

Although it is suggestive that this hemopoietic potentiality resides within the population of small round cells called "marrow lymphocytes" in this paper, further studies are necessary to exclude the endothelial cells or reticular cells as precursor cells of the regenerating cells. For instance, one could speculate that the dilution of label in the marrow lymphocytes was due to proliferation of these cells to repopulate their own cell pool without having anything to do with the wave of labeled hemopoiesis. If this were true, the only alternative source would be reticular or endothelial cells that had transformed into blast cells, and would thus not be counted within the original cell population. However, this seems unlikely, because one would then have to assume also that this population of reticular and endothelial cells would diminish in size without being repopulated since the labeling intensity over these cells remained constant and hence did not indicate cellular proliferation in this cell pool. Also, the absolute number of cells of this pool appeared to remain constant.

If only one dose of hydroxyurea was given, the labeled marrow lymphocytes did not lose their label (not reticular or endothelial cells), and the wave of regenerating blast cells and their descendents was not labeled with ^{3}HTdR. This was felt to indicate that one dose of hydroxyurea eliminated only a fraction of the "committed" or actively turning over cell pool, but not enough to prevent regeneration from within the pool.

Summary

Investigation of the angio- and neuroarchitecture of the rat bone marrow indicates the delicate ecology of endothelial cells, reticular cells, bone marrow nerves, and hemopoietic cells. The angioarchitecture is characterized by a nutrient artery branching into many arterioles. The sinusoidal system can be described by a series of sinusoidal trees with a rectal sinusoid as the trunk and contortal sinusoids as branches supplied from the arterial side. The arteries and arterioles are accompanied by nerves. But many nerve fibers go straight through the parenchyma and by sinusoids. A hypothesis is advanced that fine regulation of hematopoietic proliferation and cell release is performed via nerve fibers which may be sensitive to minute pressure changes (pressure from hemopoietic growth and from blood flow), particularly in an organ which is protected from external pressure through a rigid bony capsule.

The renewal characteristics of the cellular elements of the stroma (endothelial cells, reticular cells, marrow lymphocytes) were studied by the "complete ^{3}HTdR labeling method." It was shown that the renewal of these elements slows down considerably during the first 4–6 weeks of life. After that time, these cells may divide under normal conditions perhaps not more than once a year.

If cytotoxic stimuli are applied to the marrow (radiation, nitrogen mustard, hydroxyurea) it can be shown that endothelial cells and reticular cells do not participate in the hematopoietic regenerative activity of the marrow. However, the small round cell of the marrow, called marrow lymphocyte, appears to be able to respond to those stimuli with a burst of proliferative activity.

If hydroxyurea was given four times 6 hours apart (500 mg./kg. body weight per injection), a wave of labeled blast cells and subsequently differentiated myelocytic, erythropoietic and megakaryocytic cells was found. It may be that these were derived from the labeled marrow lymphocyte population, which thus would have under normal steady state conditions all characteristics of a "resting stem cell pool."

REFERENCES

1. Bohne, F., Haas, R. J., Fliedner, T. M., and Fache, I.: Slowly proliferating cells initiating bone marrow recovery from aplasia caused by hydroxyurea in rats. In preparation.

2. Bond, V. P., Fliedner, T. M., and Archambeau, J. O.: Mammalian radiation lethality, a disturbance in cellular kinetics. New York and London, Academic Press, 1965.

3. Bradley, T. R. and Metcalf, D.: The growth of mouse bone marrow cells in vitro. Aust. J. exp. Biol. med. Sci. 44:287, 1966.

4. Bradley, T. R., and Summer, M. A.: Stimulation of mouse bone marrow colony growth in vitro by conditioned medium. Aust. J. exp. Biol. med. Sci. 46:607, 1968.

5. Bruce, W. R., and Meeker, B. E.: Dissemination and growth of transplanted isologous murine lymphoma cells. J. Nat. Cancer Inst. 32:1145–1159, 1964.

6. Calvo, W.: The innervation of the bone marrow in laboratory animals. Amer. J. Anat. 123:315–328, 1968.

7. Calvo, W., and Haas, R. J.: Die Histogenese des Knochenmarkes der Ratte. Nervale Versorgung, Knochenmarkstroma und ihre Beziehung zur Blutzellbildung. Z. Zellforsch. 95:377–395, 1969.

8. Calvo, W., and Forteza-Vila, J.: On the development of bone marrow innervation in newborn rats, as studied with silver impregnation and electron microscopy. Amer. J. Anat. 126:355–372, 1969.

9. Carmena, A., Lucarelli, G., Carnevali, C., and Stohlman, F.: Regulation of erythropoiesis XIX: Effect of hypoxia on erythropoiesis in the newborn animal. Proc. Soc. Exp. Biol. Med. 121:652, 1966.

10. Cronkite, E. P., Bond, V. P., Fliedner, T. M., and Killmann, S.: The use of tritiated thymidine in the study of hemopoietic cell proliferation. Ciba Foundation Symposium on Haemopoiesis. 70. London: Churchill, 1960.

11. Fliedner, T. M., Sandkühler, S., and Stodtmeister, R.: Die Knochenmarkstruktur bei Ratten nach Bestrahlung mit schnellen Elektronen. Z. Zellforsch. 43:195–205, 1955.

12. Fliedner, T. M., Stodtmeister, R., and Sandkühler, S.: Untersuchungen über die Gefäßarchitektonik des Knochenmarkes der Ratte. Z. Zellforsch. 45:328–338, 1956.

13. Fliedneı, T. M., Bond, V. P., and Cronkite, E. P.: Structural, cytologic and autoradiographic (H^3-thymidine) changes in the bone marrow following total body irradiation. Amer. J. Path. 38:599–623, 1961.

14. Fliedner, T. M., Haas, R. J., and Stehle, H.: Die ^{3}H-Thymidinmarkierung aller Zellkerne neugeborener Ratten. Acta Histochem. Suppl. VIII:231, 1968.

15. Fliedner, T. M., Haas, R. J., Stehle, H., and Adams, A. C.: Complete labeling of all cell nuclei in newborn rats with H^3-thymidine. A tool for evaluation of rapidly and slowly proliferating cell systems. Lab. Invest. 18:249, 1968.

16. Forteza-Vila, J.: Fine structure of bone marrow stroma. In preparation.

17. Garcia, J. F., and van Dyke, D. C.: Response of rats of various ages to erythropoietin. Proc. Soc. Exp. Biol. Med. 106: 585–588, 1961.

18. Haas, R. J., Stehle, H., and Fliedner, T. M.: Autoradiographische Untersuchungen über schnell und langsam proliferierende Zellsysteme des neonatalen Knochenmarkes. Schweitz Med. Wschr. 97:1472, 1967.

19. Haas, R. J., Stehle, H., and Fliedner, T. M.: Autoradiographic studies on rapidly and slowly proliferating cell systems in neonatal bone marrow. Helv. Med. Acta. 34: 54, 1967.

20. Haas, R. J., Fliedner, T. M., and Stehle, H.: Cytokinetic analysis of slowly renewing bone marrow cells after administration of nitrogen mustard. *In* Proceedings. Vienna, International Atomic Energy Agency, 205, 1968.

21. Haas, R. J., Bohne, F., and Fliedner, T. M.: On the development of slowly-

turning-over cell types in neonatal rat bone marrow. Blood 34:791–805.

22. Haas, R. J., Bohne, F., Harriss, E. B., and Fliedner, T. M.: Cytokinetic analysis of slowly proliferating bone marrow cells during recovery from radiation injury. Rad. Res. (in press).

23. Lajtha, L. G.: Utilisation of formate ^{14}C for synthesis of DNA by human bone marrow cells in-vitro. Nature 174:1013, 1954.

24. Lucarelli, G., Porcellini, A., Carnevali, C., Carmena, A., and Stohlman, F.: Fetal and neonatal erythropoiesis. Ann. N.Y. Acad. Sci. 149:544–559, 1968.

25. McCulloch, E. A., and Till, J. E.: The sensitivity of cells from normal mouse bone marrow to gamma radiation in-vitro and in-vivo. Rad. Res. 16:822–832, 1962.

26. Metcalf, D., and Stanley, E. R.: Quantitative studies on the stimulation of mouse bone marrow colony growth in-vitro by normal human urine. Aust. J. Exp. Biol. Med. Sci. 47:453–466, 1969.

27. Morse, B. S., Renricca, N. J., and Stohlman, F.: The effect of hydroxyurea on differentiated marrow erythroid precursors. Proc. Soc. Exp. Biol. Med. 130:986, 1969.

28. Stohlman, F., Jr., ed.: Kinetics of cellular proliferation. New York, Grune & Stratton, 1959.

29. Swan, H. T., Reisner, H., Jr., and Silverman, M.: The effect of various metabolites on the growth of marrow cells in-vitro. Blood 10:735–752, 1955.

30. Zamboni, L., and Pease, D.: The vascular bed of red bone marrow. J. Ultrastruct. Res. 5: 65–85, 1961.

Experience with Injured and Implanted Bone Marrow: Relation of Function to Structure

By William H. Crosby

Some years ago at Walter Reed Army Medical Center, Knospe et al. undertook a study of the reaction of bone marrow to local irradiation.[1] When the femur of the rat was exposed to 4000 R and examined 2 to 4 days later the marrow was in a state of disarray. Hemopoiesis was wiped out and there was phagocytosis of the debris of the hemopoietic tissue. Integrity of the sinusoids was retained. Within 2 weeks hemopoietic activity was completely restored. It has been well established that a single dose of wholebody irradiation in excess of 1000 R is capable of destroying all hemopoiesis in the body. Transfusion of hemopoietic cells from unirradiated marrow can spare some animals thus irradiated. If 1000 R total body irradiation produces total destruction of hemopoiesis, it would seem reasonable to conclude that 4000 R local irradiation must produce local destruction of hemopoiesis in the locally irradiated femur. Restoration of hemopoiesis at 2 weeks cannot result from hemocytoblasts rising from the dead. Reestablishment of hemopoiesis must result from stem cells moving from uninjured areas into the blasted areas via the blood stream. It is significant that at 2 weeks as well as at 2 days the sinusoidal structure of the marrow was preserved.

When the irradiated tibia was examined 3 months after exposure to 4000 R the marrow once again was "aplastic." There was no hemopoiesis and the sinusoidal structure had disappeared. This suggests an essential relation between the marrow's microcirculation and its hemopoietic function. Without the unique sinusoids of the marrow there can be no hemopoiesis. The experiment demonstrating a "biphasic aplasia" indicates that the marrow consists of two organs, one dependent on the other—the hemopoietic organ depending upon the adventitial organ. Each consists of cells with finite life spans. Therefore each must be constantly replaced from a substrate of stem cells. Irradiation with 1000 R destroys hemopoietic stem cells, but cells of this sort are present in the blood and when the injury is local the damage can be repaired by circulating hemopoietic stem cells. The adventitial structure is not destroyed by 1000 R. On the other hand, the injury inflicted by 4000 R does destroy the stem cells of the adventitia. The damage is not immediately evident because the life span of the cells which comprise the sinusoids probably can be counted in weeks rather than hours. Two weeks after injury by 4000 R the microcirculation remains relatively intact and hemopoiesis can depend upon it. But as the adventitial cells die one by one and cannot be replaced, the structure of this specialized microcirculation gradually decays until it finally can no

---

From the Blood Research Laboratory, New England Medical Center Hospitals and the Department of Medicine, Tufts University School of Medicine, Boston, Massachusetts.

These studies were supported with funds from U.S. Public Health Service Grant No. AM124402 from the National Institute of Arthritis and Metabolic Disease and Atomic Energy Commission Contract No. AT(30–1) 3808.

longer provide the anatomic basis for hemopoiesis and the second phase of irradiation aplasia becomes evident.

Now we have a basis for two varieties of aplastic disorders of the marrow, one arising from an injury or defect of the hemopoietic cells (example, the marrow 2 days after irradiation injury) and the other arising from an abnormality of the adventitia or microcirculation (example, the marrow 2 months after irradiation injury).

The next experiments of Knospe challenged this concept of an essential interrelation between hemopoiesis and the microcirculation.[2] At 3 months after irradiation of the rat's tibia when the second aplasia was firmly established the tibial cavity was opened and thoroughly reamed out. Then a block of unirradiated marrow tissue lifted from the contralateral femur was placed in the cavity. This implant of marrow "took" and grew in the irradiated bone. Hemopoiesis reestablished itself, demonstrating that it was not injury to the bone or its haversian canals which caused the second phase of aplasia following irradiation. As a control Knospe demonstrated that mechanical disruption of the irradiated marrow without replacement from the other femur resulted only in fibrosis without restoration of the sinusoidal microcirculation and without hemopoiesis.

The experiments of Weinstein [3] confirmed those of Knospe. He demonstrated also that mixing of normal hemopoietic marrow with fibrotic postirradiation marrow resulted in the establishment of islands of hemopoiesis and fibrosis. The two sorts of tissue, at least in rats, seemed not to be agressive enough to eliminate one another.

After these successful transplantations of marrow Tavassoli attempted to implant autogenous marrow into other extramedullary sites.[4] He found that implants would establish themselves not only in the spleen—another sinusoidal organ—but also in kidney, liver, muscle, omentum, and subcutaneously. The latter is the most useful site because the animals can be easily used for repetitive experiments. It has also been observed that vascularity of the "host" organ modifies the eventual structure of the implant. In poorly vascularized tissues the implant develops a heavy shell of bone.[5] In richly vascular organs such as spleen, the implant may develop no bony shell at all.[6]

The bit of implanted marrow tissue removed from the bony cavity does not itself survive. Tavassoli demonstrated this by examining implants at intervals. The richly cellular hemopoietic tissue (Fig. 1A) removed from the tibia to a subcutaneous site soon loses all of its hemopoietic cells. This defoliation occurs so swiftly as to suggest that the hemopoietic cells simply crawl away, ameboid-fashion. There is no evidence of phagocytosis to suggest wholesale destruction by the reticular cells which remain in situ at 24 hours after the implantation (Fig. 1B).

At 24 hours the implant is not only divested of its hemopoietic elements, but surrounding it there has developed a rather intense proliferation of small capillaries which not only encompass the implant but pierce it through. Vascularization of the implant develops early but with ordinary capillaries, not sinusoids. Now the cells of adventitial framework, remaining after the hemo-

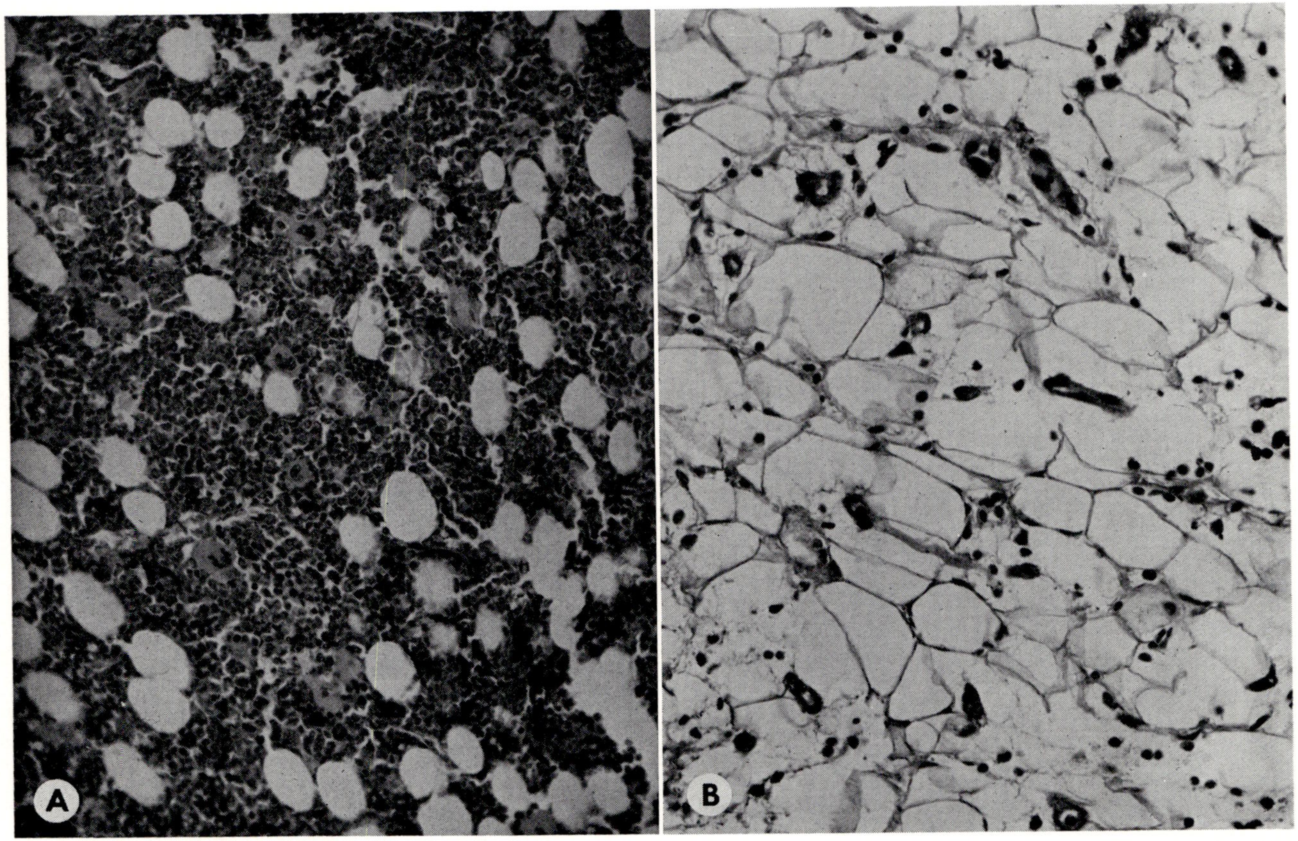

FIG. 1—(A). Cellular marrow removed en bloc from rat's tibia. (H & E × 400.) (B) Autogenous implant of rat's tibial marrow recovered from subcutaneous site after 24 hours. Only the advential scaffolding and empty sinusoids remain. (H & E × 425.)

poietic cells depart, undergo an immediate proliferation. Within 2 days the implant is overgrown with a primitive cell resembling a reticulum cell (Fig. 2A). In this dense cellular matrix streaks of osteoid appear which rapidly become cancellous bone (Fig. 2B). The implant is now 8 days old and the first sinusoidal vessels appear in the reticulum cell's network. Two days later the sinusoidal pattern is profuse and the sinusoids become greatly dilated occupying all of the space of the interstices of the cancellous bone. And now a few tiny islands of hemopoietic cells appear (Fig. 3A). These rapidly enlarge causing the dilated sinusoids to become narrow (Fig. 3B). Erosion of the cancellous bone results in narrowing of the spicules and then they disappear leaving the reconstituted implant, a fragment of actively hemopoietic marrow surrounded by a thin shell of lamellar bone (Fig. 4).

The same sequence of events occurs during repair of the marrow within the tibia following removal of the tissue for an implant: proliferation of primitive cells, formation of osteoid, bone, sinusoids, then hemopoiesis and resorption of the cancellous bone. It has been pointed out that the sequence is a recapitulation of the ontogeny of marrow development in the fetus.[7] The marrow implant, like implanted splenic tissue,[8] is totally reconstituted from primitive cells. The primitive mesenchymal cells which proliferate to restore transplanted spleen and marrow resemble each other microscopically, but genetically they contain somewhat different information. Those of the spleen are predetermined to the restoration of splenic sinusoids, white pulp, and penicillar arteries. Those of the marrow seem to be the precursor of osteoblasts which form the cancellous and lamelar bone of the developing implant, the osteoclasts which resorb the cancellous bone, and the reticuloendothelial cells which arrange themselves as marrow sinusoids.[7] Whether or not these cells may also give rise to the hemopoietic tissue of these implants is not evident. It is not essential that they should be. Experiments with local irradiation show that even after 10,000 R the hemopoietic marrow of the irradiated tibia is acutely restored and remains functional until the sinusoids deteriorate.[1] It is evident that all hemopoietic stem cells are destroyed by such irradiation. It is equally evident that the temporary restoration of this marrow depends upon hemopoietic stem cells which enter the irradiated marrow via the circulation from unirradiated parts of the body.

Tavassoli implanted tissue of other organs and found that there is no functional survival of liver or kidney, for example. Skin, like spleen and marrow, also survives as we know from experience with skin grafting. Tavassoli noted that these successful implants become quickly surrounded and penetrated by a complex of capillaries.[9] The tissues which fail do not provoke this vascular reaction. The nature of the stimulus is not known. It is of further interest that renal medullary tissues, while it does not survive with recognizable tubules, does form bone and finally a nodule which contains functional hemopoietic elements. Cortical tissue does not give rise to bone marrow.[10]

In summary, it is evident that the transplantation experiments reiterate the requirement of hemopoiesis for the unique sinusoidal structure which is characteristic of the marrow's microcirculation. The quality of environment which

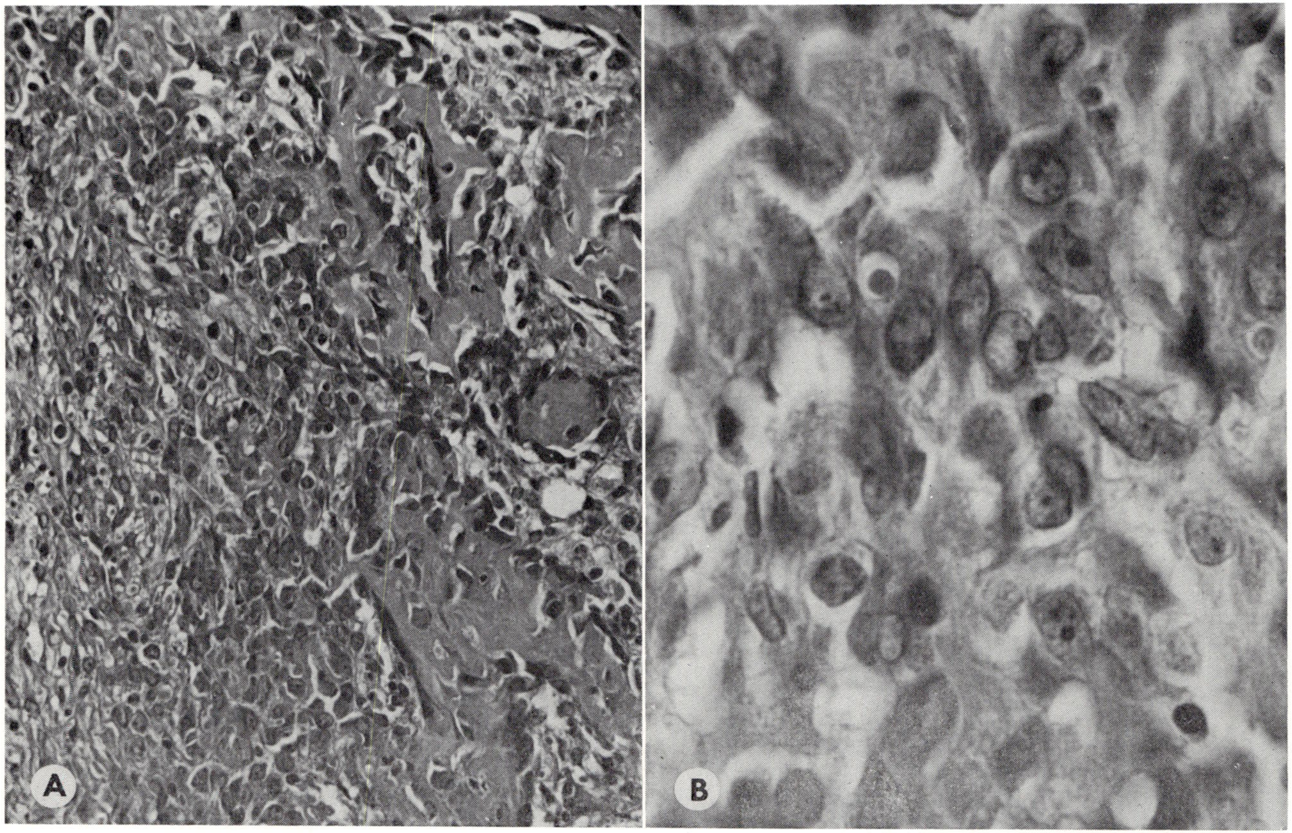

Fig. 2—(A.) Autogenous subcutaneous implant of rat's tibial marrow at 6 days. In the center is the dense proliferation of primitive-appearing cells. On the right is the beginning of cancellous bone formation. (H & E $\times$ 500.) (B.) Detail of the primitive-appearing "reticulum cells" which evidently give rise to cancellous bone and later to marrow sinusoids. (H & E $\times$ 2000.)

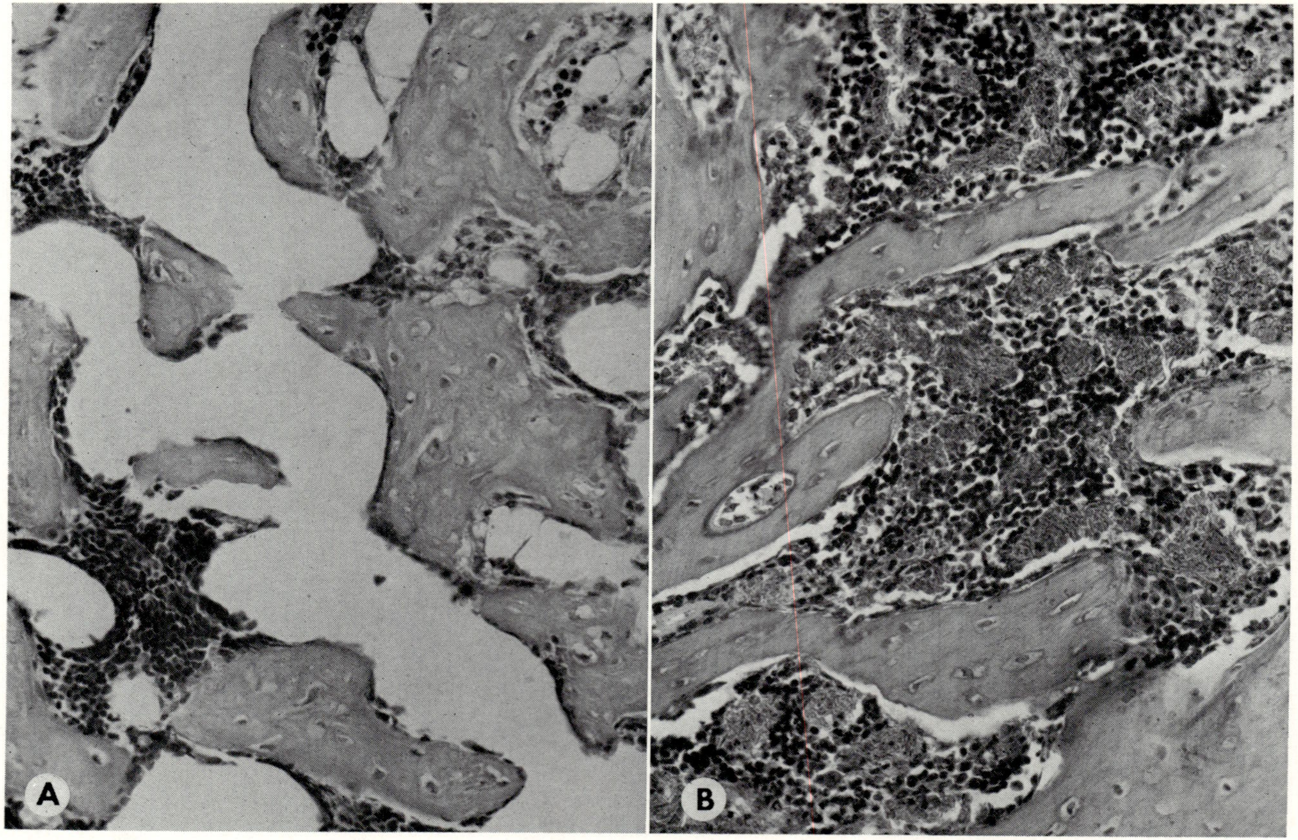

FIG. 3—(A.) Autogenous subcutaneous implant of rat's tibial marrow at 10 days. The primitive cells are gone. At 7 or 8 days the first marrow sinusoids appeared and have now expanded to fill the interstices of the cancellous bone. (Sinusoids are empty because the blood cells were washed away during preparation of the section.) The accumulations of dark staining cells are the first islands of hemopoiesis. (H & E × 425). (B.) Autogenous implant is spleen of rat's tibial marrow at 29 days. The streaks of cancellous bone are narrow, the sinusoids (filled with packed red cells) are smaller, and the hemopoietic marrow occupies the intervening space. (H & E × 425.)

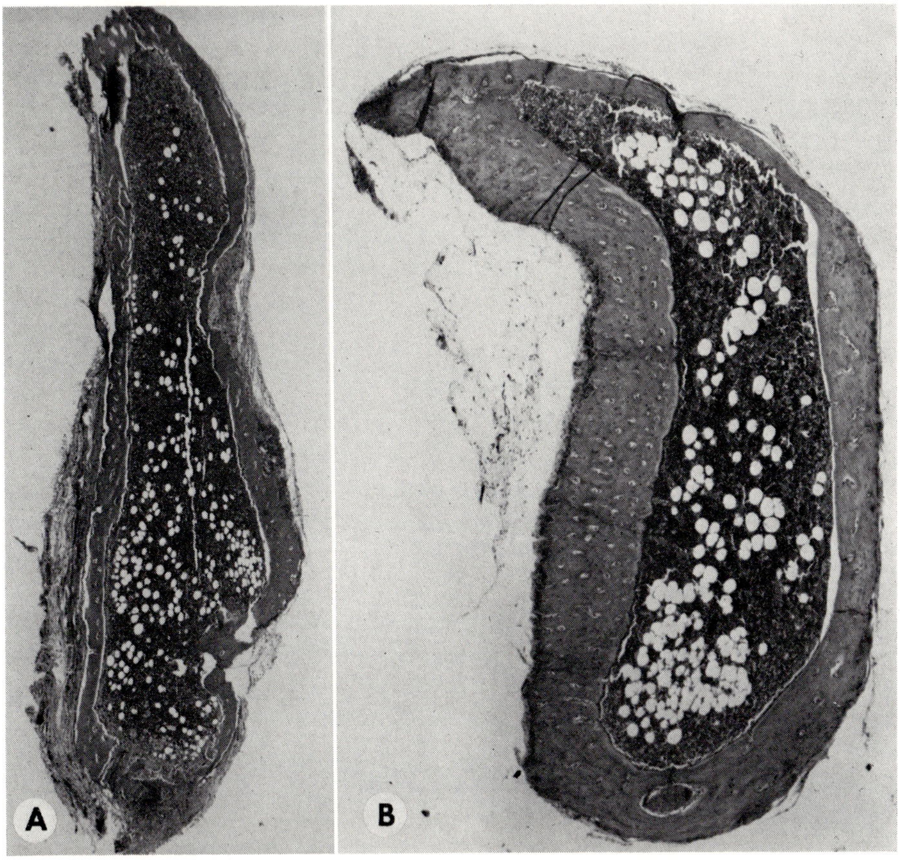

FIG. 4—(A.) Autogenous subcutaneous implant of rat's tibial marrow at 35 days. The cancellous bone has disappeared entirely from the marrow. The implant is encased in a thin bony cortex. The presence of mitotic figures indicate that these reconstituted marrow implants are proliferating hemopoietic tissue. (H & E × 25.) (B.) Autogenous subcutaneous implant of rat's tibial marrow at 400 days. The implants evidently remain a permanent part of the animal's hemopoietic organ.

these vessels furnish may be a function of blood flow which is provided—or withheld—during phases of hemopoiesis. The shell of bone which rigidly determines the volume of the marrow must stifle almost completely pulsitile flow and prevents the marrow's expanding with the proliferation of hemopoiesis; the volume of every cell which leaves the marrow must be replaced by incoming blood, by proliferation of other cells or by fat. The residual thin crust of bone on the surface of the marrow implants (Fig. 4) speaks, teleologically, of a need for this rigid constraint.

Aplastic Anemia

The dependence of hemopoiesis upon the sinusoidal structure of the marrow indicates that aplastic anemia may come about in two ways at least:

inadequacy of the hemopoietic cells and inadequacy of the sinusoidal microcirculation. Knospe's experiment provides an example of both.[1] Immediately following 4000 R of localized irradiation there is localized aplasia, a consequence of destruction of the hemopoietic cells. The marrow recovers, repopulated by stem cells which come in the circulating blood. Later aplasia reappears, this time when the microcirculation deteriorates to an extent that can no longer support hemopoiesis.

Clinically we can distinguish these two varieties of the aplastic state of the marrow. Myelosuppressive agents, such as nitrogen mustard as used in the treatment of Hodgkin's disease, can destroy normal hemopoietic cells and cause aplasia. Hemopoietic cells also would seem to be primarily at fault in the aplastic disease related to chloramphenicol idiosyncrasy, in the aplasia which is sometimes encountered as a preliminary to acute granulocytic leukemia, and in Fanconi's hereditary aplastic disease. In all three there is evidence of chromosomal aberration. The acute aplastic reaction associated with viral infection also seems to be a result of destruction of hemopoietic cells.

Alteration of the marrow's adventitia and microcirculation is probably the basis for the normal involution whereby the hematopoietic marrow in our extremities becomes fatty—aplastic—as we mature. Evidently this is a reversible state and the marrow may become hemopoietic again, for example, when chronic hemolytic disease develops. Some patients with aplastic marrow are found to have small hyperplastic, hyperactive areas of hemopoiesis.[11] Normal hemopoietic elements are hard at work in these areas, while in adjacent areas there is no activity. A reasonable explanation would be that in the aplastic areas the adventitial microcirculation is inadquate to sustain hemopoiesis. It is conceivable that the involutionary process, which causes the active marrow in our extremities to become inactive as we mature, may extend beyond the limits defined by normal growth and maturation.

We suspect that the normal involution of marrow in the extremities is not a function of its location, blood supply or temperature. My associate, Tavassoli, removed blocks of marrow from the tibia of a rabbit and implanted them under the skin of the abdomen. They underwent the same recapitulation as described above for the marrow of rats: proliferation of primitive cells, cancellous bone formation, appearance of sinusoids, and then active hemopoietic marrow. The implants at 5 weeks seemed the same, resembling Figure 3B. But when the implant was taken from the proximal tibia it was completely cellular at 8 weeks whereas the fatty marrow from the distal tibia of the same animal once gain was fatty.[12] The recapitulation of the fatty implant's development extended beyond cellularity to involution. This would indicate that the involution of the implant, like the rest of the recapitulation of its fetal development, is genetically determined.

REFERENCES

1. Knospe, W. H., Blom, J., and Crosby, W. H.: Regeneration of locally irradiated bone marrow. I. Dose dependent, long-term changes in the rat, with particular emphasis upon vascular and stromal reaction. Blood 28:398, 1966.

2. Knospe, W. H., Blom, J., and Crosby, W. H.: Regeneration of locally irradiated

bone marrow. II. Induction of regeneration in permanently aplastic medullary cavities. Blood 31:400, 1968.

3. Weinsten, M. B., and Crosby, W. H.: Bone marrow injury and repair, irradiation and mechanical disruption. Acta Haemat. (Basel) 40:55, 1968.

4. Tavassoli, M., and Crosby, W. H.: Transplantation of marrow to extramedullary sites. Science 161:54, 1968.

5. Tavassoli, M., Maniatis, A., and Crosby, W. H.: Studies on marrow histogenesis. I. The site of choice for extramedullary marrow implants. Proc. Soc. Exper. Biol. Med. 133: 878, 1970.

6. Chiari, O. M.: Vorläufige Mittellung über Knochenmarkstransplantation. Munchen Med. Wschr. 59:2502, 1912.

7. Amsel, S., Maniatis, S., Tavassoli, M., and Crosby, W. H.: The significance of intramedullary cancellous bone formation in the repair of bone marrow tissue. Anat. Rec. 164:101, 1969.

8. Perla, D.: The regeneration of autoplastic splenic transplants. Amer. J. Path. 12:655, 1936.

9. Tavassoli, M., and Crosby, W. H.: The fate of fragments of liver implanted in ectopic sites. Anat. Rec. In press.

10. Tavassoli, M.: Unpublished observation.

11. Hashimoto, M.: Pathology of bone marrow. Acta Haemat. (Basel) 27:193, 1962.

12. Tavassoli, M. and Crosby, W. H.: Histogenesis of the bone marrow: a comparison of fatty and red marrow. Submitted to Science.

DISCUSSION

George Brecher, M.D.: I am delighted to have been asked by Dr. Stohlman to open the discussion of what has become a new area of intense investigation. One of the striking features of this new field is that many of the facts of regeneration after mechanical depopulation have been known for some time but it is only recently (when these facts were reexamined from the point of view of local origin of regenerating marrow versus seeding from circulating or transfused cells and when the possibility of therapeutic utilization of these facts has been considered) that these long known facts have suddenly become highly provocative and intriguing. Most of us who have worked in radiobiology and regeneration of bone marrow after marrow supression by x-radiation, have taken it for granted that marrow regeneration is entirely a function of the stem cells already committed to the production of hemopoetic descendents and that vascular tissue to support the proliferation of such stem cells is always available. This appeared logical in radiation injury because connective tissue was thought to be relatively radioresistant as was bone. After mechanical depopulation, however, it appears that restoration of the hemopoietic marrow is delayed until some connective tissue proliferation can take place which is followed by bone formation and formation of bone marrow. It is tempting to view this seqence as Dr. Patt has indicated as a repetition of fetal bone and bone marrow formation and to suppose, particularly in view of Knospe's experiments, that one might short circuit the undesirable bone formation by local injection of marrow from the contralateral femur. Preliminary results suggest, as indicated by Dr. Patt, that the cycle of bone formation preceding marrow regeneration may be indeed short circuited in this fashion. I suspect, however, that the situation may be very much more complex than this simple view indicates. Dr. Patt has published extensive documentation of the fact that radiation with as little as 1000 R prior or after depopulation delays marrow regeneration for at least a month at which time there is extensive bone formation almost filling the marrow cavity as well as fibrous tissue. However, the entire process of widening of the haversian canals, which begins within a few days of depopulation and which may provide the initial origin of connective tissue organizing the clot in the unirradiated femur, is delayed for 2 to 3 weeks in the depopulated and irradiated femur. The clot which begins to be organized within 2 or 3 days after simple depopulation does not become organized for at least 2 weeks in the irradiated and depopulated femur. Knospe's experiment which succeeded in restoring a marrow which had become fibrotic after 4000 R by depopulation and local injection of marrow may therefore be dependent on the prior recovery of the lining cells of the haversian canals from radiation injury. A composite view of these experiments might be that after depopulation of the marrow by mechanical means the cells of the haversian canals must initiate connective tissue, bone and marrow formation. X-radiation injures these cells and delays the initial phase of regeneration but eventually leads to similar sequences of bone and marrow formation as seen in the depopulated marrow without radiation. With a high enough dose of radiation, however, the fibrotic process predominates, although the recovery of the mesenchymal elements in the haversian canals is sufficient to support marrow regeneration after renewed depopulation. Unfortunately, this does not answer the question as to why seeding could not take place once the continuity of the vasculature is reestablished. Very likely we have not yet explored all the possible temporal combinations of depopulation with transfusion or local marrow injection. With an optimal sequence we may yet succeed in the therapeutic goal of restoring normally functioning marrow by a combination of mechanical removal and repopulation from matched donors.

CORNELIUS ROSSE; M.D.: Cytokinetic studies of the lymphoid cell compartment of guinea pig bone marrow support the findings of Dr. Fliedner. After repeated 4-hourly inspection of ^{3}H-thymidine for a period of 5 days, around 15 per cent of small lymphocytes remained unlabeled in the marrow in addition to endothelial cells, reticulum cells, and plasma cells. By 5 days all other cell types were 100 per cent labeled. This indicates that a fraction of marrow small lymphocytes in the guinea pig is long-lived. Everett and Caffrey found, however, in the rat that 100 per cent of small lymphocytes were labeled in the marrow by the fifth day under the same ^{3}H-thymidine injection schedule. This is at variance with Dr. Fliedner's findings, the explanation for which could possibly be found in different criteria of cell identification.

Reference: N. B. Everett, and R. W. Caffrey: Radioautographic studies of bone marrow small lymphocytes. *In* Yoffey, J. M. (Ed.): The Lymphocyte in Immunology and Haemopoiesis. London, Edward Arnold, 1967, p. 108.

E. A. McCULLOCH M.D.: I should like to comment on Dr. Fliedner's tentative identification as "small round cells" the cells of origin of marrow repopulation. In the mouse, it is now established that pluripotent stem cells may be recognized as colony-forming units (CFU). Recently, technics (velocity sedimentation) have been developed that separate cells on the basis of size. Analysis of the marrow by velocity sedimentation shows that CFU are among the larger cells of the marrow. CFU are clearly separated from the small cells that would be identified as lymphocytes. I believe it is now established that CFU are *not* small lymphocytes.

The Renal Erythropoietic Factor (Erythrogenin) and Erythropoietin (ESF)

By ALBERT S. GORDON AND ESMAIL D. ZANJANI

ALTHOUGH THE KIDNEY is now recognized as the principal site of ESF production, the processes underlying ESF biogenesis have only recently been under scrutiny. Three possible mechanisms have been suggested: (1) The ESF is produced as such in the kidney much in the way that other glycoprotein hormones (e.g., those of adenohypophyseal origin) would be, in sequential steps from basic building blocks to the final more complex biologically active substance, (2) A pro-ESF is fabricated by the kidney which requires activation and/or protection by a serum protein to form the ESF.[1,2] (3) The kidney produces a factor, chemically unrelated to the ESF, which acts upon a serum protein to generate the functional ESF. Work performed in our laboratory for the past several years [3,4] has supported the validity of the latter mechanism for the biogenesis of the ESF. A few features of this system and an example of its possible relation to a clinical case will be described in this paper.

THE CONCEPT OF THE RENAL ERYTHROPOIETIC FACTOR (REF) OR ERYTHROGENIN

In 1966 our laboratory reported the extraction of a factor from the kidneys of hypoxic rats which, when administered intraperitoneally alone, was not erythropoietically active but which, when injected intravenously alone (unpublished) or when incubated with normal rat serum in vitro, engendered the production of a principle capable of stimulating erythropoiesis in test polycythemic mice.[4] This activity, termed the Renal Erythropoietic Factor (REF) and more recently erythrogenin,[4] was found to be present in the light-mitochondrial and microsomal fractions of rat kidney. Studies of the kinetics of the erythrogenin-serum system suggested that erythrogenin is an enzyme which converts a serum substrate (erythropoietinogen) into the active form of the ESF.[6] That it is actually the ESF which is generated in the erythrogenin-serum incubation system is supported by the finding that neutralization of the erythropoiesis stimulating activity occurs upon addition of anti-ESF globulin to the incubation mixture.[7]

A modification of the original procedure [5] is now used in the preparation of erythrogenin extracts with reproducible activity (Fig. 1).

It should be noted that the so-called "light-mitochondrial" fraction, from which the active erythrogenin extracts are derived when viewed with the electron microscope, revealed few if any mitochondria, a number of lysosomes but predominantly microbodies or more specifically the peroxisomes of deDuve.

From the Laboratory of Experimental Hematology, Department of Biology, Graduate School of Arts and Science, New York University, New York, New York.

Supported by grants from the National Heart Institute of the National Institutes of Health, United States Public Health Service (2 RO1–HE03357–11, 12, 13).

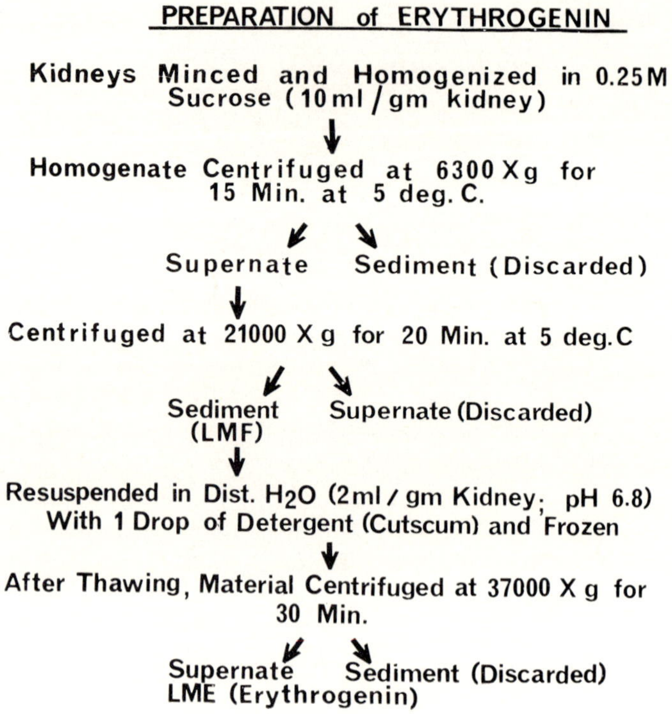

FIG. 1—Method of preparation of erythrogenin from kidney tissue.

More purified and more highly active fractions prepared by the method of Leighton et al.[8] showed progressively greater concentrations of peroxisomes and fewer lysosomes.

Confirmation of the existence of erythrogenin in extracts of the light-mitochondrial fraction of kidneys has been made by McDonald et al.,[9] Schooley et al.,[10] Malgor and Fisher,[11] as well as by Adamson, Keighley, Mirand, and Whitcomb (personal communications).

The inability of some laboratories to extract erythrogenin from the kidney may be attributable to several causes. Thus, a too vigorous initial homogenization of the renal tissue may result in the release of factors which inactivate the generated ESF.[12–14] A second possibility concerns the use in the procedure of distilled water with a pH of 5 and below that is destructive to erythrogenin.[15] Another possible cause arises from the inability of assays employed by some workers [16] to consistently detect ESF activity in the vicinity of 0.05 to 0.1 IRP units. The activity of most of our preparations, in the quantities generally administered, falls within this range.

FEATURES OF THE ERYTHROGENIN-SERUM SYSTEM

Methods

1. Dialysis of normal rat serum is carried out against 100 volumes of 0.005 M. disodium ethylenediamine tetra-acetate (EDTA) for 24 hours at 4 C. Neu-

tralization of the EDTA with 1N NaOH is performed before the serum is used. Redialysis of the dialysand is done immediately thereafter against 100 volumes of cold deionized water for 24 hours. The dialyzed serum (pH 6.8–7.2) is stored in frozen form until used.

2. The erythrogenin-serum incubation system contains equal volumes of dialyzed or undialyzed serum and the erythrogenin preparation. The incubations are conducted in a water bath left exposed to air shaken at 37 C. Saline is added to bring the incubation mixtures to constant volumes. Reactions are terminated at the end of the incubation period by plunging the reaction flasks into ice water. The preparations are administered to the assay mice immediately after incubation.

3. Per cent of radioiron incorporation into peripheral red cells of ex-hypoxic polycythemic mice is used as the criterion for erythropoiesis-stimulating effects of the preparation tested. In some cases, the RBC radioiron incorporation values are converted into equivalent IRP units of the ESF. The assay mice are made polycythemic by subjection to 0.40 atmospheres of air in a special decompression chamber, 19 hours/day for a total of 219 hours. Five or 6 CF1 female mice (23–28 Gm.) are employed to assay each preparation. The test mice are injected intraperitoneally with the entire dose on day 3 post-hypoxia which corresponds to the first day of the assay schedule. These mice receive 0.5 μc. $^{59}FeCl_3$ intravenously on day 3 (day 5, post-hypoxia). Radioiron incorporation values are obtained for whole blood at 48 hours after isotope injection. The blood volume is estimated to be 7.35 per cent of the body weight.[17] For assays of the erythrogenin-serum mixtures each mouse receives 2 ml. of the incubation fluid. In some experiments, the preparations are assayed in transfusion-induced plethoric mice.[18]

Results

Figure 2 demonstrates, in the present experiment, that intraperitoneal injections of erythrogenin (derived from the kidney of rats exposed to 0.40 atmospheres of air for 19 hours) alone or of normal rat serum alone failed to augment erythropoiesis. However, erythropoiesis-stimulating activity was generated upon incubation of erythrogenin for 10 minutes with untreated normal rat serum. Upon extending the incubation time to 60 minutes, a loss of erythropoiesis-stimulating activity was noted. This inactivating reaction which appears to proceed simultaneously with the ESF-generating reaction, but at a slower rate, may be prevented by dialyzing the serum used in the incubation against EDTA. Thus, incubation for 10 minutes of erythrogenin with EDTA-dialyzed serum led to the generation of 100 per cent more activity than noted with undialyzed serum (0.96 IRP units vs. 0.48 IRP units). Figure 2 also shows that whereas erythropoietic activity had disappeared after 60 minutes when undialyzed serum was employed, mixtures containing erythrogenin and dialyzed serum showed more activity at 60 than at 10 minutes.

These experiments suggest that the inactivation mechanism developing in the erythrogenin-serum substrate system requires for its operation a cation or

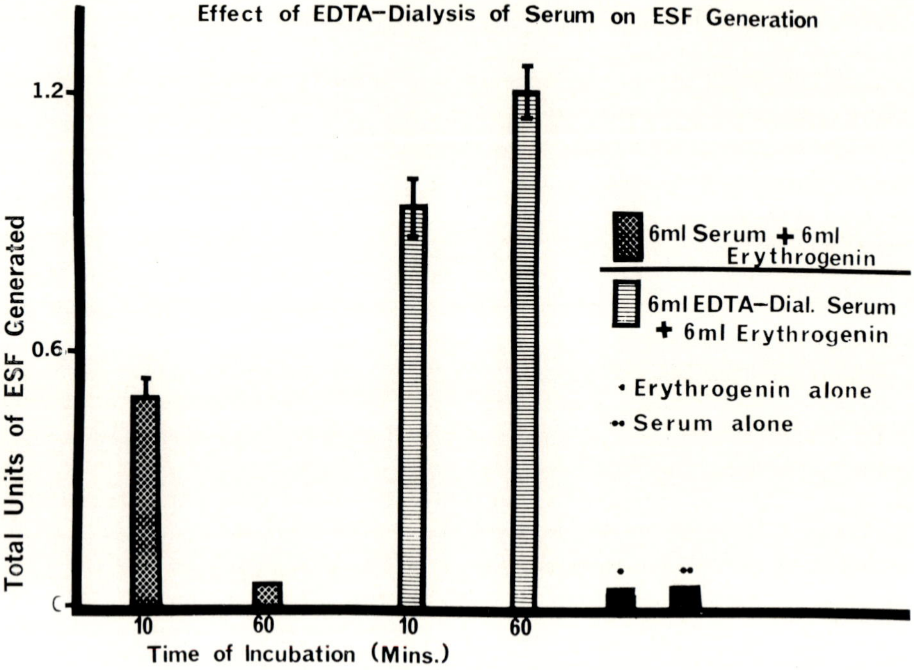

FIG. 2—Comparison of ESF production in incubation mixtures containing erythrogenin and dialyzed serum versus mixtures containing erythrogenin and undialyzed serum.

cations present in serum. It is unlikely that the inactivating agent is a component of serum since incubation of the ESF preparations with undialyzed serum leads to potentiation or no influence on the erythropoietic activity of the hormone rather than its destruction.[17–19] It would seem more probable that the erythrogenin extracts also contain a factor which in the presence of a serum-borne ion(s) inactivities the ESF generated during incubation. Further evidence that the erythrogenin preparations do not harbor this ion is shown by our finding that dialysis of erythrogenin extracts against EDTA leads to no increase in activity of the incubation fluid.

STUDIES WITH ANTI-ESF

Serum from rabbits immunized with ESF, previously shown to neutralize the erythropoiesis-stimulating effects of ESF both in vivo and in vitro [20–22] has been employed to demonstrate that the immunologic features of erythrogenin and the serum substrate are different from those of the product generated from their interaction, namely the ESF. The description to follow is a synopsis of the article by Schooley, Zanjani, and Gordon.[10]

Methods

Rabbits are immunized with human urinary ESF following the method of Schooley and Garcia.[23] One ml. of the antiserum from these animals abolishes the erythropoietic effects of approximately 25 IRP units of ESF. Goat antirabbit gamma globulin (GARGG) is obtained from goats receiving monthly immunizations with rabbit gamma globulin in complete Freund's adjuvant. One ml. of this antiserum which shows only one precipitation line in agar immunodiffusion, precipitates a minimum of 4 mg. of rabbit gamma globulin. This immune serum is employed to antagonize the effects of the anti-ESF.

In the incubation procedures employed here, 10 ml. of hypoxic rat erythrogenin was incubated with 10 ml. of EDTA-dialyzed normal rat serum at 37 C. for 45 minutes. Precipitates appearing after antibody additions were removed by centrifugation. Antibody added was thoroughly mixed and allowed to react for 15 minutes before centrifugation. All reaction mixtures, without immune precipitates, were diluted to 26 ml. and kept temporarily in ice water until administered to assay mice.

The erythropoiesis stimulating effects of the various incubation mixtures were tested in both ex-hypoxic polycythemic and in transfusion-induced plethoric mice. The method for preparing the ex-hypoxic assay mice has been described above. To obtain plethorized mice, female C_3H mice are given 2 consecutive intraperitoneal injections of 1 ml. of packed isologous red cells and 5 days later are employed for the assay. At 56 hours after administration of the test samples, radioiron is administered intravenously; 72 hours later, the radioiron content of 0.5 ml. of blood is determined. The results for both types of assay are expressed as the per cent of RBC ^{59}Fe incorporation in the calculated blood volume. In all experiments described in this phase of the study, 2 ml. of the preparations to be assayed were given intraperitoneally. As assay animals, 10–12 mice served for each material tested.

Results

The assay data are shown for transfusion-induced plethoric mice in Figure 3 and for ex-hypoxic polycythemic mice in Figure 4. Although the type of result obtained was similar in both assay systems, the RBC radioiron incorpo-

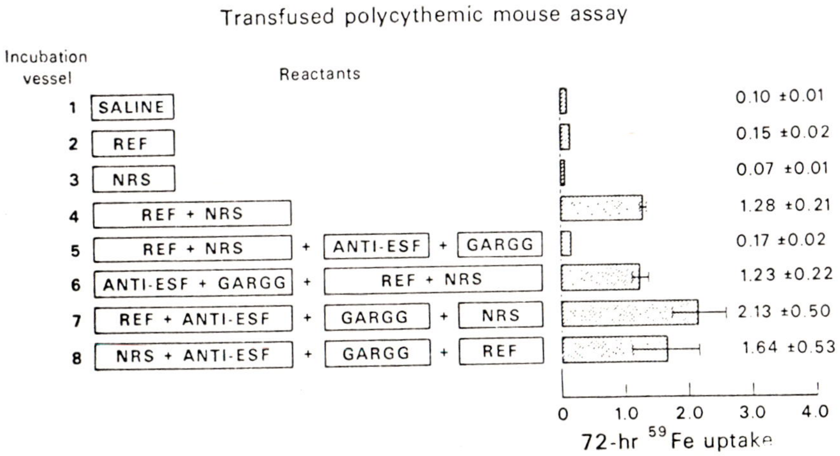

FIG. 3—Influence of anti-ESF on erythropoietic activity of erythrogenin, normal serum and mixtures containing both erythrogenin and serum. Transfusion-induced plethoric mice were used as assay animals (taken from Schooley et al.[10]).

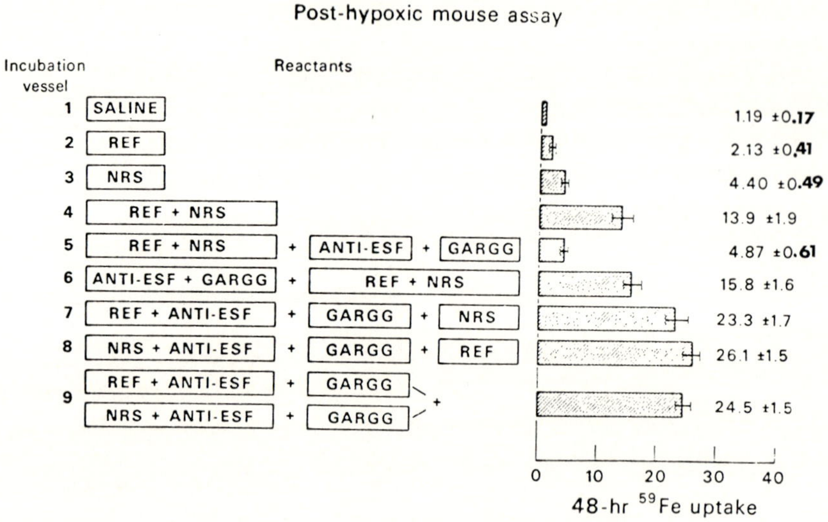

FIG. 4—Influence of anti-ESF on erythropoietic activity of erythrogenin, normal serum and mixtures containing both erythrogenin and serum. Hypoxia-induced polycythemic mice were used as assay animals (taken from Schooley et al.[10])

ration values were approximately tenfold greater in the ex-hypoxic mice. The various substances in the system are shown within the boxes and the order of their addition is indicated from left to right in the figures. Little or no erythropoietic activity was noted when erythrogenin (vessel 2) or normal rat serum (vessel 3) was given alone to assay mice. When erythrogenin was incubated with NRS, however, significant erythropoiesis stimulating activity was detectable in both assay systems (vessel 4). If anti-ESF (sufficient to antagonize 12.1 IRP units of ESF) was added to the mixture after incubation (vessel 5), the generated erythropoietic activity was neutralized. The excess of anti-ESF which had not reacted with ESF (and probably the anti-ESF which had combined with ESF) was removed from the incubation mixture by adding goat antirabbit gamma globulin (GARGG) before administration to the assay mice. It is clear that the quantities of anti-ESF and GARGG selected in each case were adequate since their addition before the incubation of erythrogenin and NRS did not influence the intensity of the measured erythropoietic response (cf. vessels 6 and 4). This demonstrates that the anti-ESF was not introduced into the assay mice after reaction with GARGG, a result which confirms our previous finding[7] that the ESF was produced in the reaction vessel and not in the assay animal.

The data for vessels 7 and 8 of Figure 3 and vessels 7, 8, and 9 of Figure 4 show that the production of ESF by erythrogenin and serum is not changed if the anti-ESF is added to erythrogenin and to NRS and then removed by GARGG before incubation of the 2 reactants.

It must be emphasized that a significant erythropoiesis stimulating influence of the erythrogenin-serum incubation mixtures was seen not only in the ex-hypoxic polycythemic but in the transfusion-induced plethorized mice as well.

Positive effects with similar incubation fluids in transfusion-induced plethoric mice have also been reported previously.[7] These results are at variance with the note by Erslev and Kazal [16] that no effects on erythropoiesis are exerted by such incubation mixtures in mice rendered plethoric by transfusion. Since transfusion-induced plethora in mice is associated with a virtual disappearance of nucleated erythroid elements in the blood-forming organs, these workers have suggested that the product of the erythrogenin-serum reaction is not exerted on stem cells but possibly on members of the nucleated erythroid series. The positive effects reported in the present article in hypertransfused as well as in ex-hypoxic polycythemic mice render their interpretation untenable.

The present experiments strongly support the concept that active ESF is generated in vitro when erythrogenin is incubated with serum. The cleavage of a part of the protein substrate in a manner similar to that of the renin-angiotensin II system may be involved in this reaction. The inability of anti-ESF to inactive erythrogenin or the serum substrate is in accord with this hypothesis and suggests that antigenic sites, probably on the serum substrate, become available to react with anti-ESF only after the action of erythrogenin.

It is interesting to speculate that a pro-ESF, erythropoietically inactive (erythropoietinogen) circulates normally in the plasma and that its conversion into active ESF is dependent not only on the availability of this precursor substance but also on erythrogenin, the production and/or release of which is triggered by hypoxia. In possible support of this concept are observations by Lowy and Keighley [24] that fractions, with similar chemical properties, can be obtained from human urine, some with high ESF activity from severely anemic patients and others with little or no ESF properties from normal subjects.

Relations of the Erythrogenin-Serum Reaction Concept to a Clinical State

The involvement of the erythrogenin-serum reaction in the biogenesis of ESF under normal and hypoxic conditions in animals suggests that clinical dyscrasias in erythropoiesis may exist that are associated with or even caused by changes in erythrogenin and substrate production and activity. Thus it is conceivable that certain clinical states of altered erythropoiesis may arise from a lack or excess of these factors or possibly from abnormalities in their basic chemical structure and activity. We intend to investigate this possibility in certain of the polycythemic and aplastic anemic states in man as well as in some of their apparent animal counterparts, e.g., murine polycythemia of viral origin [25] as well as the anemias in mice of genetic origin.[26]

The literature contains numerous examples of tumor states in man associated with erythrocytosis. These include some diseases of the kidney,[27–30] cerebellar hemangioblastoma,[31,32] uterine fibromas,[33] pheochromocytoma,[34] adrenal adenoma,[35] virilizing ovarian tumors,[36] and hepatomas.[37–40] Increased amounts of the ESF have been reported in the serum, plasma or urine and in the tissues or cyst fluid in some of these states. It has been suggested that increased production of the ESF by the tumor may represent the mechanism for the accompanying erythrocytosis.

We present here a case of hepatocellular carcinoma in man associated with erythrocytosis.[41] Attempts were made to determine whether or not alterations in the ESF-erythrogenin-serum substrate levels had occurred in this state.

Case Report

A 39-year old Negro male entered the Mount Sinai Hospital in New York City for the first time with a 5-week history of right-sided chest pain, cough, and hemoptysis. His liver was enlarged 3 cm. below the right costal margin. The spleen was not palpable. X-ray studies revealed numerous small masses in both lung fields and fluid at the right base. Intravenous pyelograms and aortogram were negative for renal tumor. A biopsy specimen of liver exhibited active cirrhosis; however, a scintiscan of this organ using colloidal ^{198}Au revealed a greatly reduced uptake over the entire lateral portion of the right lobe. At this time his hemoglobin level was 21.6 Gm. per cent; hematocrit, 64 per cent; RBC count, $7.9 \times 10^6/mm.^3$; and reticulocytes, 4.4 per cent. His white cell count, platelet numbers, leucocyte alkaline phosphatase, and blood chemistries were within the normal range.

Over the next two months, he was hospitalized on three occasions because of reaccumulation of pleural fluid but several thoracenteses did not reveal tumor cells. Although metastatic carcinoma remained his principal diagnosis, the primary site was not evident. At the end of his second month of illness, he was readmitted in severe respiratory distress, clinically cyanotic for the first time, and expired within 8 hours. Throughout his illness he maintained elevated hemoglobin and hematocrit values without an associated increase in his platelet or white cell count.

Autopsy Findings

The major pathology occurred in the liver and lungs. The liver was enlarged, weighing 2850 Gm. and the right lobe almost completely replaced by a tumor mass 18 cm. in diameter. The rest of the liver was finely nodular. There was a macroscopic invasion of the hepatic veins and inferior vena cava by tumor tissue. Both lungs were studded with firm nodules up to 3 cm. in diameter and the medium to small pulmonary arteries showed tumor emboli. Microscopically, the nontumorous portion of the liver contained thick fibrous bands extending from portal tracts to central veins, delimiting parenchymal nodules of varying size. The tumor consisted of bizarre anaplastic cells, which in some areas were continuous with parenchymal cords. Hepatoma cells in the lung were generally in the form of nests and sheets and some were surrounded by brown, bile-staining material. The kidneys showed glomerular congestion, but were not otherwise unusual. The spleen weighed 150 Gm. and contained no foci of extramedullary hematopoiesis. The bone marrow was moderately cellular and showed increased erythroid activity.

Methods

1. WHOLE TISSUE HOMOGENATES: One of the tumorous and one of the nontumorous portions of the liver were used for the extraction of the light-mito-

chondrial fraction, and the remaining parts were prepared as follows: The tissues were frozen overnight at -20 C., cut into smaller pieces while in the frozen state and homogenized in cold 0.02 M. phosphate buffer (pH 7) (2 ml./Gm. tissue) with a Waring blender. The mixture was frozen, thawed and centrifuged at 10,000 × g for 20 minutes. The supernatant fluid was filtered through glass wool and frozen again until used. A portion of the renal tissue was subjected to similar procedures.

2. LIGHT-MITOCHONDRIAL EXTRACTS: The following procedure [3,4] was used for the remaining portions of the liver, kidney, and liver tumor. The tissue, not permitted to freeze, was minced and homogenized in 0.25 M. sucrose solution (10 ml./Gm. tissue) in a Potter-Elvehjem homogenizer. The homogenate was centrifuged at 6300 × g for 10 minutes and the precipitate discarded. The supernatant was recentrifuged at 21,000 × g for 20 minutes. The sediment containing the light-mitochondrial fraction was resuspended by homogenization in cold distilled water (2 ml./Gm. tissue) in the presence of a drop of detergent and the mixture frozen for 48 hours. After thawing, the mixture was centrifuged at 37,000 × g for 30 minutes and the supernatant fluid (light-mitochondrial extract) was stored frozen until used.

3. INCUBATION TECHNIC: All incubations were carried out at 37 C. for 1 hour in a water bath incubator shaker as described above for the rat studies. In those cases in which the capacity of whole tissue homogenates to function as the substrate for erythrogenin was being studied, light-mitochondrial extracts from hypoxic rat kidneys, with known erythrogenin activity, were used in the incubations. These extracts were prepared from the kidneys of rats exposed to 0.40 atmospheres of air for 19 hours according to the procedure outlined above.

Incubation mixtures established to determine the effect of whole tissue homogenates on the activity of exogenous ESF contained 6 ml. of the homogenate and 0.6 IRP units of human urinary ESF dissolved in 6 ml. saline. These mixtures were also incubated at 37 C. for 1 hour.

4. BIOASSAY METHOD: The erythropoietic activity of plasma as well as all other preparations was determined in hypoxia-induced polycythemic mice. In these experiments, each mouse in the group received the entire dose of the test material as a single intraperitoneal injection on day 3 post-hypoxia. In cases where the activity of an incubation mixture was being determined, 2 ml. of the incubation mixture was injected. All other samples were administered as a 1 ml. dose. On day 5 the mice were given 0.5 μc. of radioiron via tail vein and the per cent red blood cell radioiron incorporation values determined 48 hours later.

Results

Table 1 indicates the presence of considerable erythrogenin activity in the kidneys of a patient with hepatoma. This activity was equivalent to that generally detectable in normal kidneys.[4] In addition, highly significant amounts of ESF were found in the patient's plasma, averaging 0.35 IRP units/ml.

From Table 1A it is evident that erythropoiesis stimulating activity was absent in whole homogenates and in light-mitochondrial extracts of the non-tumorous portion of the liver of the patient. Moreover, light-mitochondrial ex-

TABLE 1—*Effects of Liver, Tumor, and Kidney of Patient L.A. on Erythropoiesis (taken from Gordon et al.*[41])

Material Tested	% RBC-^{59}Fe Uptake ± S.E.M. *	IRP units of ESF
NHS† alone	1.00 ± 0.30	N.D.‡
Rat erythrogenin¶ alone	1.99 ± 0.34	N.D.
Human erythrogenin (normal) alone	1.63 ± 0.28	N.D.
Human erythrogenin (L.A.) alone	0.66 ± 0.12	N.D.
Plasma of L.A. alone	9.60 ± 1.16	0.13
Rat erythrogenin + NHS	9.14 ± 2.37	0.12
Human erythrogenin (normal) + NHS	7.01 ± 1.66	0.08
Human erythrogenin (L.A.) + NHS	6.53 ± 1.24	0.05

* Standard error of the mean.
† EDTA-dialyzed normal human serum.
‡ Nondetectable activity.
¶ Hypoxic rat erythrogenin.

tracts of the nontumorous portions of the liver did not generate ESF when incubated with normal human serum. In addition, incubation of the whole homogenate of the nontumorous portion of the liver with rat erythrogenin failed to engender production of the ESF. These results imply that nontumorous human liver tissue is unable to function as a substrate for erythrogenin nor can light-mitochondrial extracts of liver generate the ESF during incubation with normal serum.

Similarly, erythropoiesis stimulating activity was lacking in both the whole homogenate and in extracts of the light-mitochondrial fraction of the hepatoma. Note in Table 1A, however, that although the light-mitochondrial extract of the liver tumor did not evoke ESF production when incubated with normal

TABLE 1A—*Hepatic Erythropoietic Activity*

Material Tested	% RBC-^{59}Fe Uptake ± S.E.M.*	IRP units of ESF
LME† from liver (nontumorous) alone	0.87 ± 0.10	N.D.
LME from liver (nontumorous) + NHS	1.92 ± 0.46	N.D.
Whole homogenate of liver (nontumorous) alone	2.13 ± 0.92	N.D.
Whole homogenate of liver (nontumorous) + rat erythrogenin	1.86 ± 0.62	N.D.
Whole homogenate of liver (normal rat) + rat erythrogenin	2.16 ± 0.56	N.D.
LME from liver tumor alone	0.66 ± 0.11	N.D.
LME from liver tumor + NHS	1.72 ± 0.31	N.D.
Whole homogenate of liver tumor alone	1.61 ± 0.56	N.D.
Whole homogenate of liver tumor + rat erythrogenin	9.87 ± 1.41	0.13

* Standard error of the mean.
† Light-mitochondrial extract.

TABLE 1B—*Hepatic Inhibition of ESF Activity*

Material Tested	% RBC-^{59}Fe Uptake ± S.E.M.*	IRP units of ESF
0.10 Units of ESF† alone	10.88 ± 1.12	
0.10 Units of ESF + whole homogenate of liver (normal rat)	5.16 ± 1.09	0.04
0.10 Units of ESF + whole homogenate of liver (nontumorous, L.A.)	4.33 ± 0.89	0.03
0.10 Units of ESF + whole homogenate	11.14 ± 1.93	0.13

* Standard error of the mean.
† International units of ESF.

human serum, incubation of a whole homogenate of the liver tumor mass with rat erythrogenin resulted in the formation of highly significant quantities of the ESF. It would appear, therefore, that the hepatoma contains substrate for the erythrogenin but possesses no "erythrogenin-like" activity.

Of additional importance was the finding (Table 1B) that whereas incubation of whole homogenates of nontumorous portions of the liver with human urinary ESF resulted in a loss of approximately 50 per cent of ESF activity, addition of a similar weight of homogenate derived from the liver tumor to the same amount of urinary ESF was not attended by any inactivation of the hormone.

Discussion

An elevated level of plasma ESF accompanied the erythrocytosis in this patient with hepatoma. McFadzean et al.[42] also reported an increase in the quantity of ESF in the plasma of 3 to 4 patients with hepatocellular carcinoma and concomitant erythrocytosis. Although our patient displayed cyanotic symptoms terminally, these were not observed during the earlier course of his illness. If, however, the ESF produced was triggered largely as a result of supervening hypoxia, this would not explain the specific observation of increased substrate activity in only the liver tumor. Polycythemia vera is probably excluded by the absence of elevated platelet or white cell counts, normal leucocyte alkaline phosphatase activity, and the presence of a normal spleen at autopsy.

Several possibilities may be advanced for the elevated quantities of ESF in the plasma and the associated erythrocytosis in this patient. Since the ESF has been detected in tumor homogenates or cystic fluids in subjects with hypernephroma, cerebellar hemangioblastoma, and pheochromocytoma, it is conceivable that the tumors in these instances may actually be the sites of ESF origin. Despite the relatively high levels of ESF found in the plasma of the presently studied patient, no activity could be detected in the tumor homogenates. This conforms with other findings [42] that no ESF activity was present in tumor homogenates from 2 patients with hepatocellular carcinoma associated with erythrocytosis. Neither did light-mitochondrial extracts of the normal regions of liver nor the tumor display a capacity to generate ESF when incubated with normal serum. It was of interest to note, however, the ability of the whole tumor homogenate, but not that of nontumorous liver tissue, to function effectively as a substrate for rat erythrogenin.

Experiments performed in rabbits [43] and rats [44] suggest that the liver may be a site of ESF destruction. Of significance in the present study was the finding that although, as has been shown for normal liver,[12] incubation of homogenates of the nontumorous portion of liver with the ESF led to considerable inactivation of the hormone, homogenates of the hepatoma failed to alter the activity of added ESF. Since the entire right lobe of the liver in this patient was found to be almost completely replaced by the tumor, which lacked ESF-inactivating ability, decreased destruction could conceivably account, at least in part, for the increased quantities of circulating ESF.

It seems possible, therefore, that the heightened levels of plasma ESF in this patient may be attributed to an augmented production by the hepatoma of the substrate for erythrogenin (the erythrogenin was found to be normal in this patient). Whether or not a diminished capacity of the diseased liver to inactivate the ESF in vivo was also in part responsible for the increased quantity of circulating ESF is not clear.

Experiments are under way to determine the nature of the ESF-inactivating factor(s) present in normal liver and absent in the hepatoma as well as the identity of the substance in the hepatoma which has the capacity to serve as a substrate for erythrogenin.

SUMMARY

A renal erythropoietic factor (REF or erythrogenin) exists in extracts of light-mitochondrial and microsomal fractions of mammalian kidneys. Although not generally erythropoietically active when administered intraperitoneally alone, incubation of erythrogenin with normal serum in vitro results in the generation of erythropoiesis stimulating activity. When undialyzed serum is employed, loss of erythropoiesis stimulating activity occurs in the incubation mixtures. This inactivation phenomenon appears to depend on a component associated with the erythrogenin preparations and ions existent in serum.

Antibody effective in neutralizing the ESF is incapable of inactivating either erythrogenin or the serum substrate. This inability of anti-ESF to combine with erythrogenin or the serum substrate suggests that antigenic sites, most likely on the serum substrate, are enabled to react with anti-ESF only after the action of erythrogenin.

A case of hepatocellular carcinoma associated with erythrocytosis in man is described. In seeking the origin of the erythrocytosis, the ability of the nontumorous and tumorous portions of the liver to function either as erythrogenin or as substrate was determined. No erythrogenin-like activity was detectable in either the tumorous or nontumorous lobes of the liver. However, homogenates of the tumor, but not of the nontumorous portions, were capable of serving as substrate for erythrogenin. Although homogenates of nontumorous hepatic tissue inactivated exogenous ESF, this destructive influence was not apparent in homogenates of the tumor tissue. It is suggested that the erythrocytosis in this case of hepatocellular carcinoma may be a consequence of additional substrate furnished by the tumor for the erythrogenin-serum substrate system. Whether or not a decreased inactivation of endogenously produced ESF also contributes to the erythrocytotic state remains to be determined.

REFERENCES

1. Kuratowska, Z., Lewartowski, B., and Lipinski, B.: Chemical and biological properties of an erythropoietin generating substance obtained from perfusates of isolated anoxic kidneys. J. Lab. Clin. Med. 64:226, 1964.

2. Kuratowska, Z.: The renal mechanism of formation and inactivation of erythropietin. Ann. N.Y. Acad. Sci. 149:128, 1968.

3. Gordon, A. S., Cooper, G. W., and Zanjani, E. D.: The kidney and erythropoiesis. *In* Miescher, P. A. (Ed.): Seminars in Hematology, Vol. IV, No. 4. New York, Grune & Stratton, 1968, p. 337.

4. Gordon, A. S., and Zanjani, E. D.: Some aspects of erythropoietin physiology. *In* Gordon, A. S. (Ed.): Regulation of Hematopoiesis. New York, Appleton-Century-Crofts, 1970. In press.

5. Gordon, A. S., Zanjani, E. D. Mirand, E. A., and Back, N.: Biogenesis of erythropoietin. *In* Back, N., Martini, L., and Paoletti, R. (Eds.): Pharmacology of Hormonal Polypeptides and Proteins. New York, Plenum Press, 1968, p. 553.

6. Zanjani, E. D., Contrera, J. F., Gordon, A. S., Cooper, G. W., Wong, K. K., and Katz, R.: The renal erythropoietic factor (REF). III. Enzymatic role in erythropoietin production. Proc. Soc. Exp. Biol. Med. 125: 505, 1967.

7. Zanjani, E. D., Schooley, J. C. and Gordon, A. S.: Neutralization of *in vitro* generated erythropoietic activity by anti-erythropoietin. Life Sci. 7:505, 1968.

8. Leighton, F., Poole, B., Beaufay, H., Baudhin, P., Coffey, J. W., Fowler, S., and de Duve, C.: The large-scale separation of peroxisomes, mitochondria, and lysosomes from the livers of rats injected with Triton WR-1339. Improved isolation procedures, automated analysis, biochemical and morphological properties of fractions. J. Cell Biol. 37:482, 1968.

9. McDonald, T. P., Zanjani, E. D., Jordan, T., Lange, R. D., and Gordon, A. S.: Studies of the renal erythropoietic factor using a hemagglutination inhibition system. Proc. Soc. Exp. Biol. Med. 132:533, 1969.

10. Schooley, J. C., Zanjani, E. D., and Gordon, A. S.: Immunologic aspects of the biogenesis of erythropoietin. Blood. In press.

11. Malgor, L. A., and Fisher, J. W.: Influence of testosterone and hypoxia on erythropoietin production in isolated perfused dog kidneys. Amer. J. Physiol. In press.

12. Erslev, A. J., and Kazal, L. A.: Inactivation of erythropoietin by tissue homogenates. Proc. Soc. Exp. Biol. Med. 129:845, 1968.

13. Fisher, J. W., Hatch, F. E., Roh, B. L., Allen, R. C., and Kelley, B. J.: Erythropoietin inhibitor in kidney extracts and plasma from anemic uremic human subjects. Blood 31:440, 1968.

14. Kuratowska, Z.: The renal mechanism of the formation and inactivation of erythropoietin. Ann. N.Y. Acad. Sci. 149:128, 1968.

15. Contrera, J. F., and Gordon, A. S.: Erythropoietin: Production by a particulate fraction of rat kidney. Science 152:653, 1966.

16. Erslev, A. J., and Kazal, L. A.: Renal Erythropoietic Factor: Lack of effect on hypertransfused mice. Blood 34:222, 1969.

17. Weintraub, A. H., Gordon, A. S., and Camiscoli, J. F.: Use of the hypoxia-induced polycythemic mouse in the assay and standardization of erythropoietin. J. Lab. Clin. Med. 62:743, 1963.

18. Garcia, J. F., and Schooley, J. C.: Dose-response relationships for erythropoietin given in serum or in saline. Proc. Soc. Exp. Biol. Med. 120:614, 1965.

19. Gardner, E., Jr., Wright, C-S., Lewis, J. P., and Moores, R. R.: Potentiation of purified erythropoietin with serum proteins. Brit. J. Haemat. 13:317, 1967.

20. Schooley, J. C., and Garcia, J. F.: Immunological studies of human urinary erythropoietin. Proc. Soc. Exp. Biol. Med. 109:325, 1962.

21. Garcia, J. F., and Schooley, J. C.: Immunologic neutralization of various erythropoietins. Proc. Soc. Exp. Biol. Med. 112:712, 1963.

22. Schooley, J. C., Garcia, J. F., Cantor,

L. N., and Havens, V. W.: A summary of some studies on erythropoiesis using anti-erythropoietin immune serum. Ann. N.Y. Acad. Sci. 149:286, 1968.

23. Schooley, J. C., and Garcia, J. F.: Some properties of serum obtained from rabbits immunized with human urinary erythropoietin. Blood 25:204, 1965.

24. Lowy, P. H., and Keighley, G.: The fractionation of erythropoietin from anemic and normal human urine. Ann. N.Y. Acad. Sci. 149:54 1968.

25. Mirand, E. R.: Murine viral-induced polycythemia. Ann. N.Y. Acad. Sci. 149:486, 1968.

26. Russell, E. S.: Abnormalities of erythropoiesis associated with mutant genes in mice. *In* Gordon, A. S. (Ed.): Regulation of Hematopoiesis. New York, Appleton-Century-Crofts, 1970. In press.

27. Forssell, J.: Polycythemia in a patient with hypernephroma. Nord. Med. 30:1415, 1946.

28. Damon, A., Holub, D. A., Melicor, M., and Uson, A. C.: Polycythemia and renal carcinoma: Report of ten new cases, two with long hematologic remission following nephrectomy. Amer. J. Med. 25:182, 1958.

29. Hewlett, J. S., Hoffman, G. S., Senhauser, D. A., and Battle, J. D.: Hypernephroma with erythrocythemia. Report of a case and assay of the tumor for erythropoietic-stimulating substance. New Eng. J. Med. 262:1058, 1960.

30. Lange, R. D., and Gallagher, N. I.: Clinical and experimental observations on the relationship of the kidney to erythropoietin production. *In* Jacobson, L. O., and Doyle, M. (Eds.): Erythropoiesis. New York, Grune & Stratton, 1962, p. 361.

31. Waldmann, T. A., Levin, E. H., and Baldwin, M.: The association of polycythemia with a cerebellar hemangioblastoma. The production of an erythropoiesis stimulating factor by the tumor. Amer. J. Med. 31:318, 1961.

32. Hennessy, T. G., Stern, W. E., and Herrick, S. E.: Cerebellar hemangioblastoma: Erythropoietic activity by radioiron assay. J. Nucl. Med. 8:601, 1967.

33. Hertko, E. J.: Polycythemia (erythrocytosis) associated with uterine fibroids and apparent surgical cure. Amer. J. Med. 34:288, 1963.

34. Bradley, J. E., Young, J. D., and Lentz, G.: Polycythemia secondary to pheochromocytoma. J. Urol. 86:1, 1961.

35. Mann, D. L., Gallagher, N. I., and Donati, R. M.: Erythrocytosis and primary aldosteronism. Ann. Intern. Med. 66:335, 1967.

36. Kepler, E. J., Dockerty, M. B., and Priestly, J. T.: Adrenal-like ovarian tumor associated with Cushing's snydrome (so-called masculinovoblastoma, luteoma, hypernephroma, adrenal cortical carcinoma of the ovary). Amer. J. Obstet. Gynec. 47:43, 1947.

37. McFadzean, A. J. S., Todd, D., and Tsang, K. C.: Polycythemia in primary carcinoma of the liver. Blood 13:427, 1958.

38. Donati, R. M., McCarthy, J. M., Lange, R. D., and Gallagher, N. I.: Erythrocythemia and neoplastic tumors. Ann. Intern. Med. 58:47, 1963.

39. Lehman, C. J., Erslev, A. J., and Myerson, R. M.: Erythrocytosis associated with hepatocellular carcinoma. Amer. J. Med. 35:439, 1963.

40. Brownstein, M. H., and Balliard, H. S.: Hepatoma associated with erythrocytosis: Report of eleven new cases. Amer. J. Med. 40:204, 1966.

41. Gordon, A. S., Zanjani, E. D., and Zalusky, R.: A possible mechanism for the erythrocytosis associated with hepatocellular carcinoma in man. Blood. 35:151, 1970.

42. McFadzean, A. J. S., Todd, D., and Tso, S. C.: Erythrocytosis associated with hepatocellular carcinoma. Blood 29:808, 1967.

43. Jacobsen, E. M., Davis, A. K., and Alpen, E. L.: Relative effectiveness of phenylhydrazine treatment and hemorrhage in the production of an erythropoietic factor. Blood 11:937, 1956.

44. Prentice, T. C., and Mirand, E. A.: Effects of acute liver damage plus hypoxia on plasma erythropoietin content. Proc. Soc. Exp. Biol. Med. 94:231, 1957.

DISCUSSION

E. A. Mirand, M.D.: Our laboratory has shown for man, baboons, and rodents the existence of extra-renal erythropoietin. Therefore, it is important that one has to postulate the existence of an extra-renal site for erythrogenin.

In response to Dr. Gurney's question as to whether or not one sees a high level of a plasma factor in a polycythemic state, we can say that in viral polycythemia, the plasma factor as well as erythrogenin was not elevated in this disease. However, it is important to emphasize, that we have shown that the viral polycythemia was not erythropoietin dependent.

Oxygen Delivery by Abnormal Hemoglobins: Effects on Erythropoietin Production

By JOHN W. ADAMSON

ABNORMAL HEMOGLOBULINS WITH ERYTHROCYTOSIS

THE AVAILABILITY OF OXYGEN for cellular metabolism has long been recognized as the primary factor responsible for the regulation of erythropoiesis.[1] Specifically, some site within the kidney probably behaves as an oxygen sensor and controls erythropoiesis in the normal state by modulating the production or release of erythrogenin and, thereby, erythropoietin.[2,3] Defects in the oxygen delivery system classically increase erythropoietin production which, in turn, increases the effective production of red cells when the target organ is intact and responsive. Recently, alterations in oxygen delivery, apparently mediated by specific biochemical abnormalities of the hemoglobin molecule itself, have received a good deal of attention by a number of workers. Beginning with the clear description by Charache and co-workers of the association of familial erythrocytosis with hemoglobin Chesapeake [4] a total of 5 abnormal hemoglobins have been completely characterized and shown to be associated with erythrocytosis [5–9] (Table I). Hemoglobins Ypsilanti and J Capetown probably will be added to this list; however, the amino acid substitution of Ypsilanti is uncertain,[10] and, while J Capetown is substituted at the same site as hemoglobin Chesapeake [11] (position 92 of the alpha chain) and is characterized by increased oxygen affinity,[12] it is not yet clear that affected family members have significant erythrocytosis.[11,13] What is certain, however, is that a number of other such hemoglobins exist and merely await complete description. As reviewed by Perutz [14,15] and Weatherall,[16] the element common to all of these mutant hemoglobins is the fact that the amino acid substitutions are at or spatially near the FG segment and, thus, these substitutions are in the $\alpha_1\beta_2$ contact region, a location which would teleologically enhance their influence on configurational and functional changes in the hemoglobin molecule.

Recently, studies have appeared which relate erythropoietin production to erythropoiesis in members of families bearing abnormal hemoglobins associated with erythrocytosis.[17] These studies were conducted on members of families having either hemoglobin Yakima or Rainier. In each case, as shown in Figure 1, the whole blood oxygen-hemoglobin dissociation curve was shifted to the left of normal. In addition, there was a loss of the normal sigmoid shape of the dissociation curve. The P_{50}'s, or characteristic partial pressure of oxygen at which 50 per cent of the hemoglobin is saturated, were 12.0 (37 C.), 12.8 (38 C.), and 26 mm. Hg. (38 C.) for hemoglobins Yakima,[6] Rainier, and A,[17] respectively. Affected members of these families were characterized by having a stable but elevated red cell mass and hematocrit. Clinically, it appeared that a

This work was supported in part by U.S. Public Health Service Research grant nos. GM–15253, 5–RO1–HE–06242 and Clinical Research Center Facility grant no. FR–37.

TABLE 1—*Mutant Hemoglobins Associated with Erythrocytosis*

HEMOGLOBIN	SUBSTITUTION
Chesapeake	α^{92} Arginine ⟶ Leucine
Yakima	β^{99} Aspartic acid ⟶ Histidine
Kempsey	β^{99} Aspartic acid ⟶ Asparagine
Rainier	β^{145} Tyrosine ⟶ Histidine
Hiroshima	β^{143} Histidine ⟶ Aspartic acid

new equilibrium state had been reached, the increased amount of circulating hemoglobin presumably compensating for the increased oxygen affinity. If this were indeed the case, one might anticipate that erythropoietin production, as reflected by excretion, would be normal. It would follow, then, that altering the oxygen-carrying capacity by lowering the hemoglobin level would set in motion mechanisms to correct the state of tissue hypoxia; specifically that

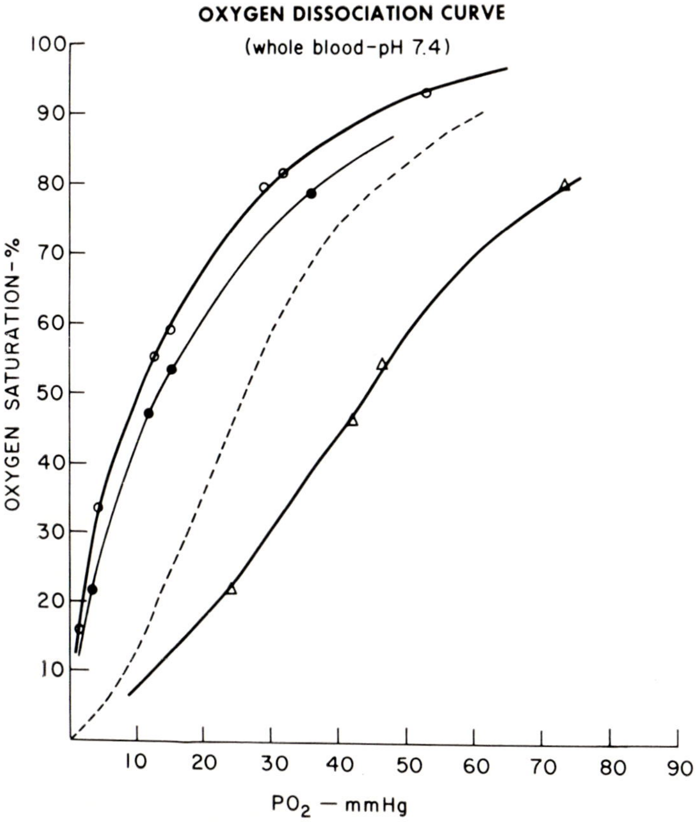

FIG. 1—Whole blood oxygen dissociation curves for hemoglobins Yakima (○), Rainier (●) and Seattle (Δ) compared to the normal adult human hemoglobin (dotted line).

measurements of both erythropoietin and marrow production would increase.

To test this relationship, measurements of circulating reticulocytes [18] (corrected for hematocrit to provide an index of marrow production [19]) and urinary erythropoietin excretion [20] were obtained for 3 days preceding and 5 days following an acute erythropheresis. The exchange phlebotomies were designed to lower the red cell mass by approximately 20 per cent. This would make the stimulus comparable between these individuals and normal subjects previously studied in the same manner. The results of these experiments are shown in Figures 2A and 2B. Baseline levels of erythropoietin output in the subject with hemoglobin Yakima averaged 2.4 Standard B units per day, within the laboratory's normal range of approximately 4 ± 2 units per day.[20,21] Following phlebotomy there was a clear rise in the reticulocyte index as well as an increase in erythropoietin output with a peak value of 12.9 units per day. Figure 2B depicts similar data obtained on the proposita with hemoglobin Rainier. Baseline erythropoietin excretion in this individual averaged 3.0 units per day, again clearly within the normal range. As in the first study, the response to phlebotomy was manifested by an increase in erythropoietin excretion and reticulocyte production, the latter confirming the physiologic significance of the hypoxic stress. At their peaks, the erythropoietin response to phlebotomy was equivalent to that seen in normal subjects in whom the hematocrit had been lowered to the range of 30 to 34 per cent. Figure 3 shows the average baseline and acute phlebotomy response data plotted simultaneously for normal subjects and for individuals having hemoglobins Rainier or Yakima. In addition, the average erythropoietin excretion is shown for the propositus with hemoglobin Yakima, whose hematocrit had been stabilized at a level of 47.5 per cent. Urinary erythropoietin averaged 9.2 units per day in this individual, over 2 standard deviations above the normal mean.

Thus, in the conditions of erythrocytosis associated with abnormally functioning hemoglobin, the curve relating erythropoietin response to phlebotomy is shifted to the right of normal—yet another example of a physiologic "shift," in this instance reflecting the adequacy of oxygen delivery to the site of the renal sensor at the induced hematocrit level.

Abnormal Hemoglobins with Anemia

If the concept of changes in hemoglobin structure altering oxygen delivery in vivo is a valid one, it should be possible to extend these observations to the other side of the normal hematocrit/erythropoietin response curve—in other words, document a "shift to the left" of the curve. Recently published studies by Finch and colleagues have shown this to be the case.[22] The molecular prototype of this model is hemoglobin Seattle, whose oxygen dissociation curve is shown in Figure 1. This hemoglobin, which has glutamic acid substituted for alanine at position 76 of the beta chain, is characterized by a markedly decreased oxygen affinity, moderate shortening of the red cell survival and a chronic, stable anemia. The P_{50} of whole blood in hemoglobin Seattle is 43.0 mm. Hg. (38 C.). Hematologic data for the proposita with hemoglobin Seattle and her 2 affected sons are presented in Table 2. Reflecting

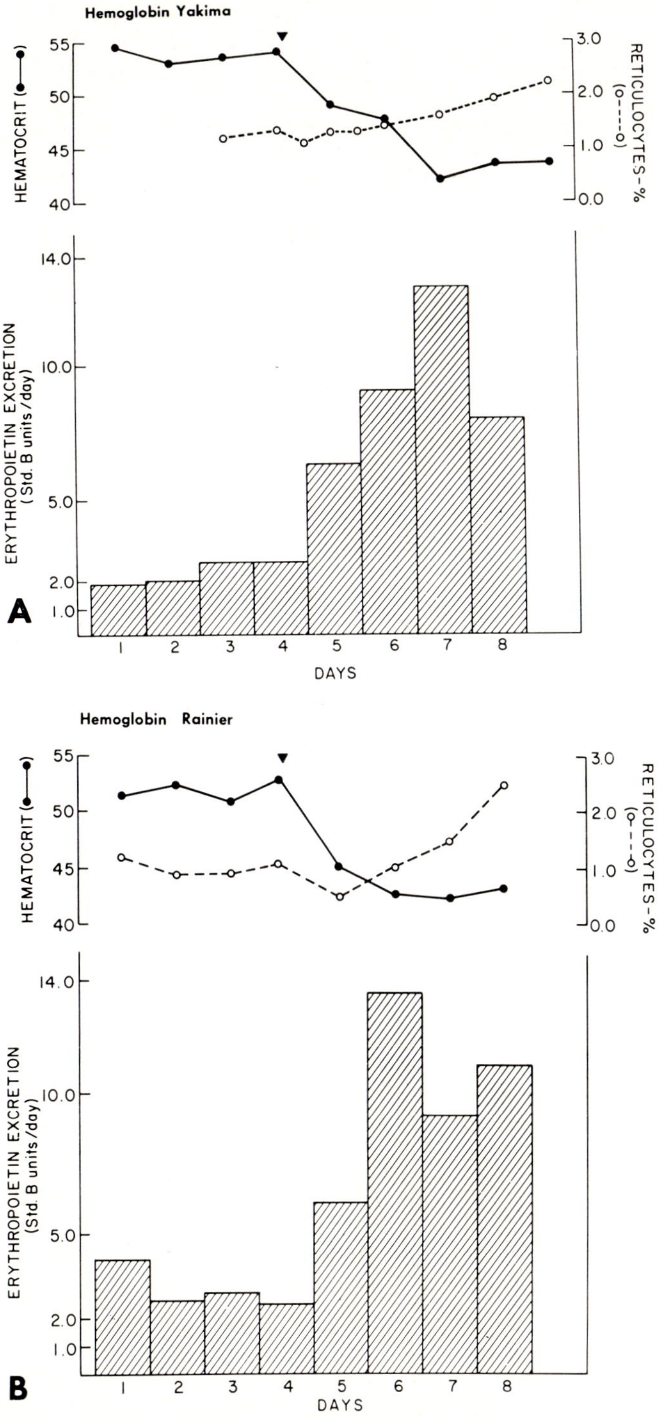

FIG. 2—Daily observations of hematocrit, corrected reticulocyte count, and erythropoietin excretion in subjects with hemoglobin Yakima and Rainier. (From Adamson et al.[17])

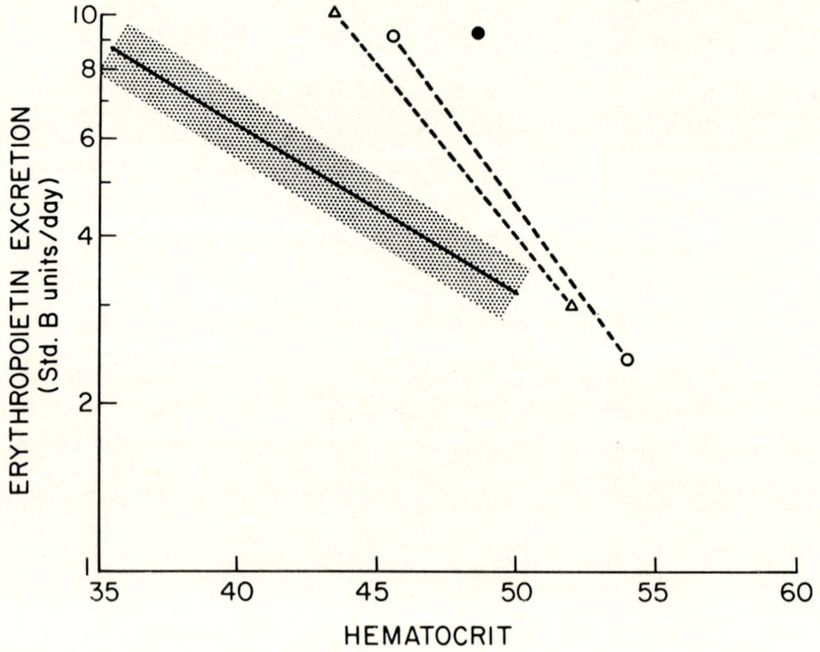

FIG. 3—Average erythropoietin excretion pre- and post phlebotomy in subjects with hemoglobin Yakima (○, ●) and Rainier (Δ). Shaded area represents results obtained in 12 normal subjects similarly studied. (From Adamson et al.[17])

total erythropoiesis, both the rate of reticulocyte production and iron kinetic studies were in agreement, yielding values of about twice normal. Previous studies have shown that in the absence of any specific limitation to proliferation, the marrow is able to increase erythroid productivity to 2 to 3 times normal within two weeks' time and to an even greater degree when presented with a more chronic stress.[23] Consequently, while there is a shortened red cell survival in subjects with hemoglobin Seattle, it is not of sufficient magnitude to explain the degree of anemia seen in the affected family members. Insight into the mechanism of the anemia was provided by demonstrating that erythropoietin excretion was within normal limits, indicating no appreciable

TABLE 2—*Hemoglobin Seattle*

SUBJECT	AGE/SEX	HCT.	RETIC (%)	PIT*	ESF Units/day
M.S.	33/F	32.5	2.0	1.21	1.5
R.S.	10/M	24.0	1.7		3.8
S.S.	12/M	25.0	2.1		3.3

*Plasma iron turnover (PIT) as mgm/100 ml whole blood/day.

deficiency in oxygen supply to the renal sensor. At hematocrit values similar to these subjects, as indicated in Figure 4, normal control subjects would have erythropoietin excretion levels on the order of 10 to 20 units per day. Figure 4 summarizes our experience relating hematocrit to erythropoietin excretion in normal subjects made chronically anemic. The data comprising this figure were gathered from 3 individuals who had been made anemic by phlebotomy and allowed to stabilize for a period of 2 to 3 weeks prior to returning to normal and one patient with chronic iron deficiency anemia who was serially studied during a course of iron replacement therapy. Thus, the relationship of erythropoietin excretion to hematocrit in these individuals differs from the curve of erythropoietin response to acute blood loss [21] by being displaced to the left. The erythropoietin values observed in the subjects with hemoglobin Seattle, Yakima, and Rainier are shown in Figure 5 and compared to normal. Because of the low hematocrit levels in hemoglobin Seattle, phlebotomy and transfusion experiments were not conducted and, thus, no data were obtained characterizing erythropoietin excretion over a wide hematocrit range.

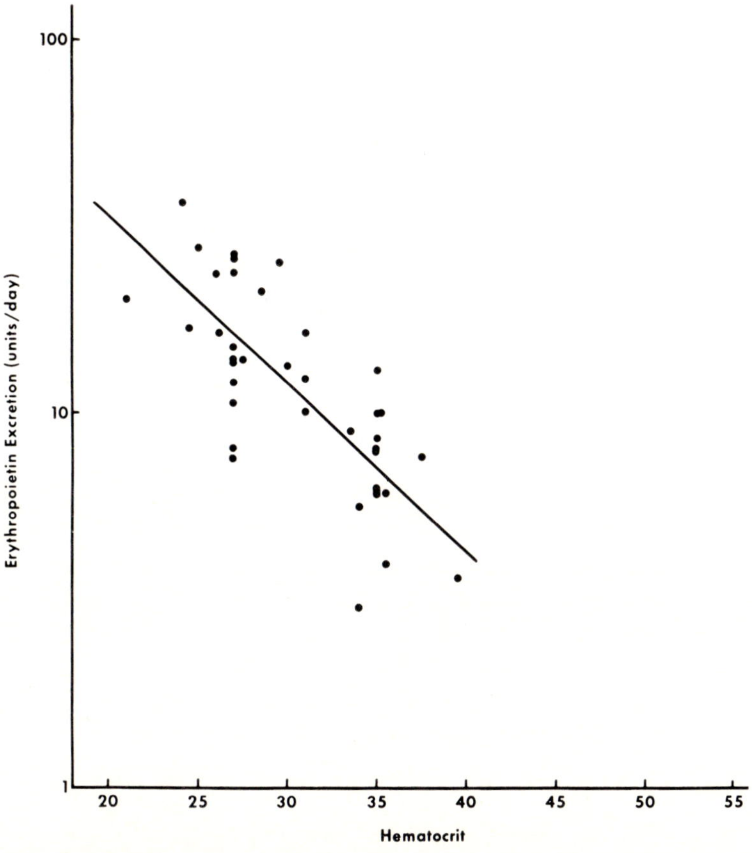

FIG. 4—Relationship of erythropoietin excretion to hematocrit in 4 subjects studied daily while recovering from phlebotomy-induced or iron deficiency anemia.

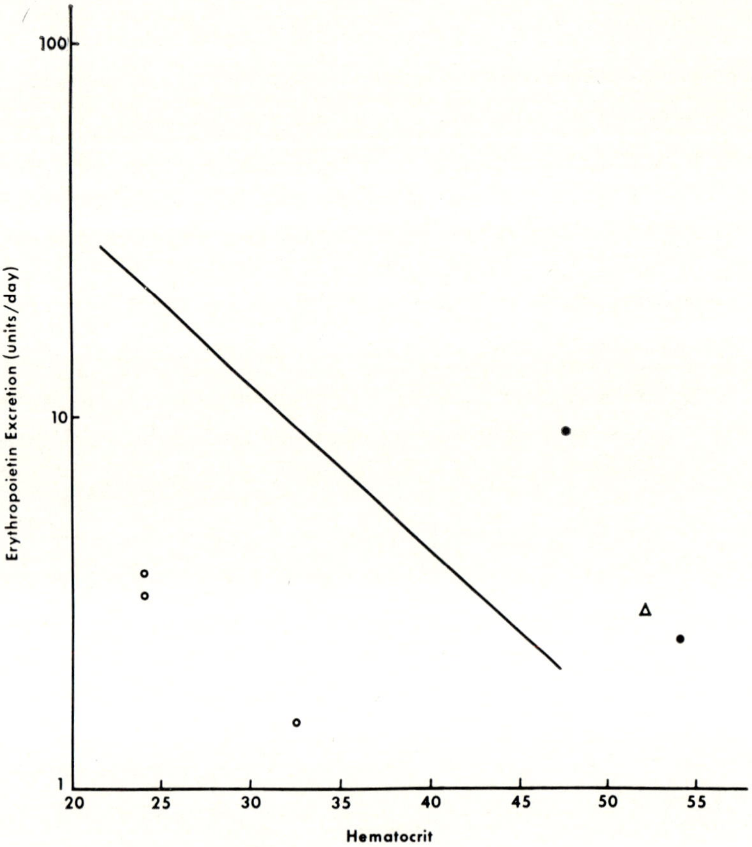

FIG. 5—Erythropoietin excretion for normal subjects and subjects with hemoglobins Seattle (○), Yakima (●), and Rainier (Δ).

CONSIDERATIONS

In these cases it is useful to consider those factors contributing to oxygen delivery to the site within the kidney which regulates erythrogenin production. While this may be exceedingly complex, there are some interesting observations which contribute to the understanding or a definition of what may constitute "compensation" or normalization of oxygen delivery. Figure 6 is a composite of the hemoglobins studied which relates the oxygen content of the blood to partial pressure of oxygen. For the purposes of this figure, hemoblobins Rainier and Yakima may be considered to behave virtually identically. Fundamental to understanding the situation in humans is the fact that the mixed venous oxygen tension in a wide variety of mammalian species, including man, is virtually identical. Parer and co-workers clearly demonstrated this fact in mammals having markedly different oxygen affinities, oxygen consumption rates, and whose body weights varied by over two thousand fold.[24] Included in these studies were observations by Dawson and Evans of 2 groups of sheep whose characteristic hemoglobins have quite different oxygen affinity.[25] In each instance there appeared to be a matching of oxygen affinity and carrying capacity to provide oxygen delivery. In examining the similarities between previous studies [26] and the observations to date on humans having abnormally functioning hemoglobins, one must make the assumption that in the resting state the arterial-venous oxygen difference is within the normal range (4 to 5 volumes per cent) and that the cardiac output is normal. Indeed, this has been shown to be the case in family members bearing hemoglobin Yakima. If one makes these assumptions for the mutant hemoglobins considered here, then there is an area depicted in the figure common to all; that area is the range of the normal mixed venous partial pressure of oxygen. With these mutant hemoglobins of man, and in a constant state, oxygen delivery is related directly to oxygen affinity. Factors which lower the level of hemoglobin, such as phle-

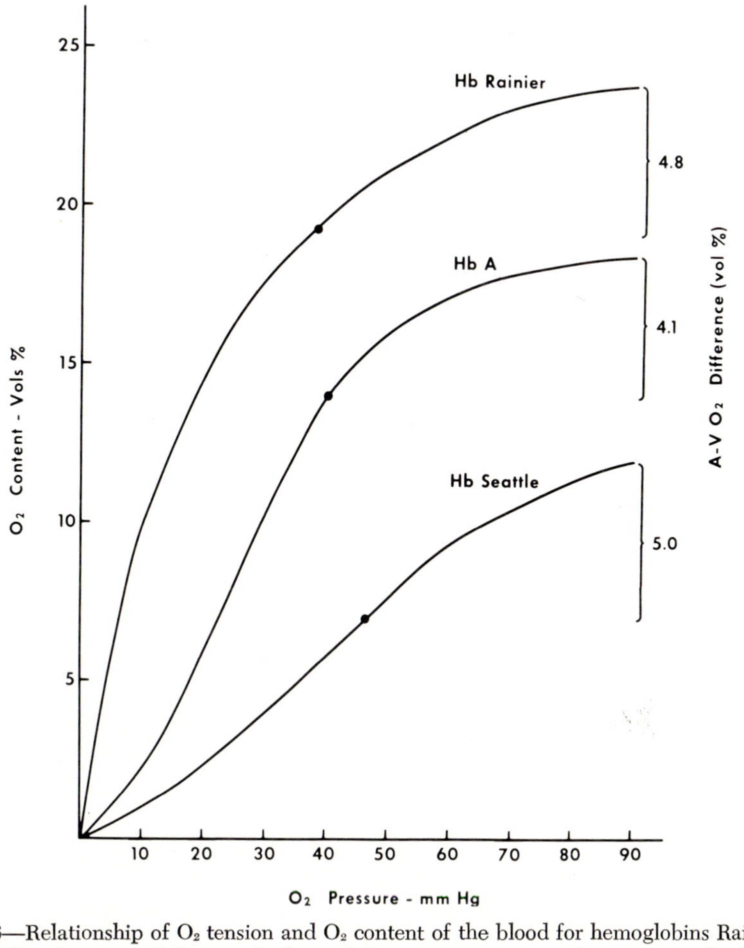

FIG. 6—Relationship of O_2 tension and O_2 content of the blood for hemoglobins Rainier, A, and Seattle. The A-V_{O_2} difference for hemoglobin Rainier is assumed to be virtually the same as that for hemoglobin Yakima[6]; the value for hemoglobin A is from Bartels et al.[28] The dots (●) represent either the known or calculated mixed venous O_2 tension for each hemoglobin.

botomy, would thereby decrease the oxygen content of the blood and the oxygen carrying capacity. Maintenance of a normal or constant tissue oxygen supply would necessitate at least a normal arterial-venous oxygen difference and, therefore, would produce a lowering of the mixed venous partial pressure of oxygen. Physiologically, this would be viewed as an hypoxic stress, sensed as such by the kidney, and be accompanied by increased erythropoietin production.

Although no direct studies have been conducted on members of the family bearing hemoglobin Kansas, extrapolation of known data, provided that cardiac output and oxygen consumption were normal, produces similar results as with the hemoglobins reported here. Although the oxygen affinity curve of hemoglobin Kansas is shifted markedly to the right of normal, there is suboptimal hemoglobin oxygen saturation while breathing room air, producing cyanosis in the propositus.[27] Based on the observation that the hemoglobin was only 60 per cent saturated under resting conditions, one could predict that in order to achieve a normal mixed venous oxygen tension, a hemoglobin level of approximately 13 to 14 Gm. per cent would be necessary. This indeed was the situation.

In summary, then, the following seem consistent: (1) In the resting, stable state, oxygen affinity and carrying capacity appear matched to provide adequate delivery of oxygen to tissues, both in a variety of mammalian species as well as in man with mutant hemoglobins which are functionally abnormal. (2) The adequacy of the oxygen delivery system appears best correlated with the mixed venous oxygen tension and the excretion of erythropoietin. (3) Further insight into this relationship may be provided by the direct correlation of erythropoietin production to the mixed venous partial pressure of oxygen or, more specifically, to the partial pressure of oxygen in renal venous blood.

REFERENCES

1. Grant, W. C., and Root, W. S.: Fundamental stimulus for erythropoiesis. Physiol. Rev. 32:449, 1952.

2. Gordon, A. S., Cooper, G. W., and Zanjani, E. D.: The kidney and erythropoiesis. Seminars Hemat. 4:337, 1967.

3. Gordon, A. S.: Hormonal relations to erythropoiesis. XII Congr. Int. Soc. Hemat., 288, 1968.

4. Charache, S., Weatherall, D. J., and Clegg, J. B.: Polycythemia associated with a hemoglobinopathy. J. Clin. Invest. 45:813, 1966.

5. Jones, R. T., Osgood, E. E., Brimhall, B., and Koler, R. D.: Hemoglobin Yakima. I. Clinical and biochemical studies. J. Clin. Invest. 46:1840, 1967.

6. Novy, M. J., Edwards, M. J., and Metcalfe, J.: Hemoglobin Yakima. II. High blood oxygen affinity associated with compensatory erythrocytosis and normal hemodynamics. J. Clin. Invest. 46:1848, 1967.

7. Reed, C. S., Hampson, R., Gordon, S., Jones, R. T., Novy, M. J., Brimhall, B., Edwards, M., and Koler, R. D.: Erythrocytosis secondary to increased oxygen affinity of a mutant hemoglobin, hemoglobin Kempsey. Blood 31:623, 1968.

8. Stamatoyannopoulos, G., Yoshida, A., Adamson, J. W., and Heinenberg, S.: Hemoglobin Rainier (β^{145} tyrosine $\rightarrow$ histidine): Alkali resistant hemoglobin with increased oxygen affinity. Science 159:741, 1968.

9. Hamilton, H. B., Iuchi, I., Miyaji, T., and Shibata, S.: Hemoglobin Hiroshima (β^{143} histidine $\rightarrow$ aspartic acid): A newly identified fast moving beta chain variant associated with increased oxygen affinity and compensatory erythremia. J. Clin. Invest. 48: 525, 1969.

10. Glynn, K. P., Penner, J. A., Smith, J. R., and Rucknagel, D. L.: Familial erythrocytosis: Description of three families, one with hemoglobin Ypsilanti. Ann. Intern. Med. 69:769, 1968.

11. Botha, M. C., Beale, D., Isaacs, W. A., and Lehmann, H.: Hemoglobin J Capetown—α_2 92 argenine → glutamine β_2. Nature (London) 212:792, 1966.

12. Lines, J. G., and McIntosh, R.: Oxygen binding by haemoglobin J-Cape Town (α_2 argenine → glutamine). Nature (London) 215:297, 1967.

13. Jenkins, T., Stevens, K., Gallo, E., and Lehmann, H.: A second family possessing hemoglobin J Capetown. S. Afr. Med. J. 42:1151, 1968.

14. Perutz, M. F., Kendrew, J. C., and Watson, H. C.: Structure and function of haemoglobin. II. Some relations between polypeptide chain configuration and amino acid sequence. J. Molec. Biol. 13:669, 1965.

15. Perutz, M. F., and Lehmann, H.: Molecular pathology of human hemoglobin. Nature (London) 219:902, 1968.

16. Weatherall, D. J.: Polycythemia resulting from abnormal hemoglobins. New Eng. J. Med. 280:604, 1969.

17. Adamson, J. W., Parer, J. T., and Stamatoyannopouplos, G.: Erythrocytosis associated with hemoglobin Rainier: Oxygen equilibria and marrow regulation. J. Clin. Invest. 48:1376, 1969.

18. Brecher, G., and Schneiderman, M.: A time-saving device for the counting of reticulocytes. Amer. J. Clin. Path. 20:1079, 1950.

19. Giblett, E. R., Coleman, D. H., Pirzio-Biroli, G., Donohue, D. M., Motulsky, A. G., and Finch, C. A.: Erythrokinetics: Quantitative measurements of red cell production and destruction in normal subjects and patients with anemia. Blood 11:291, 1956.

20. Adamson, J. W., Alexanian, R., Martinez, C., and Finch, C. A.: Erythropoietin excretion in normal man. Blood 28:354, 1966.

21. Adamson, J. W.: The erythropoietin/hematocrit relationship in normal and polycythemic man: Implications of marrow regulation. Blood 32:597, 1968.

22. Stamatoyannopoulos, G., Parer, J. T., and Finch, C. A.: Implications of a hemoglobin with decreased oxygen affinity. New Eng. J. Med. 281:915, 1969.

23. Coleman, D. H., Stevens, A. R., Dodge, H. T., and Finch, C. A.: Rate of blood regeneration after blood loss. Arch. Intern. Med. (Chicago) 92:341, 1953.

24. Parer, T. T., and Metcalfe, J.: Oxygen transport by blood in relation to body size. Nature (London) 215:653, 1967.

25. Dawson, T. J., and Evans, J. V.: Effect of hemoglobin type on the cardiorespiratory system of sheep. Amer. J. Physiol. 209:593, 1965.

26. Parer, J. T., Jones, W. D., and Metcalfe, J.: A quantitative comparison of oxygen transport in sheep and human subjects. Resp. Physiol. 2:196, 1967.

27. Reissmann, K. R., Ruth, W. E., and Nomura, T.: A human hemoglobin with lowered oxygen affinity and impaired heme-heme interaction. J. Clin. Invest. 40:1826, 1961.

28. Bartels, H., Beer, R., Fleischer, E., Hoffheinz, H. J., Krall, J., Rodenwald, G., Wenner, J., and Witt, I.: Bestimmung von Kurzschlussdurchblutung und Diffusionskapazität der Lunge bei Gesunden und Lungenkranken. Pflüger. Arch. Ges. Physiol. 261:99, 1955.

The Hemagglutination-Inhibition Assay for Erythropoietin: A Progress Report

By Robert D. Lange, Ted P. McDonald, Tommye Ann Jordan, Frank E. Trobaugh, Jr., Arthur L. Kretchmar, and Amoz I. Chernoff *

Introduction

Immunochemical methods are potentially valuable as a means of assaying erythropoietin (ESF). An essential element of the technic is the production of a suitable antiserum. Recent evidence from our laboratory indicates that antibodies to erythropoietin are, at the least, of two types.[1] The first type of antibody causes neutralization of the biologic activity of the hormone and the second brings about hemagglutination of red blood cells (RBC) sensitized to ESF. ESF preparations were found to inhibit the hemagglutination reaction. Since the amount of test material which inhibits the reaction can be compared to the amount of standard which causes the same degree of inhibition, the quantity of ESF in test materials can be calculated from a calibration curve.

This paper reports a number of experiments that were performed to measure the sensitivity, reproducibility, specificity, and applicability of the hemagglutination-inhibition (HAI) technic for the assay of erythropoietin.

Methods and Materials

The methods used for immunization, hemagglutination and biologic assay of ESF have been described.[1–3] ESF preparations for immunization, cell sensitization, and laboratory standards were prepared by diethylaminoethyl column chromatography. The International Reference Preparation of ESF was obtained from the Medical Research Council, National Institute of Medical Research, London.

Rabbits were immunized by a variety of methods. The procedure currently in use in our laboratory employs denatured antigen complexed with methylated rabbit albumin as previously outlined[3] and represents a modification of Plescia's immunization technic.[4]

The HAI test method has been presented.[1–3] The microtiter system of Sever,[5] based on the original technic introduced by Takatsy[6] was utilized. Human group O RBC are now used and the phosphate buffer is employed without the addition of normal rabbit serum.

The biologic assay of ESF was made indirectly by measuring the incorpora-

From the University of Tennessee Memorial Research Center and Hospital, Knoxville, Tennessee and Presbyterian-St. Luke's Hospital, Chicago, Illinois.

Supported in part by United States Public Health Service grant No. HE 10567–04 from the National Heart Institute, contract no. AT–(40–1)–3547 from the Atomic Energy Commission, and United States Public Health Service Research grant CA–04144 awarded by the National Cancer Institute to Presbyterian-St. Luke's Hospital.

* Dr. Chernoff is recipient of Research Career Award No. 5-K6-GM-3756 from National Institutes of General Medical Sciences.

tion of ^{59}Fe in RBC of mice made polycythemic in silicone rubber membrane enclosures.[7,8]

The sera of normal blood bank donors were obtained from the Knoxville Blood Center. The pathologic sera were obtained from patients under study in Presbyterian-St. Luke's Hospital, Chicago, and the University of Tennessee Memorial Research Center and Hospital, Knoxville. The usual clinical and laboratory criteria were used to make diagnoses.

Results

Table 1 contains a list of some of the ESF preparations which have been found to inhibit the hemagglutination of red blood cells sensitized with ESF. The degree of agreement between the biologic units measured in the polycythemic mouse assay and immunochemical units determined by the HAI technic should be noted. Both plasma and urine ESF sources were assayable by the HAI method.

The agreement between determinations performed with dilutions of an ESF extract made from human urine and similar dilutions of an ESF rich serum are shown in Figure 1. Both sources had the same reaction slope and the inverse correlation between the amount of ESF present and the quantity of material which first allowed 3 plus hemagglutination is readily apparent.

Results of experiments performed by adding known amounts of a laboratory standard ESF preparation to normal human serum and measuring the quantity recovered are recorded in Figure 2, which indicates an almost quantitative recovery of the added hormone.

The accuracy and reproducibility of the HAI method were tested by performing 10 replicate determinations of the amount of ESF present in each of 8 dilutions of the same laboratory standard. The quantity of ESF which first allowed 3 plus hemagglutination was found to be the same in each

TABLE 1—*List of Extracts Causing the Inhibition of Hemagglutination of ESF Sensitized Red Blood Cells*

Extract Number	Source	ESF/ml. i *	ESF/ml. b †
IRP ‡	Human urine	0.25	0.20
		0.50	0.50
26 B_3	Human urine	1.27	3.56
29 B_3	Human urine	3.50	3.50
86 B_3	Human urine	0.44	1.74
91 B_3	Human urine	9.50	10.36
94 B_3	Human urine	10.93	11.40
95 B_3	Human urine	7.00	11.00
115 B_3	Human urine	1.75	2.08
Step III	Sheep plasma	3.50	4.60

*i = immunochemical units.
† b = biologic units.
‡IRP = International Reference Preparation of erythropoietin.

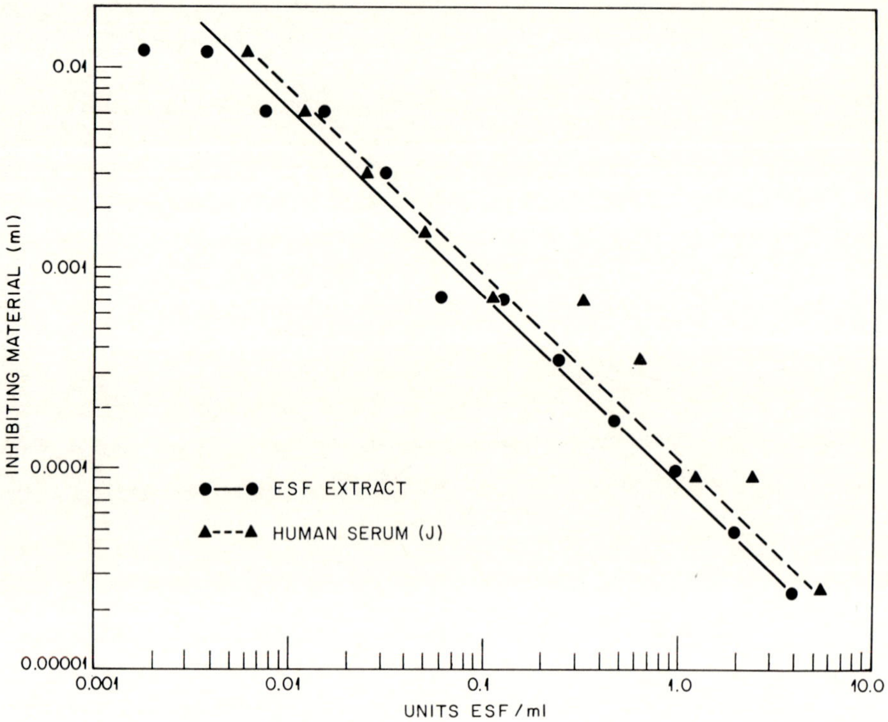

FIG. 1—Hemagglutination-inhibition of anti-ESF 127S with serial dilutions of an ESF extract and serum from a patient with refractory anemia. The volume of the inhibiting material represents an equivalent of the original starting material which was serially diluted. The line of best visual fit for each test material was drawn. Reproduced by courtesy of the American Journal of Medical Technology.[2]

instance. By the HAI technic, 1.76 milli-immunochemical units (mi) of ESF per ml. could be detected ± 0.1 mi. per ml. The maximum precision was 8 per cent. In another test, the immunochemical units of ESF in 36 different sera from normal volunteers and patients were determined and then redetermined after each specimen had been divided into 2 parts and independently coded. Good agreement was present if the independently measured results did not vary by more than the amount of ESF present in one doubling dilution. In only 5 of 108 samples did one of the three determinations vary by more than the amount of material present in one doubling dilution. The other 2 values were either the same or were different by less than the amount of test material present in one doubling dilution.

The results of 3 sets of measurements of the ESF present in the serum of 254 blood donors, made by comparing them to 3 different ESF preparations, are given in Table 2. As shown, 95 per cent of the values fell in the range of 7.5 to 60 mi. ESF per ml.

The sera of 14 patients with a primary anemia were used to compare the biologic and immunochemical levels of ESF (Fig. 3). As shown, at both high and low levels of activity, a fair degree of correlation was present.

TABLE 2—*Erythropoietin Levels in 254 Blood Donors Measured by the Hemagglutination-Inhibition Technic*

ESF/ml. mi.*	Number of Donors at Each Level			
	Assay 1 †	Assay 2 ‡	Assay 3 ¶	Totals
3.0	9	0	0	9
7.5	32	0	0	32
15.0	54	2	0	56
30.0	45	22	26	93
60.0	1	41	18	60
120.0	0	2	2	4

* Milli-immunochemical units.
† Performed using laboratory standard 29 B_3 as a control.
‡ Performed using laboratory standard 88 B_2 as a control.
¶ Performed using laboratory standard 119 A_2 as a control.

The relation between the hematocrit measurement and the units of ESF determined by the HAI technic are shown in Figure 4. The sera of 254 blood donors and 78 patients were used in these determinations. The sera of only 4 blood donors were found to have greater than 60 mi. of ESF per ml. Of the patients, the relation between the hematocrit and the units of erythropoietin was reasonably good in 65 instances. In 13 instances, however, the correlation was poor, i.e., 5 patients had a hematocrit of less than 40 per cent but 60 or less mi. ESF per ml. and 8 patients had a hematocrit of greater than 40 per cent with more than 60 mi. ESF per ml.

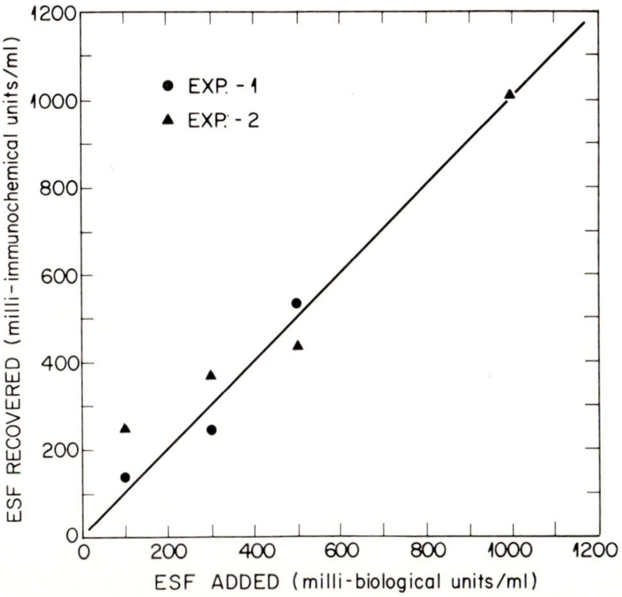

FIG. 2—Determination of erythropoietin added to serum obtained from a hematologically normal volunteer using the HAI technic.

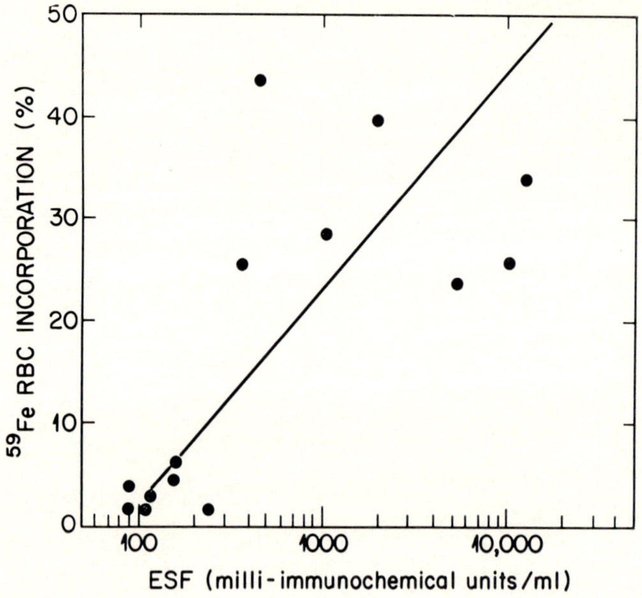

FIG. 3—Relation between results obtained by the assay of sera of patients with primary anemia by biologic and immunochemical methods. The best visual fit line was drawn.

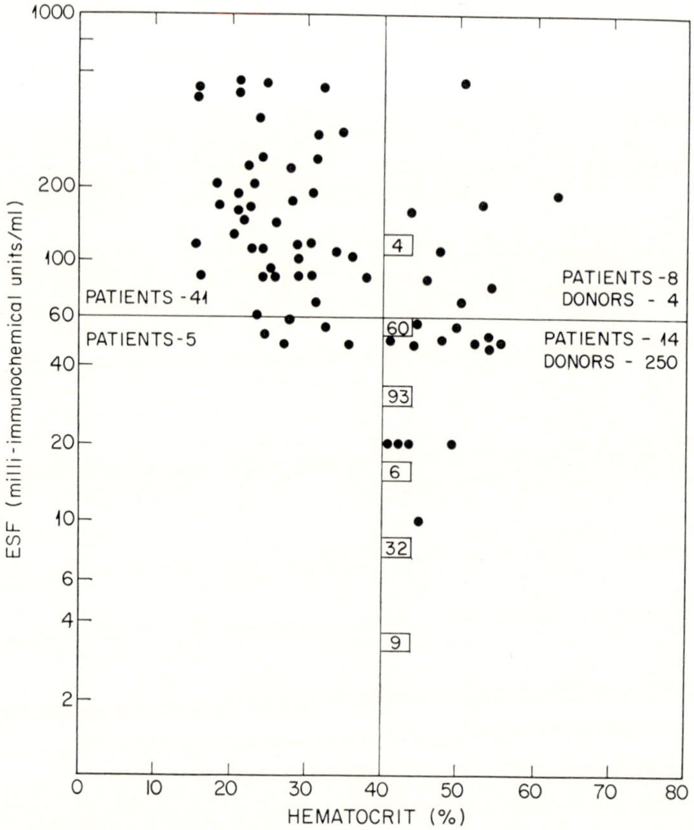

FIG. 4—Comparison between hematocrit values and immunochemical units of erythropoietin. The numbers enclosed in boxes represent the number of sera of blood donors which assayed immunochemically at that level.

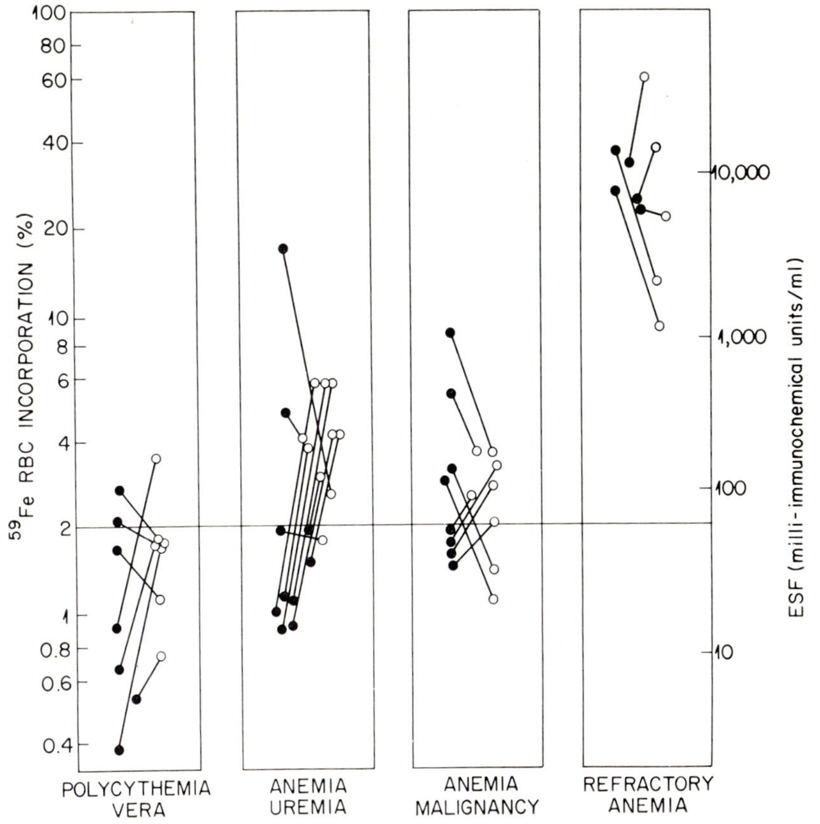

Fig. 5—The correlation between the biologic and immunochemical determinations of ESF activity in the sera of patients with polycythemia vera, the anemia of uremia, anemia associated with malignancy, and refractory anemia. The per cent ^{59}Fe incorporation value is represented by a solid circle and is connected to an open circle which gives the immunochemical determination of the same serum. The horizontal line represents the upper 95 per cent confidence limits of normal serum values for ESF, determined by HAI (60 mi. per ml.) and for the ^{59}Fe red blood cell incorporation of saline injected mice.

In a group of patients with polycythemia vera, refractory anemia, and anemia associated with uremia or malignancy, the mi. ESF per ml. was correlated to the RBC ^{59}Fe incorporation caused by the sera in the polycythemic mouse assay. The results are shown in Figure 5. It should be noted that, in general, the sera of uremic patients assayed at higher levels of ESF by the immunochemical technic than by the biologic system.

Discussion

Immunochemical methods are being used with increasing frequency for measuring the levels of peptide hormones and proteins in serum or body fluids. Radioimmunoassays are currently employed to measure serum levels of insulin, growth hormone, adrenocorticotrophic hormone, chorionic gonadotropin, and others.[9] The HAI technic has also been applied to quantitating proteins and peptides.[10–13]

As shown by the results presented in this communication, the immunochemical determination of erythropoietin is sensitive and reproducible. It was found to be more precise than the mouse assay, in which Adamson[14] has estimated a variation of ± 25 per cent in the difference of means of response between groups of mice given the same ESF preparation. By making 8 dilutions of a given sample, a precision in the order of 8 per cent can be obtained, a figure which contrasts with the 20 per cent precision Baechler et al.[15] found in the immunoassay of human chorionic gonadotropin. To obtain these results careful attention must be given to certain details of the technic.[2] Among the conditions which are critical are: (1) the concentration of tannic acid; (2) the concentration of sensitizing antigen; and (3) the potency of the reference standard. Since no international reference preparation for immunochemical units exists, the number of immunochemical units is necessarily related to materials standardized to biologic units. Dukes et al.[16] have succinctly pointed out the difficulties in bioassaying erythropoietin preparations and have raised the question as to whether or not one needs more than a single standard, since erythropoietin may exist in more than one molecular species.

A fair degree of correlation was found between the biologic and immunologic determination of ESF in extracts and sera of patients with primary anemias. It would not have been surprising, however, if major differences had occurred. Borth[17] has referred to the WHO Expert Committee in Biological Standardization report which stated in 1968: [18] "Although where highly purified hormones are used, a systematic correlation between the results of biologic assays and immunoassays might be expected, this is not necessarily so when, for reasons of stability or scarcity, less pure materials must be employed . . . Since available hormone antigens are not always ideal in this sense, the problem is that of deciding on a unit of immunoreactivity that can be distinguished from the unit of biological activity." That different methods of testing may give results that vary many fold for the same sample was also pointed out by Borth.[17] He found that the mean concentration of human chorionic gonadotrophin in 23 serum samples was 191 ib. (international biological units) per ml. by the rat test, 64 ib. per ml. by the ovarian hyperaemia test, 16.9 ii. (international immunochemical units) per ml. by the complement fixation test and 49.1 ii. per ml. by the hemagglutination-inhibition test. ESF correlation studies have an added difficulty for it is frequently difficult to obtain enough serum to assay the material in more than 5 mice. While the polycythemic mouse provides a very good biologic assay system, more than 5 mice are required before one can develop confidence in the results.

The determination of ESF levels in the serum of blood donors has been discussed.[1,3] The results reported in this study again indicate the presence of erythropoietin in normal serum which, by inference, is evidence for the importance of the hormone in the regulation of normal erythropoiesis; this point has been in doubt due to the inability of the bioassay methods to detect routinely the presence of ESF in normal serum.[19–21] A reasonably good correlation was found between the hematocrit determination and mi. ESF in the sera of 254 blood donors and in 65 of the pathologic sera. In 13 instances a similar correlation was not present. In the 5 patients with a hematocrit of less than 40 and 60 or less mi. ESF per ml. the diagnoses were as follows: carcinoma of the breast, cirrhosis, Hodgkin's disease, anemia of uremia, and cholelithiasis. Of the 8 patients with a hematocrit of greater than 40 per cent and more than 60 mi. ESF per ml., 2 had polycythemia vera, and 1 stress erythrocytosis. The other 5 patients had: leukemia, congestive heart failure (2), the nephrotic syndrome, and diabetes. Adamson [14] has previously reported a relation between the hematocrit and the excretion of ESF in the urine of normal volunteers and patients with hematologic disorders. The low ESF values found by both biologic and immunochemical methods for polycythemic patients and the high ESF values found in the patients with refractory anemia were to be expected. The elevated immunochemical ESF results in the patients with the anemia associated with uremia were anomalous since the anticipated low levels were found by the mouse assays.[22] Whether or not these results represent truly false values for the HAI test in the presence of an elevated BUN or inhibition of the biologic expression remains to be tested. Recently a patient with sickle cell anemia and a hemoglobin of 5 Gm. per cent was found to have 1.1 units ESF per ml. by the HAI test but no ESF activity by the polycythemic mouse assay. Further studies are underway to ascertain the cause of these discrepant results.

Many more specimens must be tested both biologically and by HAI before the HAI method can be considered standard, but the method is promising, and deserves further evaluation in other laboratories.

Summary

1. The hemagglutination-inhibition technic for the assay of erythropoietin has excellent reproducibility.

2. As little as 1.8 ± 0.1 milli-immunochemical units or erythropoietin can be detected with a precision of 8 per cent.

3. Erythropoietin was quantitatively recoverable when added to normal serum; on serial dilution, extracts of erythropoietin and erythropoietin rich human serum demonstrated the same reaction slope.

4. Normal human serum has up to 60 milli-immunochemical ESF units per ml.

5. When related to hematocrit determinations, the hemagglutination-inhibition test showed reasonably good correlation; in patients with refractory and iron deficiency anemias, the ESF unitage determined biologically and immunochemically appeared to be positively correlated.

6. There was discordance between the biologic and immunochemical assays for ESF in the sera of patients with the anemia of uremia.

ACKNOWLEDGMENTS

The advice of Dr. T. Mitchell and the excellent technical assistance of Miss Carol Drinkard and Mrs. Pauline Lotkin is gratefully acknowledged. We also wish to thank Mr. T. Foster of the Knoxville Blood Center who made available the sera from blood donors.

REFERENCES

1. Lange, R. D., McDonald, T. P., and Jordan, T.: Antisera to erythropoietin: Partial characterization of two different antibodies. J. Lab. Clin. Med. 73:78, 1969.

2. Jordan, T. A., McDonald, T. P., and Lange, R. D.: The hemagglutinating properties of an antiserum to erythropoietin. Amer. J. Med. Techn. 35:595, 1969.

3. Lange, R. D.: Immunological studies on erythropoietin. *In* Fisher, J. W. (Ed.): Hormones in the Kidney, II. London, Academic Press. In press.

4. Plescia, O. J., Braun, W., and Palczuk, N. C.: Production of antibodies to denatured deoxyribonucleic acid (DNA). Proc. Nat. Acad. Sci. U.S.A. 52:279, 1964.

5. Sever, J. L.: Application of a microtechnique to viral serological investigations. J. Immun. 88:320, 1962.

6. Takatsy, G.: Kiseral Orvostud bes. 2: 393, 1950. Cited in Sever, J. L., *ibid* 88:320, 1962.

7. Lange, R. D., Simmons, M. L., and Dibelius, N. R.: Polycythemic mice produced by hypoxia in silicone rubber membrane enclosures: A new technique. Proc. Soc. Exp. Biol. Med. 122:761, 1966.

8. McDonald, T. P., and Lange, R. D.: Erythropoietin bioassays utilizing silicone rubber membrane enclosures. J. Lab. Clin. Med. 70:48, 1967.

9. Yalow, R. S., and Berson, S. A.: General principles of radioimmunoassay. *In* Hayes, R. L., Goswitz, F. A., and Murphy, Beverly E. Pearson. (Eds.): Radioisotopes in Medicine In Vitro Studies. Oak Ridge, Tenn., U.S. AEC Division of Technical Information Extension, 1968.

10. Read, C. H., and Bryan, G. T.: The immunological assay of human growth hormone. Recent Progr. Hormone Res. 16:187, 1960.

11. Arquilla, E. R., and Stavitsky, A. B.: The production and identification of antibodies to insulin and their use in assaying insulin. J. Clin. Invest. 35:458, 1956.

12. Gold, E. R., and Fudenberg, H. H.: Chromic chloride: A coupling reagent for passive hemagglutination reactions. J. Immun. 99:859, 1967.

13. Gusdon, J. P., Jr.: Improved hemagglutination-inhibition assay: Clinical application to measurement of human placental lactogen. Obstet. Gynec. 33:397, 1969.

14. Adamson, J. W.: The erythropoietin/hematocrit relationship in normal and polycythemic man: Implications of marrow regulation. Blood 32:597, 1968.

15. Baechler, C., Borth, R., and Menzi, Annette: Immunoassay of human chorionic gonadotrophin: Diurnal viriation in urinary antigen excretion. Acta Endocr. (Kobenhavn) 52:199, 1966.

16. Dukes, P. P., Hammond, D., and Shore, Nomie A.: Comparison of erythropoietin preparations yielding different dose-response slopes in the exhypoxic polycythemic mouse assay. J. Lab. Clin. Med. 74:250, 1969.

17. Borth, R.: A terminology for units of biological and immunochemical activity. Acta Endocr. (Kobenhavn) 60:1, 1969.

18. WHO Expert Committee on Biological Standardization: WHO Techn. Rep. Ser. 384:8, 1968.

19. Stohlman, F., Jr., and Brecher, G.: Humoral regulation of erythropoiesis. IV. Relative heat stability of erythropoietine. Proc. Soc. Exp. Biol. Med. 95:797, 1957.

20. Mirand, E. A., Weintraub, A. H., Gordon, A. S., Prentice, T. C., and Grace, J. T., Jr.: Erythropoietic activity of untreated and deproteinized normal human plasma. Proc. Soc. Exp. Biol. Med. 118:823, 1965.

21. Cotes, Mary: Personal communication.

22. Gallagher, N. I., McCarthy, J. M., and Lange, R. D.: Observations on erythropoietic-stimulating factor (ESF) in the plasma of uremic and nonuremic anemic patients. Ann. Intern. Med. 52:1201, 1960.

APPENDIX

A. Hemagglutination (HA) technic: The hemagglutination tests are performed in titration plates * by the method of Sever,[5] except that "U" plates are used.

* Microtitering Kit, Cooke Engineering Company, Alexandria, Virginia.

The sensitized cells for HA are prepared as follows:

1. Human group O red blood cells, collected in heparin, are washed 3 times in 0.9 per cent NaCl, centrifuged, and resuspended in saline to a concentration of approximately 20 per cent v/v.
2. Equal volumes of cell suspension and 0.005 per cent tannic acid † solution are mixed and placed in a 37 C. water bath for 15 to 30 minutes or until frank agglutination of the cells occurs.
3. The tanned cells are washed 3 times with 0.075 M. phosphate buffered saline, pH 7.2 and are resuspended to a concentration of approximately 20 per cent, v/v. Suspensions of these cells remain usable for ESF coating over a period of 16 to 24 hours when kept at 4 C.
4. The tanned red blood cells are mixed with an equal volume of diluted sensitizing antigen and incubated for one hour at room temperature. The optimum condition for the ESF antigen-RBC conjugation in our laboratory is a ratio of antigen to cells of 1 unit ESF per 12 ml. of a 20 per cent v/v RBC suspension.
5. An equal volume of 1 per cent formalin is mixed with the tanned, sensitized cell suspension and the mixture is incubated at room temperature for an additional one hour, after which the mixture is kept at 4 C. for 24 hours.
6. The cells are washed 3 times in phosphate buffered saline and are resuspended to a final concentration of 2 per cent v/v in phosphate buffered saline.

All test materials (antigens or antisera) are inactivated by heating at 56 C. for 45 minutes, cooled to room temperature, and absorbed for 30 minutes at room temperature with about 0.2 ml. of triply washed, unsensitized group O red blood cells for each ml. of test material. The cells are removed by centrifugation.

Phosphate buffered saline is used as a diluent for both the HA and HAI tests.

The HA titer is determined by making serial dilutions of antiserum in "U" plates using 0.025 ml. volumes of phosphate buffered saline as a diluent. A volume of 0.025 ml. sensitized, tanned, formalinized cells is added to all dilutions. After addition of antiserum and cells, the plates are covered with plastic tape and the contents are mixed. The HA patterns are read after at least 4 hours' incubation at room temperature. The titer of the antiserum is expressed as a reciprocal of the highest dilution showing 3 plus HA (Fig. 6 rows A-D).

B. Hemagglutination-inhibition technic:

1. Doubling dilutions of test materials and standards (0.025 ml.) are made serially with phosphate-buffered saline in microtiter plates.
2. Antiserum (0.025 ml.) is added and, after mixing, the combination is incubated at room temperature for 30 minutes. A high titered, hemagglutinating antiserum is made up to a concentration of one-half of its previously determined titer with phosphate-buffered saline to allow for dilution with test substances.
3. Sensitized cells (0.025 ml.) are added, the plates sealed, mixed, and kept at room temperature for 12 hours.
4. The HAI titer is read as the amount of material which first permits 3 plus agglutination by the antiserum.
5. The units of ESF in the test material are calculated by comparing the amount of test material which causes inhibition to the amount of standard ESF, which causes the same degree of inhibition. The units of ESF are calculated by the following formula:

$$A \times A^1 = B \times B^1$$

A = Units ESF in standard.
A^1 = Amount of standard allowing 3 plus HA.
B = Units of ESF in test material.
B^1 = Amount of test material allowing 3 plus HA.

See Figure 6, rows E-H, for examples of HAI tests for sera and extracts.

† Mallinckrodt Chemical Works, St. Louis, Missouri.

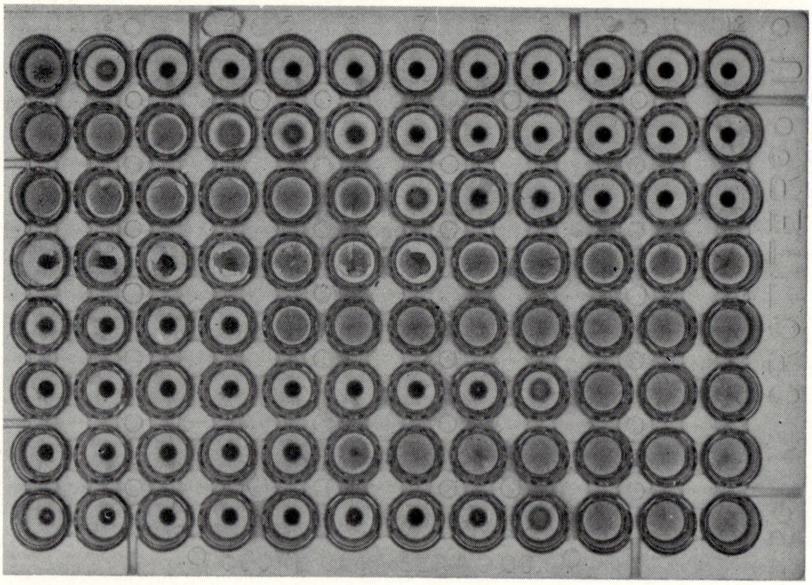

FIG. 6—Hemagglutination and hemagglutination-inhibition reactions:

A. Rows A-D: Hemagglutination titers of 4 anti-ESF sera.

Row	Antiserum	Last well showing 3+ HA	Titer
A	AS* 121 (1/10)	1	1:20
B	AS-UT9 (1/10)	2	1:40
C	AS 101 (1/10)	5	1:320
D	AS 132 (1/10)	11	1:20,480

B. Rows E-H: Hemagglutination-inhibition reactions for 2 human sera and 2 ESF extracts.

Row	Material	First well showing 3+ HA	m-i u ESF/ml.†
E	Normal human serum	5	40
F	Patient "J"	10	1400
G	ESF lot 44	6	90
H	ESF lot 91	10	1400

The following types of HA reactions should be noted:

Negative—Row A, wells 3–12
1+ —Row B, well 5
2+ —Row B, well 4
3+ —Row B, wells 1 & 2
4+ —Row D, wells 1–7

*AS = antiserum.
†m-i u = milli-immunochemical units.

The Erythropoietic Effect of Androgens

By Clifford W. Gurney, W. Fried and A. Fessler

An inquiry into the effect of androgens upon erythropoiesis receives impetus from two directions. In the first place, a sex difference in the hemoglobin levels and red cell masses is commonly recognized in a wide variety of species of mammals. In the second place, androgens have been shown to accelerate red cell production in some patients. Kennedy et al.[1] have described plethora in some women receiving testosterone as therapy for carcinoma of the breast, and androgens have become standard therapy of patients with refractory anemia since the reports of beneficial effects in some children and adults by Shahidi and Diamond[2] and Gardner and Pringle.[3]

When, as a consequence of the availability of improved assay systems for the demonstration and quantitation of erythropoietin[4] the regulation of erythropoiesis came under intensive scrutiny, it was logical to direct attention to a study of the mechanisms by which androgens influenced red cell production. Over a decade ago, many investigators first sought a direct action of androgen on the bone marrow, but early results were at best equivocal. Almost simultaneously, 3 different groups of workers reported the presence on bioassay of elevated levels of erythropoietin in the plasma of animals after testosterone administration.[5,6,7]

Figure 1 records the result of one experiment in which young adult female CF1 mice were given 2 daily subcutaneous injections of 2.5 mg. testosterone proprionate in sesame oil. Control animals received sesame oil alone. On the same days, each animal received .4 ml. of washed red cells, intraperitoneally, an amount sufficient to raise the hematocrit to the range of 59 to 64 per cent. Four days after the second injection of oil and red cells, each animal received .5 μc. of $^{59}FeCl_3$ intravenously.

Figure 1 indicates that red cell formation is virtually eliminated in the control animals, but a striking degree of erythropoiesis, as indicated by red cell incorporation into newly formed red cells, is present in the recipients of testosterone. Because this response can be eliminated by injection of anti-erythropoietin antibody,[8,9] it is reasonable to attribute this erythropoietic effect of androgens to production in recipient animals of erythropoietin.

Two features of this experiment are worthy of emphasis. First, the radioactive iron is administered 4 days after the second injection of androgen. This is in contrast to the optimal time of 48 hours for administration of radioactive iron to demonstrate the erythropoietic stimulating properties in polycythemic mice of a single injection of erythropoietin. It therefore appears that the production of erythropoietin in response to androgen is slow to be activated.[10] Such a slow response is quite different from the prompt elaboration of erythropoietin when hypoxia, rather than androgen, is the stimulus. Secondly,

---

These investigations were supported in part by the Atomic Energy Commission and in part by United States Public Health Service Grant AM 1 CA–11609.

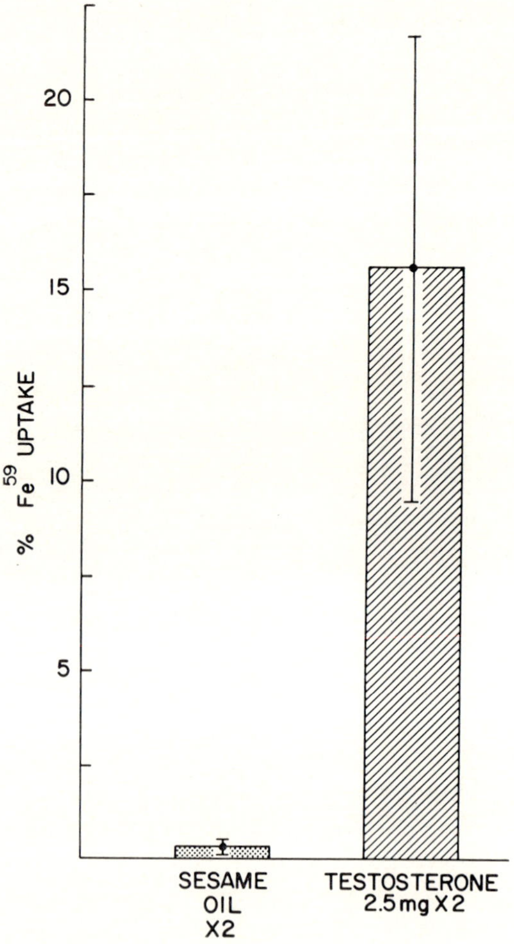

FIG. 1—The effect of androgen on ^{59}Fe uptake in plethoric mice.

the animals were rendered only moderately plethoric. It has been demonstrated that large degrees of plethora dampen the erythropoietic effect of androgen.[11]

Figure 2 shows the results obtained when different groups of mice, given increasing volumes of transfused blood, are used to demonstrate the erythropoietic effect of 2 daily injections of 2.5 mg. of testosterone. It will be noted that very slight plethora is sufficient to dampen the effect of androgens. Non-transfused littermates in this experiment had hematocrits ranging between 46 and 49 per cent. Sesame oil injected control mice had average ^{59}Fe uptakes of approximately 5 per cent when hematocrits were 50–51 per cent, and under .4 per cent when hematocrits were 55 per cent and higher.

The amounts of androgen employed in these experiments are disturbingly large. If, however, a dose-response curve to androgens is obtained in mice receiving 2 daily injections of only .2 ml. of packed cells, it is possible to demonstrate an erythropoietic effect with smaller amounts of testosterone. In Figure 3 it is noted that as little as .1 ml. of testosterone, administered daily for 2 days, increases the iron incorporation in mildly plethoric mice, and progressively greater radioiron uptake is observed when larger amounts of androgen are administered. In these experiments the radioiron was injected 96 hours after the second dose of testosterone, so as to give the androgen ample time to exert its erythropoietic effect.

Figure 4 illustrates a comparison of hypoxia and testosterone as stimuli for production of erythropoietin in normal CF1 female mice. In the instance of hypoxia, the maximum titer of plasma erythropoietin can be very high if the stimulus is intense but the feature to be noted here is the prompt rise to maximum level, followed by a quick decline to barely detectable degrees of elevation after 48 hours. The plasma concentration of erythropoietin achieved after a single dose of testosterone as great as 5 mg. often does not exceed ½ unit, but of particular interest is the delay between injection

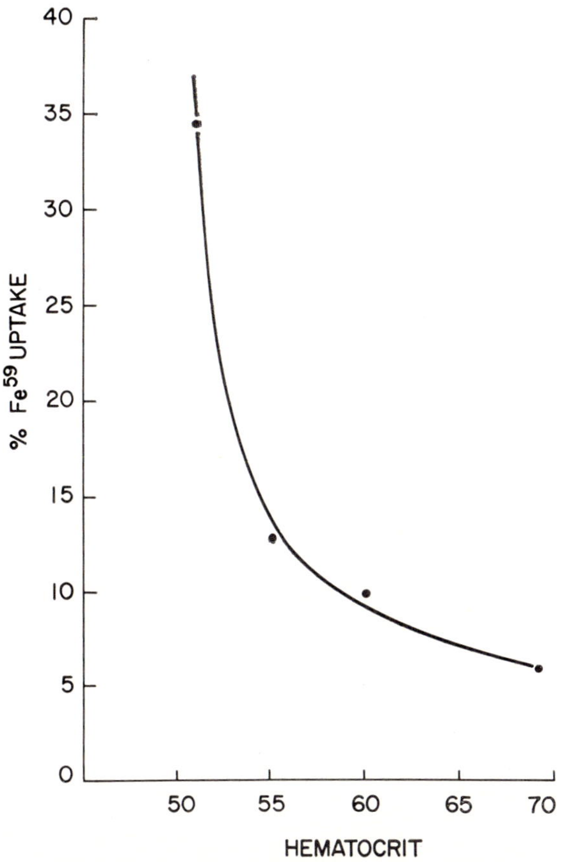

FIG. 2—The dampening effect of increasing plethora on androgen-stimulated erythropoiesis.

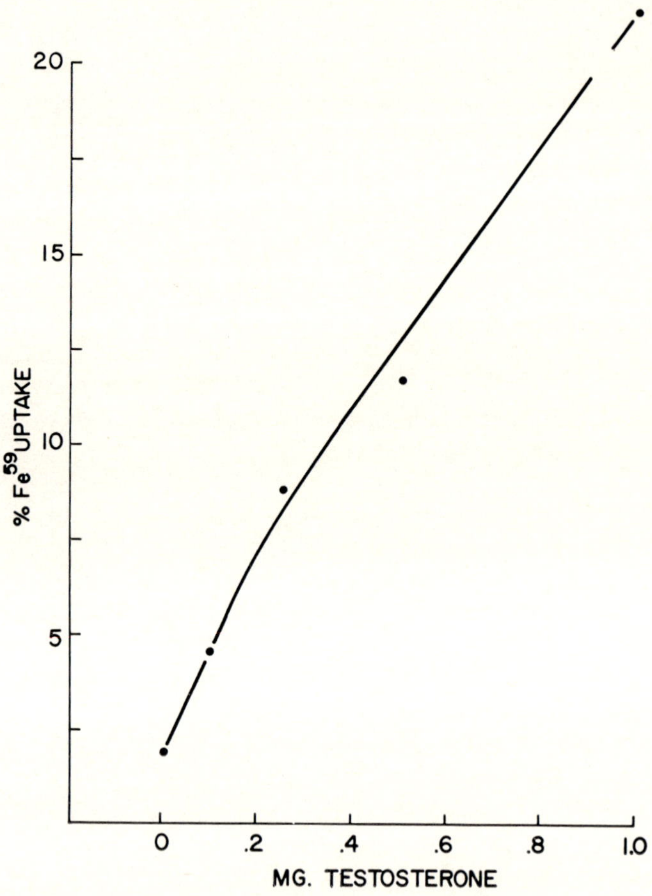

FIG. 3—Dose-response curve of the erythropoietic effect of androgens in mildly plethoric mice.

of the androgen and the response. The maximal concentration to be achieved is not observed until the fourth day after the injection of androgen. The reasons for this delay are at this time not known.

Gordon et al. have presented evidence that the kidneys produce Renal Erythropoietic Factor, an enzyme which they believe converts an inactive plasma precursor of erythropoietin into the active erythropoietic hormone, and it is REF rather than erythropoietin that increases in the kidney after administration of testosterone.[12] Although anephric man [13] and animals subjected to a hypoxic stress can produce erythropoietin,[14] and testosterone potentiates the effect of hypoxia in the production of erythropoietin in animals with intact kidneys,[5] Fried and Kilbredge recently showed that the production of erythropoietin in hypoxic anephric rats is not enhanced if the animals are given androgens.[15] Thus, although erythropoietin can be produced in the absence of the kidneys, it would appear that renal tissue may be responsible for most androgen-induced production of erythropoietin in the rat. In anephric man, however, androgen therapy may increase plasma erythropoietin titers.[16]

There is a marked difference in the degree to which male and female mice respond to androgens. Figure 5 summarizes the results obtained when young male and female CBA mice of the same size are made moderately plethoric with .8 ml. of packed mouse red cells given intraperitoneally. Although hematocrits are comparable, being in the low 60s, the degree of suppression of erythropoiesis observed in females, as indicated by ^{59}Fe incorporation into newly formed red cells, cannot be obtained in the males. Conversely, the erythropoietic stimulus obtained by the injection of androgens is substantially less in the male than that observed in the female. That endogenous production of androgens in the male mice is somehow responsible, at least in part, for this sex difference in response to administered androgen is suggested by the disappearance of this sex difference in animals investigated 4 months after surgical removal of the gonads.[17]

Androgens are profound anabolic agents and, for this reason, the possibility of a direct erythropoietic effect on the bone marrow also has received much attention. Reisner has described increased concentration of erythroblasts in bone marrow tissue cultures in response to testosterone[18] and Jacobson et al. describe the increased uptake of tritiated thymidine by marrow cultures of some, but not all, children with a variety of illnesses when testosterone was added to the culture.[19] However, McDonald has not been able to confirm the existence of a direct effect of testosterone on the incorporation of ^{59}Fe by cultured bone marrow, and he was not able to demonstrate any enhancement of the effect of erythropoietin on marrow cultures by androgen.[20] Recently the enhancement of hemoglobin synthesis by chick blastoderms has been

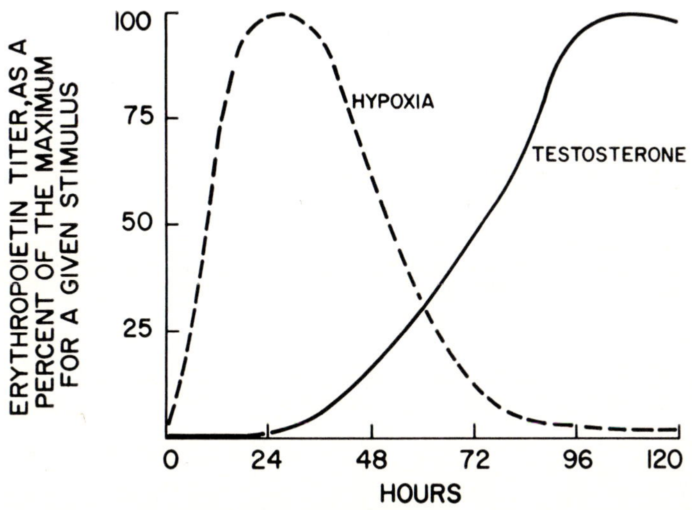

FIG. 4—Change in erythropoietin titer with time, after continuous hypoxia or single injection of androgen.

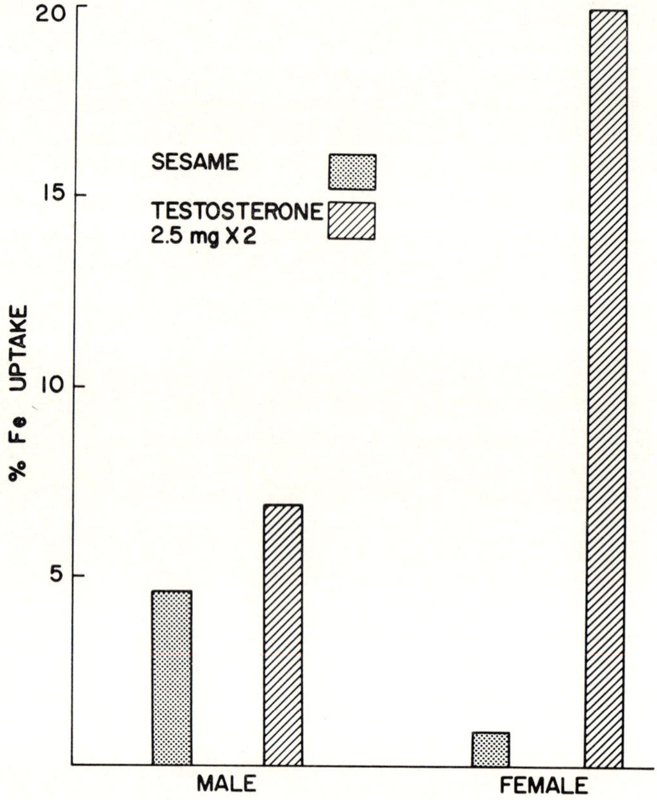

FIG. 5—Effect of testosterone on male and female CBA mice.

described in response to the addition of steroid metabolites.[21] This observation is of special interest because some of the delay in appearance of erythropoietin after the injection of testosterone in mice might be accounted by the time required to convert the androgen to one or more active metabolities.

We have recently conducted a series of investigations on the effects of a synthetic antiandrogen, cyproterone acetate, upon erythropoiesis. This material, 1,2-methylene 6 chloro-$\Delta^{4,6}$ pregnandiene-17-ol-3, 20 dione-17-acetate was kindly supplied by the Berlin Lab., Inc. of West Germany. It has been shown to neutralize the effects of testosterone upon the levator anuses and genital glands of the castrated rat at a ratio of cyproterone to testosterone of approximately 2.5:1,[22] and the erythropoietic effect of androgens in polycythemic mice is neutralized at the same ratio.[23]

Studies were then undertaken to determine whether or not cyproterone alone had any effect on erythropoiesis. We first pretreated groups of normal female CF1 mice with either cyproterone 3 mg. per day dissolved in benzyl benzoate in castor oil, or the solvent alone. Two daily injections were given and, 48 hours after the second injection, all animals were placed in a hypoxia chamber at a setting of approximately .5 atmosphere where they remained for 12.5 hours. Immediately after removal from the chamber, all animals were

TABLE 1—*Assay, in Transfused Mice, of Plasma From*

1. Normal mice, after	BBCO* × 2	.4% + .2%
2. Normal mice, after	BBCO × 2 + 12.5 HR hypoxia	17.8% ± 7.1%
3. Normal mice, after	Cyproterone × 2 + 12.5 HR hypoxia	23.7% ± 6.1%

* Benzyl benzoate in castor oil.

bled and a single pooled plasma sample was generated from each group. This plasma was assayed in the polycythemic mouse according to the technic of DeGowin et al.,[4] each assay animal receiving .8 ml. of plasma. The results are recorded in Table 1. Results are expressed as average values per group, plus or minus one standard deviation of the mean. It will be seen that cyproterone in the amounts and timing employed here did not interfere with the ability of normal mice to produce erythropoietin in response to acute hypoxia.

Finally, we studied the effect of cyproterone on the response of polycythemic mice to erythropoietin. Normal CF1 female young adult mice were made polycythemic by 2 daily injections of 1.0 ml. of packed red cells. On the following 3 days, the mice received one daily injection of either benzyl benzoate in castor oil or 3 mg. of cyproterone in the diluent. On the day following the last injection, 2 groups of mice received 3 units of erythropoietin per mouse subcutaneously. The response to erythropoietin was determined by the technic of DeGowin et al.[4] The results are recorded in Table 2. A substantial reduction of radioiron incorporation in newly formed red cells was achieved in animals pretreated with cyproterone. Similar results have been obtained in other experiments employing different amounts of erythropoietin and different schedules of pretreatment injection of cyproterone.

Several interpretations of these results are possible. First, cyproterone might activate mechanisms leading to destruction of exogenously administered erythropoietin. Second, cyproterone could conceivably alter the ferrokinetics in the assay animals so as to reduce the fraction of the tracer dose of radioiron taken up by newly formed red cells. A third possibility is that cyproterone somehow exerts an inhibitory effect upon erythropoiesis. If an androgen antagonist does indeed inhibit erythropoiesis, it is likely such inhibition occurs in the hematopoietic tissue, since production of erythropoietin is not a factor to be considered in this experiment. It may be that the locus and mechanism of this inhibition will afford a way to study a possible direct effect of testosterone or some metabolite of testosterone upon the bone marrow.

In summary, androgens exert a stimulatory effect on erythropoiesis. There can be no doubt that this stimulation can be explained in part by increased

TABLE 2—*Effect of Cyproterone on Response of Polycythemic Mice to Erythropoietin*

BBCO* × 3		.1% ± 0 %
BBCO × 3	+E†(3u)	17.2% ± 4.9%
Cyproterone 3 mg. × 3	+E(3u)	7.9% ± 4.3%

* Benzyl benzoate in castor oil.
† Erythropoietin.

elaboration of erythropoietin. There is, however, a small amount of direct and indirect evidence consistent with the thesis that androgens or metabolic products of androgens may also have a direct effect upon erythropoiesis. Further investigations are necessary to confirm and define the importance of such a direct effect.

REFERENCES

1. Kennedy, B. J., I. T. Nathanson, D. M. Tibbetts, and J. C. Aub: Biochemical alterations during steroid hormone therapy of advanced breast cancer. Amer. J. Med. 19: 337, 1955.
2. Shahidi, N. T., and L. K. Diamond: Testosterone-induced remission in aplastic anemia of both acquired and congenital types. Further observations in 24 cases. New Eng. J. Med. 264:953, 1961.
3. Gardner, F. H., and J. C. Pringle, Jr.: Androgens and erythropoiesis. I. Preliminary clinical observations. Arch. Intern. Med. 107:112, 1961.
4. DeGowin, R., D. Hofstra, and C. W. Gurney: A comparison of erythropoietin bioassays. Proc. Soc. Exp. Biol. Med. 110:46, 1962.
5. Naets, J. P., and M. Wittek: Etude du mecanisme d'action des androgenes sur l'erythropoiese. Compt. Rend. Acad. Sci. 259:3371, 1964.
6. Mirand, E. A., A. S. Gordon, and J. Wenig: Mechanism of testosterone action in erythropoiesis. Nature (London) 206:270, 1965.
7. Fried, W., and C. W. Gurney: Erythropoietic effect of plasma from mice receiving testosterone. Nature (London) 206:1160, 1965.
8. Schooley, J. C.: Inhibition of erythropoietic stimulation by testosterone in polycythemic mice receiving anti-erythropoietin. Proc. Soc. Exp. Biol. Med. 122:402, 1966.
9. Fried, W., D. Marver, R. D. Lange, and C. W. Gurney: Studies on the erythropoietin stimulating factor in the plasma of mice after receiving testosterone. J. Lab. Clin. Med. 68: 947, 1966.
10. Fried, W., R. DeGowin, and C. W. Gurney: Erythropoietic effect of testosterone in the polycythemic mouse. Proc. Soc. Exp. Biol. Med. 117: 839, 1964.
11. Fried, W., and C. W. Gurney: Use of mile plethora to demonstrate an erythropoietic effect from small amounts of androgens. Proc. Soc. Exp. Biol. Med. 120:519, 1965.
12. Gordon, A. S., E. A. Mirand, J. Weinig, R. Katz, and E. D. Zanjani: Androgen actions on erythropoiesis. Ann. N. Y. Acad. Sci. 149:318, 1968.
13. Naets, J. P., and M. Wittek: Presence of erythropoietin in the plasma of one anephric patient. Blood 31:249, 1968.
14. Mirand, E. A., and T. C. Prentice: Presence of plasma erythropoietin in hypoxic rats with or without kidney(s) and/or spleen. Proc. Soc. Exp. Biol. Med. 96:49, 1957.
15. Fried, W., and T. Kilbredge: Effect of testosterone and of cobalt on erythropoietin production by anephric rats. J. Lab. Clin. Med. 74:623, 1969.
16. Mirand, E. A., G. P. Murphy, J. M. Steeves, J. M. Groenewald, and J. N. De-Klerk: Erythropoietin activity in anephric, allotransplanted, unilaterally nephrectomized, and intact man. J. Lab. Clin. Med. 73:121, 1969.
17. Fried, W., and C. W. Gurney: The erythropoietic response to testosterone in male and female mice. J. Lab. Clin. Med. 67:420, 1966.
18. Reisner, E. H., Jr.: Tissue culture of bone marrow. II. Effect of steroid hormones on hematopoiesis *in vitro*. Blood 27:460, 1966.
19. Jacobson, W., Sidman, R. L., and Diamond, L. K.: The effect of testosterone on the uptake of tritiated thymidine by the bone marrow of children. Ann. N. Y. Acad. Sci. 149:389, 1968.
20. McDonald, T. P.: Discussion Ann. N. Y. Acad Sci. 149:409, 1968.
21. Levere, R. D., A. Kappas, and S. Granick: Stimulation of hemoglobin synthesis in chick blastoderms by certain 5β androstane and 5β pregnane steroids. Proc. Nat. Acad. Sci. U.S.A. 58:985, 1967.
22. Berlin Lab., Inc.: Personal communication.
23. Medlinsky, J. T., C. D. Napier, and C. W. Gurney: The use of an antiandrogen to further investigate the erythropoietic effects of androgens. J. Lab. Clin. Med. 74:85, 1969.

Sheep Hemoglobins, Erythropoietin, and Genetic Regulation

By Samuel H. Boyer

One approach to the sources of differential gene action in higher organisms is through the study of natural inequalities that exist in the concentration of certain hemoglobins.[1] Such studies not only begin with the considerable advantage afforded by the preponderant synthesis of hemoglobin in older erythropoietic cells but also utilize a line of cell descent that does not synthesize hemoglobin in earlier generations. Since the younger and older erythropoietic cells can be physically separated from one another on the basis of density [2] it is, in principle, possible to distinguish transcriptional sources of quantitative inequality developing in the younger cells from translational sources occurring in the older ones. In this connection, delineation of transcriptional control has not been possible in a setting where the flux of messenger RNA from specific genes might be characterized. Among the several model systems available only one gives recognizable promise of satisfying the latter requirement. This occurs in domestic sheep [3–4] and goats [5] wherein the hemoglobin phenotype is different in health than it is in anemia. The intent of the present report is to describe some of the dimensions of this system in sheep and, in particular, to indicate those features that make it potentially useful as a model for exploring the action of erythropoietin at both cellular and molecular levels.

In healthy sheep the apparently allelic adult hemoglobins A and B form a genetic polymorphism such that three phenotypes—A, A/B, and B—are electrophoretically distinguishable.[6–7] With the advent of anemia, hemoglobin A is partially or entirely replaced by a new hemoglobin, viz., C.[3–4] In prolonged severe anemia the A phenotype of homozygous A and heterozygous A/B individuals is entirely superceded by C. Thus, in the face of profound anemia, the phenotype of A animals becomes C and that of A/B animals becomes C/B while the phenotype of B animals is unaltered.[8–9]

These changes have nothing to do with the characteristic hemoglobin of lambs.[10] The latter component is largely lost in early postnatal life and, in any case, its concentration is not appreciably influenced by anemia. Although the phenotypic alterations just described can be produced either by hemolysis or blood loss, the A → C change is particularly abrupt during phenylhydrazine induced hemolytic anemia [9,11–12] where, as illustrated in Figure 1, the older cells containing almost entirely A are destroyed and—within a period of 15 days—replaced by new cells containing largely or only C.

The long generation time and sheer size of sheep make them awkward

---

From the Division of Medical Genetics and the Clayton Laboratories Department of Medicine, The Johns Hopkins University School of Medicine and Hospital, Baltimore, Maryland.

Supported in part by grants GM 6308 and HD 02508 from the National Institutes of Health.

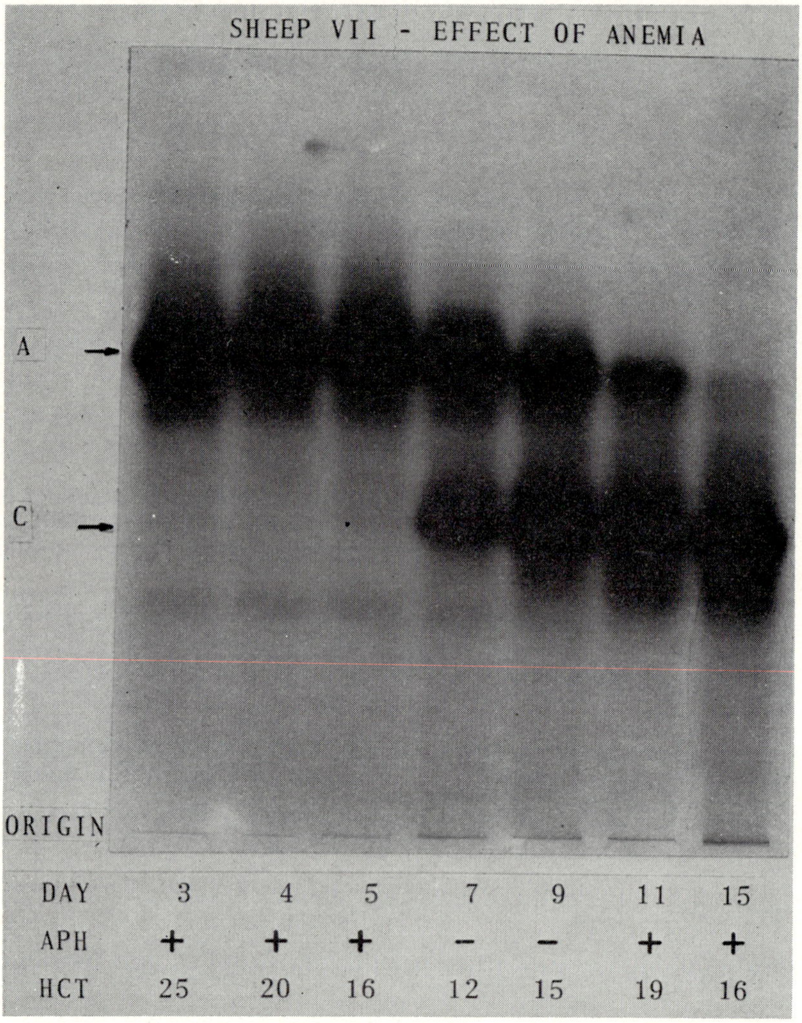

FIG. 1—Starch gel electrophoresis (anode at top) at pH 8.6 [9] of hemoglobin (stained with amido-Schwarz) from an A phenotype sheep made anemic by intramuscular administration of phenylhydrazine (APH, ~ 3mg./Kg. on days 0–5, 11–13) and showing rapid replacement of hemoglobin A by hemoglobin C with fall in hematocrit (HCT).

subjects for detailed Mendelian analysis and, thus, the genetic fine structure underlying A → C switch cannot be adduced from an analysis of matings. Consequently genetic methods cannot be used to distinguish an anemia-dependent switch in the differential activity of 2 genes, on the one hand from an anemia-dependent but extragenetic alteration in the electrophoretic mobility of hemoglobin A on the other. A distinction between these alternatives can, however, be made through analysis of protein structure. Not only are the differences between hemoglobins A. B, and C confined to their respective β chains [13] but also the multiplicity and scattered distribution of the amino acid differences between chains,[10–11] shown diagrammatically in Figure 2, leave little doubt that each β chain is the product of a distinctive structural gene and not the result of some environmental process such as partial degradation. Although the topographic relationships of the 3 β genes remain uncertain it is supposed, as the simplest explanation,[9] that (a) the genes for A-β and C-β, i.e., β^A and β^C, derive from the duplication of an ancestral gene and (b) the resulting linkage in tandem between β^A and β^C remains essentially undisturbed by recombination. Thus, it is presumed that both β^A-β^C and β^B bearing chromosomes exist whereas β^B-β^C chromosomes are, in contrast, rare or nonexistent. This presumption provides an explanation for the limitation of C to those animals that bear the gene for A-β. The argument is not without precedent and well studied instances of such inequality in the numbers and/or size of genes present on homologous chromosomes are provided, for example, by haptoglobin [14] and hemoglobin Lepore [15] variants in man.

The details of β chain sequence also provide a potential clue to the nature of the A-in-health and C-in-anemia situation. Protein sequence analysis indicates that the C-β chain has apparently lost, relative to A-β and C-β, 4 amino acid residues from its free amino terminus. The potential significance of this feature depends upon a corollary of the principle of colinearity between a gene and its protein product whereby the free amino terminus of a chain corresponds to the region of its structural gene that is believed to regulate transcription of messenger RNA (mRNA).[16] It is possible that the "four-shortened" sequence of C-β is associated with the loss of signal receptors normally used at inauguration of transcription. Carrying speculation still further, it might be supposed that the appearance of C in the course of anemia depends either on the use of new kinds of signals or new kinds of receptors.

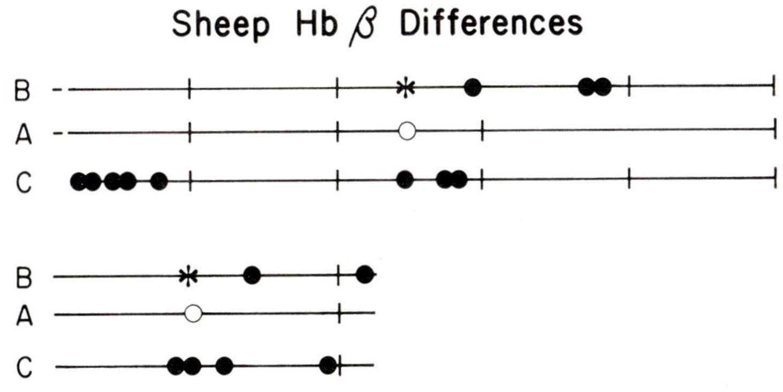

FIG. 2—Diagram of differences in the amino acid sequence of sheep A-β, B-β, and C-β chains. Diagram is based on nearly complete sequence analysis of the three chains.[9,11] Symbols indicate positions where the amino acid residue in the designated chain differs from that in the other chains. Distinctive symbols are used at single positions in cases where each chain is unique.

The shortening of C-β relative to A-β and B-β is potentially useful in still another context, viz., identification and characterization of specific mRNA. The first 14 residues of the A-β and B-β sequences are:

NH_2-Met-Leu-Thr-Ala-Glu-Glu-Lys-Ala-Ala-Val-Thr-Gly-Phe-Trp,

while the first 10 of C-β are

NH_2-Pro-Asn-Lys-Ala-Leu-Ile-Thr-Gly-Phe-Trp

Thus C-β, when examined residue-by-residue beginning at free amino terminus, is nearly always different than the other two β chains. The mRNA for C-β, when examined in an analogous fashion, should also nearly always differ from A-β and B-β in its sequence of codons and accordingly—often enough to be useful—differ in its sequence of nucleotides. This out-of-phase behavior between C-β and the other β chains should, in principle, make it possible to recognize the C-β mRNA in mixtures of α and β-mRNA derived from reticulocyte polyribosomes.[17–18] Such recognition would initially depend on the use of two sources of mRNA. The first source, reticulocytes from an anemic B homozygote, may be expected to contain mRNA for hemoglobin α and B-β. Homozygous B individuals rather than A are chosen since the latter—in the couse of burgeoning reticulocytosis—will lose the capacity for A-β, and presumably A-β mRNA, synthesis. This is of no moment; B homozygotes are an acceptable surrogate since, as already indicated above, A-β and B-β are identical in their amino-terminal portions and probably very similar in β-mRNA structure. The second source of mRNA, reticulocytes from a severely anemic animal homozygous for the β^A-β^C chromosome, should contain messengers for α and C-β. Assuming that the 5′ ends of mRNA, corresponding more-or-less to the free amino termini of polypeptide chains, can be characterized then the 5′ nucleotide beginning the α-mRNA should be common to the two sources while the remaining 5′ nucleotide should represent B-β mRNA in the case of the first source and C-β m-RNA in the case of the second. In other words, the A-B-C system of sheep provides a potential means for distinguishing α- and β-mRNA from one another and thereby an avenue for characterizing individual mRNA in a complex organism. Once the characteristic 5′ nucleotides are recognized then the resulting methodology could perhaps be applied to studies in which, for example, the flux of specific mRNA could be directly quantitated at several stages of erythropoiesis. Such optimistic projection is not, however, without flaw. The recognition scheme may be defeated (1) if insufficient quantities of mRNA are available; (2) if there is appreciable contamination with nonhemoglobin mRNA; or (3) if there exists a nontranslated portion of mRNA that is (a) antecedent to the portion coding for the amino-terminal residue, and (b) common to all hemoglobin mRNA. If conditions (a) and (b) exist then these, in themselves, would be of considerable interest and, for example, bear directly on speculation concerning differential inauguration of A-β and C-β mRNA transcription. Although the existence of these more complex conditions would diminish much of the utility of the recognition scheme for specific mRNA they might still be circumvented through successful analysis of nucleotide sequence in isolated fragments of mRNA that bear the 5′ end. Sooner or later, nucleotides corresponding to the expected codons for the amino-terminal portions of α, B-β, and C-β should be encountered.

Up to this point the tacit assumption has been made that the A→C switch primarily involves the disappearance of mRNA for A-β and the appearance of mRNA for C-β. As an alternative it was, at the outset, possible that mRNA for both β chains might be present but differentially activated in health as opposed to anemia. Distinction between these alternatives is possible as a result of the characteristic synthetic pattern that develops in a nonanemic recipient given either anemic sheep plasma [19–20] or partially purified erythropoietin preparations.[21] Synthetic patterns are obtained in samples of blood and marrow through measurement of the in vitro incorporation of radioactive amino acids into hemoglobins A and C under standardized conditions. As described elsewhere,[19] and shown in Table 1,[21] the response in peripheral blood is roughly divisible into four stages: (1) a latent stage forming the first 24 hours after in vivo challenge during which there is little or no change in synthesis; (2) a second stage, occupying the period 24–48 hours after challenge, during which the synthesis of both hemoglobins A and C increases but with little change from the relative proportions found in the control period; (3) a third stage, peaking at ~72 hours after challenge, during which the synthesis of A decreases and that of C continues to increase; and finally, (4) a recovery stage in the course of which an increase in C as a hemoglobin develops in peripheral blood and during which a slightly enhanced synthesis of C persists for several weeks time.[19–20] The second stage coincides with an increase of reticulocytes in

TABLE 1—*Effects of 7104 Units* of Human Urine Erythropoietin upon Sheep Hemoglobins (Animal #1207)†*

	Hemoglobin Synthesis (μCi. × 10^{-2}/μM total hemoglobin)						Whole Blood	
	Peripheral Blood			Marrow				
Day	A	C	%C	A	C	%C	Hct. %	Retic. %
0	14.0	0.6	4.1	1106.6	31.4	2.8	33	0.1
1	31.3	1.1	3.4	629.9	44.6	6.6	34	0.4
2	248.1	22.1	8.2	716.6	45.6	6.0	33	1.0
3	147.6	111.8	43.1	54.6	146.0	72.8	36	1.1
4	24.8	33.3	57.3	30.3	504.4	94.3	36	0.8
5	13.5	8.2	37.8	420.5	349.1	45.4	34	0.2
6	5.3	1.5	22.1	67.2	46.1	40.7	34	0.1

* 173.5 units/mg. protein.
† 173 units/Kg. sheep body weight.

Specific activities of purified hemoglobins A and C are derived, as detailed elsewhere,[19] by a standardized 3 hour, 35 C. incubation of peripheral blood and marrow samples with a mixture of ^{3}H-amino acids containing a total of 168 microcuries. Cells were obtained from a homozygous β^{A}–β^{C}/β^{A}–β^{C} sheep that received intravenously the indicated quantity of a partially purified preparation of erythropoietin from human urine on day 0. Table is taken from Thurmon et al.[21]

peripheral blood and usually with a fall in total hemoglobin synthesis by marrow (Table 1). The concordance of such events in the second stage of response is at least partially attributable to premature translocation of reticulocytes from marrow to blood by erythropoietin.[22–24] This interpretation is supported by the appearance in sheep of a dose-response relation between erythropoietin unitage employed and total second stage synthesis.[21] The observation that these second stage phenomena can be produced by conditions [19] or doses of erythropoietin [21] that do not provoke the third stage, i.e., $A \rightarrow C$, response provides independent verification for the view that erythropoietin has several distinctive dose-dependent effects.[22–25] Of immediate interest to the present argument is the ~72 hour delay in reaching the $A \rightarrow C$ stage of response. This delay, present in marrow as well as peripheral blood, strongly suggests that increased C synthesis ultimately depends on mRNA that is newly transcribed in young erythropoietic cells and does not depend on differential activation of mRNA preexisting in older cells.

The nature of the $A \rightarrow C$ switching factor has been elusive.[26] In particular it has been difficult to establish identity between this substance and erythropoietin, both of which are present in the anemic sheep plasma given to nonanemic recipients. Several biologic and physical features support the conception of identity between erythropoietin and the $A \rightarrow C$ factor. Both substances are transiently excited by exposure to high altitude hypoxia,[27] both are comparatively scant in animals with moderate anemia, and both are markedly increased when hematocrit falls below a threshold of ~11 per cent [19] to 14 per cent.[28] Moreover, a considerable portion of erythropoietin is known to be nondialyzable through cellulose membranes at neutral pH [29] and, similarly, Emilio Crosby and I (unpublished studies) have found that the $A \rightarrow C$ factor is largely retained by a Union Carbide cellulose membrane during 5-day dialysis at 4C. against 0.02M pH 7.4 phosphate buffer. Although the degree of C response developing with the use of anemic sheep plasma bears a rough correlation to the unitage of erythropoietin given,[20] and even though erythropoietin preparations from human urine can excite considerable $A \rightarrow C$ switching,[30] these observations do not in themselves establish an identity between erythropoietin and the $A \rightarrow C$ factor. Indeed, the aggregate of evidence just recounted still allows the possibility that two separate substances, i.e., erythropoietin *and* $A \rightarrow C$ factor, coexist under the conditions and the preparations described. Thurmon et al.[21] chose circumstances in which the probability of such coexistence by two separate substances was reduced to a low order. In a nonanemic A phenotype recipient sheep $A \rightarrow C$ responses obtained with anemic sheep plasma containing known quantities of erythropoietin were compared with the responses obtained, in the same animal, with known quantities of partially purified erythropoietin from human urine. Irrespective of source, an approximately linear relationship emerges between the dosage of erythropoietin and the degree of $A \rightarrow C$ switching observed. Thus, for example, the maximal effects of 7104 units of erythropoietin from human urine, shown in Table 1, on C synthesis (112×10^{-2} μCi/μM total hemoglobin) were approximately threefold greater than the maximum C synthesis (44.4×10^{-2}

μCi/μCi total hemoglobin) obtained with 4235 units of erythropoietin from anemic sheep plasma. When percentage C is used as the index of comparison an approximately twofold change in C accompanies an approximately twofold change in erythropoietin dosage. This correlation between erythropoietin dose and A → C response exists despite the use of materials that (a) originated in different species (sheep and man), (b) were assayed for erythropoietin in yet another species (mouse), and (c) had grossly different specific activities of erythropoietin such that the specific activity (erythropoietin units/mg. protein) of the human urine preparation was at least 774-fold greater than that in the sheep plasma preparation. The latter feature is perhaps the most impressive since it indicates that a comparatively strict proportionality is maintained between the biologic activities of the A → C factor and erythropoietin despite considerable differences in the quantities of other proteins present. We therefore interpret our findings as strong presumptive evidence for identity between the A → C factor and erythropoietin. The contrary interpretation, i.e., coexistence of two separate substances, seems unlikely in light of circumstances.

Despite the optimism that this provisional identification of erythropoietin as the A → C factor brings to the utility of the model certain difficulties remain. One of these difficulties is conceptual and centers on the possibility of gene-specific cloning as an explanation for the A → C phenomenon. It is possible that the A → C response is no more than an end-stage manifestation of a series of events whereby some clones of cells are precommitted to the synthesis of hemoglobin C but not A while others are precommitted to the converse situation. The limits of commitment are such that at its earliest it might develop once-and-forever in the early embryo and, at its latest, repetitively in the early stages of adult erythropoiesis. If this hypothesis is correct then the action of erythropoietin in producing an A → C response stems from specific stimulation of C-bearing clones. At a molecular level the problem becomes one of *persistent gene suppression* of β^C in one clone and β^A in another rather than one of *episodic gene switching* between β^A and β^C in all cells. If this occurs then it is unusual since cloning has not—in general—been observed among the adult hemoglobins and indeed, as reviewed elsewhere,[19] has been specifically excluded through rather precise methods for a variety of allelic and nonallelic variants. Such examples to the contrary do not, of course, gainsay the existence of A and C cloning in sheep. In any case, the dilemma should not endure very long since the bigenic or monogenic history of hemoglobin β chains present in single erythrocytes can potentially be reconstructed by any of several technics.[31–33]

Another difficulty that besets the continued dissection of the A → C model is operational and stems from an inability to provoke the A → C change in long term incubations and/or cultures of cells from nonanemic animals. Until this limitation is overcome we are left with methods that are cumbersome, require the in vivo use of enormous quantities of erythropoietin for effectiveness, and are—even then—scarcely suitable for detailed exploration of the molecular interactions between erythropoietin and A → C switching.

REFERENCES

1. Boyer, S. H.: Quantitative disparity between hemoglobins. Natural models for the analysis of the control of protein biosynthesis. J. Hered. 58:279, 1967.

2. Leif, R. C., and Vinograd, J.: The distribution of buoyant density of human erythrocytes in bovine albumin solution. Proc. Nat. Acad. Sci. U.S.A. 51:520, 1964.

3. Blunt, M. H., and Evans, J. V.: Changes in the concentration of potassium and in haemoglobin type in Merino sheep under a severe anaemic stress. Nature (London) 200:1215, 1963.

4. Van Vliet, G., and Huisman, T. H. J.: Changes in the hemoglobin types of sheep as a response to anemia. Biochem. J. 93:101, 1964.

5. Huisman, T. H. J., Adams, H. R., Dimmock, M. O., Edwards, W. E., and Wilson, J. B.: The structure of goat hemoglobins. I. Structural studies of the β chains of the hemoglobins of normal and anemic goats. J. Biol. Chem. 242:2534, 1967.

6. Harris, H., and Warren, F. L.: Occurrence of electrophoretically distinct haemoglobins in ruminants. Biochem. J. 60: xxix, 1955.

7. Evans, J. V., King, J. W. B., Cohen, B. L., Harris, H., and Warren, F. L.: Genetics of haemoglobins and blood potassium differences in sheep. Nature (London) 178: 849, 1956.

8. Beale, D., Lehmann, H., Drury, A., and Tucker, E. M.: Haemoglobins of sheep. Nature (London) 209:1099, 1966.

9. Boyer, S. H., Hathaway, P., Pascasio, F., Orton, C., Bordley, J., and Naughton, M. A.: Hemoglobins in sheep: Multiple differences in amino acid sequences in three beta-chains and possible origins. Science 153:1539, 1966.

10. Wilson, J. B., Edwards, W. C., McDaniel, M. Dobbs, M. M., and Huisman, T. H. J.: The structure of sheep hemoglobins. II. The amino acid composition of the tryptic peptides of the non-α chains of hemoglobins. Arch. Biochem. 115:385, 1966.

11. Boyer, S. H., Hathaway, P., Pascasio, F., Bordley, J., Orton, C., and Naughton, M. A.: Differences in the amino acid sequences of tryptic peptides from three sheep hemoglobin β chains. J. Biol. Chem. 242: 2211, 1967.

12. Kitchen, H., Eaton, J. W., Taylor, W. J.: Rapid production of a hemoglobin by induced hemolysis in sheep: Hemoglobin C. Amer. J. Vet. Res. 29:281, 1968.

13. Huisman, T. H. J., Reynolds, C. A., Dozy, A. M., and Wilson, J. B.: The structure of sheep hemoglobins. The amino acid compositions of the α and β chains of the hemoglobins A, B and C. J. Biol. Chem. 240:2455, 1965.

14. Smithies, O., Connell, G. E., and Dixon, G. H.: Chromosomal rearrangements and the evolution of haptoglobin genes. Nature (London) 196:232, 1962.

15. Baglioni, C.: The fusion of two peptide chains in hemoglobin Lepore and its interpretation as a genetic deletion. Proc. Nat. Acad. Sci. U.S.A. 48:1880, 1967.

16. Guest, J. R., and Yanofsky, C.: Relative orientation of gene messenger and polypeptide chain. Nature (London) 210:799, 1966.

17. Burny, A., and Marbaix, G.: Isolement du RNA messager des réticuloytes de lapin. Biochem. Biophys. Acta 103:409, 1965.

18. Laycock, D. G., and Hunt, J. A.: Synthesis of rabbit globin by a bacterial cell free system. Nature (London) 221:1118, 1969.

19. Boyer, S. H., Crosby, E. F., and Noyes, A. N.: Hemoglobin switching in non-anemic sheep. I. Mediation by plasma from anemic sheep. Johns Hopkins Med. J. 123: 85, 1968.

20. Gabuzda, T. G., Schuman, M. A., Silver, R. K., and Lewis, H. B.: Erythropoietic kinetics in sheep studied by means of induced changes in hemoglobin phenotype. J. Clin. Invest. 47:1895, 1968.

21. Thurmon, T. F., Boyer, S. H., Crosby, E. F., Shepard, M. K., Noyes, A. N., and Stohlman, F.: Hemoglobin switching in non-anemic sheep. III. Evidence for identity between the A $\rightarrow$ C factor and erythropoietin. In preparation.

22. Brecher, G., and Stohlman, Jr., F.: Reticulocyte size and erythropoietic stimulation. Proc. Soc. Exp. Biol. Med. 107:887, 1960.

23. Borsook, H., Lingrel, J. B., Scaro, J. L., and Millette, R. L.: Synthesis of haemoglobin in relation to the maturation of erythroid cells. Nature (London) 196:347, 1962.

24. Borsook, H.: Editorial. A picture of erythropoiesis of the combined morphologic and molecular levels. Blood 24:202, 1964.

25. Stohlman, Jr., F.: Humoral regulation of erythropoiesis. VII. Shortened survival of erythrocytes produced by erythropoietine or severe anemia. Proc. Soc. Exp. Biol. Med. 107:884, 1961.

26. Moore, S. L., Godley, W. C., van Vliet, G., Lewis, J. P., Boyd, E., and Huisman, T. H. J.: The production of hemoglobin C in sheep carrying the gene for hemoglobin A: Hematologic aspects. Blood 28:314, 1966.

27. Boyer, S. H., Crosby, E. F., Noyes, A. N., Kaneko, J. J., Keeton, K., and Zinkl, J.: Hemoglobin switching in non-anemic sheep II. Response to high altitude. Johns Hopkins Med. J. 123:92, 1968.

28. VanDyke, D. C., Layrisse, M., Lawrence, J. H., Garcia, J. F., and Pollycove, M.: Relation between severity of anemia and erythropoietin titer in human beings. Blood 18:187, 1961.

29. Lewis, J. P., Alford, D. A., Moores, R. R. Gardner, Jr., E., Wright, C-S., Smith, L. L., Scharnitzky, W. A.: Dialysis as a tool in the study of erythropoietin. J. Lab Clin. Med. 73:154, 1969.

30. Huisman, T. H. J., Lewis, J. P. Blunt, M. H., Adams, H. R., Miller, A., Dozy, A. M., and Boyd, E. M: Hemoglobin C in newborn sheep and goats. A possible explanation for its function and biosynthesis. Pediat. Res. 3:189, 1969.

31. Matioli, G. T., and Niewisch, H. G.: Electrophoresis of hemoglobin in single erythrocytes. Science 150:1824, 1965.

32. Marchalonis, J. J., and Nossal, G. J. V.: Electrophoretic analysis of antibody produced by single cells. Proc. Nat. Acad. Sci. U.S.A. 61:860, 1968.

33. Gitlin, D., Sasaki, T., and Vuopio, P.: Immunochemical quantitation of proteins in single cells. I. Carbonic anhydrase B, β-chain hemoglobin and γ-chain hemoglobin in some normal and abnormal erythrocytes. Blood 32:796, 1968.

Mode of Action of Erythropoietin in Polycythemic Mice III.: The Dose Response Curve

By A. L. Kretchmar, T. P. McDonald, and R. D. Lange

We have discussed, in the first two papers [1,2] of this series, the results of experiments which measured the erythropoietic response to erythropoietin (ESF) in polycythemic mice that had been treated with fluorouracil and in mice not treated with cytotoxic agents but with polycythemia induced by hypoxia or by transfusion. In this paper we report certain characteristics of the dose-response relation observed when ESF is injected into polycythemic mice and the suddenness of recovery of the ability of these animals to respond to ESF after injury to their hemopoietic system by physical agents such as drugs or x-rays.

Our discussions are based on a speculative picture of the hemopoietic system. This system is a cell renewal system.[3] There are only two general categories of cellular elements: biochemically differentiated cells, and progenitor cells. The differentiated cells arise from the progenitor cells in response to "signals" by a process which may be discussed in the context of enzyme induction.[4,5] The signals affect the progenitor cells only at "certain times" in their life cycle [5,6] and we assume that this life cycle is variable in time. Thus, progenitor cells proceed from birth by mitosis through a postmitotic time (G_1), the length (hours) of which varies systematically depending upon the number of progenitors in the population making up the cell renewal system. From G_1 the cells proceed into an S, G_2 and M phase.[4] The length of S, G_2 and M is assumed, for mathematical convenience, not to have a systematic dependence on the size of the population. In considering erythropoietic differentiation, we take the "signal" to be ESF and the "certain time" to be an interval late in G_1.[7]

We shall present, in this paper, some implications that this general picture has with regard to: (1) the relation between erythropoietic response and amount of erythropoietin to which the system is exposed, and (2) the rate of recovery of the sensitivity to erythropoietin.

Experimental Part

Early observations on the use of polycythemic mice as assay animals for erythropoietin [8] showed that hypertransfused mice were less sensitive to 0.5 units erythropoietin than were mice with hypoxia-induced erythremia. McDonald et al.[2] have pointed out that this difference in sensitivity, as well as a decline in sensitivity in mice with hypoxia-induced erythremia, is predicted from the model outlined. Indeed, a further prediction is that the dose-response

---

From the University of Tennessee Memorial Research Center and Hospital, Knoxville, Tennessee.

Supported in part by U.S.P.H.S. Research Grants RO1–AM 11894–02 and RO1–HE 10567–03 and by contract number AT–(40–1)–3547 from the U. S. Atomic Energy Commission.

relation should be different both quantitatively and qualitatively (see description of model below).

To test this prediction, female C3H mice were made polycythemic by housing them in enclosures with silicone rubber membrances as described by Lange et al.[9] The polycythemic animals were removed from the enclosures, randomly assigned to groups of 5 mice each, and housed in the usual type cages in ambient room air (approximately 20 per cent 0_2). After 4 or 7 days in room air, erythropoietin was injected in doses that ranged from 0.05 to 0.5 units and per cent ^{59}Fe incorporation in RBC measured as an index of erythropoietic response to the erythropoietin injected. The technical details have been fully described.[1,2,9]

The data are shown in Figure 1. The dose-response relation to mice 4 days

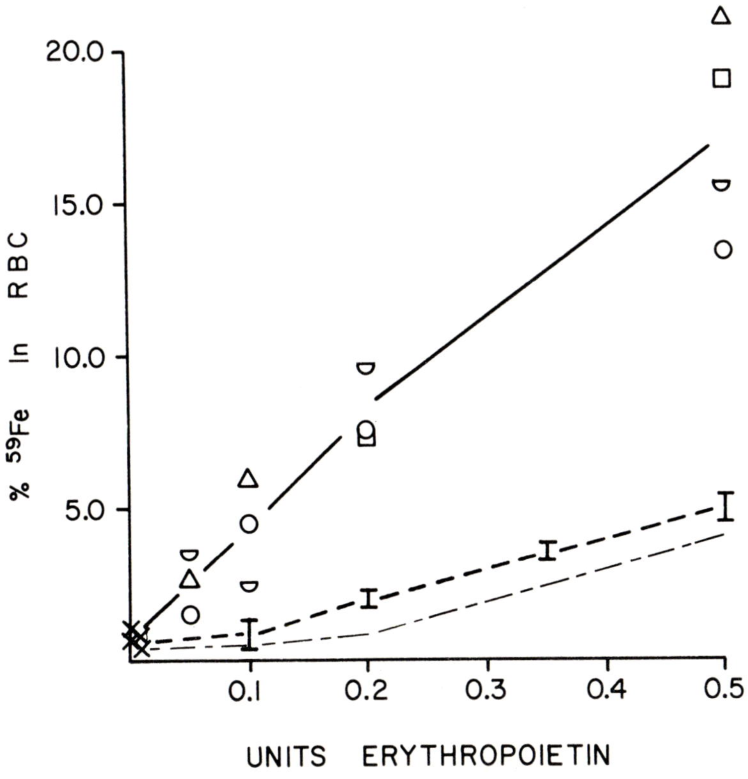

FIG. 1—Per cent of injected ^{59}Fe incorporated in RBC of polycythemic mice plotted against the amount (units) of erythropoietin administered prior to ^{59}Fe. The upper curve summarizes results from mice 4 days out of the hypoxic environment that induced the polycythemia. Symbols represent median uptake in groups of 5 mice per point; 4 separate experiments are summarized. The response shown with the triangles was determined with a different erythropoietin preparation than was used in the other 3 experiments. The middle curve shows data obtained when the same procedures were applied to mice 7 days out of the hypoxic environment. The dashed line is drawn through the average of 12 median uptakes (4 mice per group) at each point except 0.1 u. (3 medians; 5 mice per group). The vertical lines represent plus and minus-one standard error around the average of the medians. The lowest faint line represents uptake in hypertransfused mice.

out of the enclosure has been determined many times with the same and with different preparations of ESF. Quantitative differences have been noted with different preparations and from time-to-time with the same preparation of ESF. However, qualitatively the relation is consistent and is a parabolic function as shown in Figure 1. The best linear transformation for the 4 day data is to relate per cent ^{59}Fe uptake to a power of the dose of ESF, (i.e., per cent $^{59}\text{Fe}=a+bD^{\alpha}$: where a and b are constants, $\alpha=0.75$, and D is dose of ESF [T. J. Mitchell, unpublished]. The important characteristic of this function for subsequent discussion is that the dose-response relation in mice 4 days out of the enclosure does not have a "toe" but rather rises with maximum slope from the saline injected control level at the lowest doses of ESF used.

In contrast, after 7 days in ambient room air, as shown in Figure 1, mice not only have diminished response to the same dose of ESF but also the dose-response relation is qualitatively different; there is a definite "toe" with minimal slope at the lowest doses of ESF used. A similar (but somewhat lower) dose-response relation was observed in hypertransfused mice.

The suddenness with which responsiveness to ESF recovers in mice given cytotoxic doses of vinblastine is impressive.[11] Thus, uptake of ^{59}Fe after 5 units of ESF, while only 20 per cent of normal at 1 day, was within normal range at 1.6 days.[11] Our model predicts that recovery from injury caused by a cytotoxic agent that damages cells undergoing mitosis, as vinblastine does,[12] should be very rapid (see description of model below).

To test the prediction, mice were made polycythemic by housing them in membrane cages as above. On the day that they were removed from the enclosures, randomly assigned to groups of 5, and returned to ambient room air, approximately one-half the mice were injected with 50 γ vinblastine sulfate per mouse intraperitoneally. Recovery of responsiveness to ESF was tested by injecting 0.5 units ESF and measuring per cent ^{59}Fe in RBC on days 2, 3, 5 after vinblastine. Some results are shown in Figure 2. By 2 days animals already show either approximately normal response to 0.5 units ESF or no response (mice not given ESF generally had 2 to 5 per cent incorporation). At 3 days a number of animals were distinctly more responsive than normal but one still failed to respond. Five days after vinblastine the mean response was greater than in the mice not given the drug. There was no evidence of exponential recovery with time of responsiveness to ESF. Exponential recovery would be expected if differentiation occurred from an exponentially growing progenitor population which was recovering from loss of many of its members by death from the cytotoxicity of vinblastine. The response found was all or none followed by a tendency to elevated responsiveness.

The decline in the responsiveness of mice not given vinblastine (ESF only) was also predicted from our model and is the subject of another report.[2]

Description of the Model

The ideas regarding erythroid differentiation outlined in the introduction have formed the basis of a probabilistic model which has been reported by

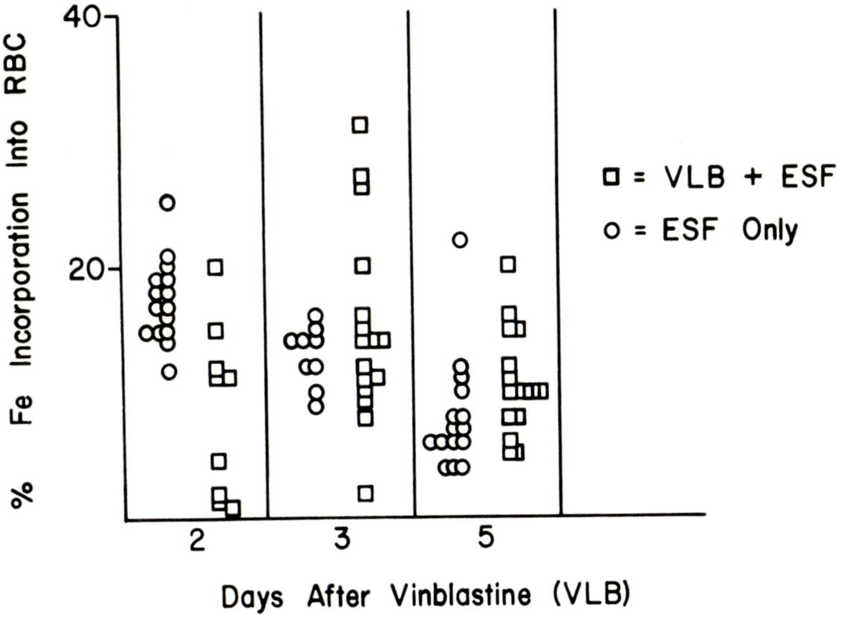

FIG. 2—Response of polycythemic mice to erythropoietin at 2, 3, or 5 days out of an hypoxic environment. Some animals were given 50 γ of vinblastine sulfate on the day they were removed from the hypoxic environment (□ = VLB + ESF). The other mice were injected with saline and served as controls (0 = ESF only). All mice were given the same amount (0.5 units) of erythropoietin.

Hoel and Mitchell.[13] Here we will illustrate the basic assumptions and give sample calculations. Some of our assumptions are highly restrictive and in this sense are unfair to the concepts outlined. We accept these restrictions in order to be able to get numerical solutions to the equations involved. The results are presented not in a spirit of offering a complete or final explanation of a very complex biologic phenomenon but rather to bring out certain predictions and implications of the idea that differentiation, in response to ESF, occurs from a progenitor population whose mean generation time varies according to the number of cells in the population. The variation is assumed to be primarily in length of G_1. Thus, when proliferation of progenitors is rapid, as in recovery of the population from injury, G_1 is short and the generation time of cells is also short; when proliferation is minimal and there are many cells in the population, G_1 is long. We assume that this is the way the size of the population is controlled, i.e., by variation in the distribution of length of G_1 for the cells in the population.

Figure 3 divides the population of progenitor cells into 2 categories. Most of the cells in the population in normal mice are in G_1. Cells in this phase are continually entering S, then G_2, and then undergoing mitosis. We assume that the cells in S, G_2 or M proceed in an orderly fashion through these phases in a time, τ, which is a random variable that has no dependence on the number of cells in the population. Thus the mean length of S, G_2, and M would be the same for rapidly proliferating and slowly growing populations. As illustrated

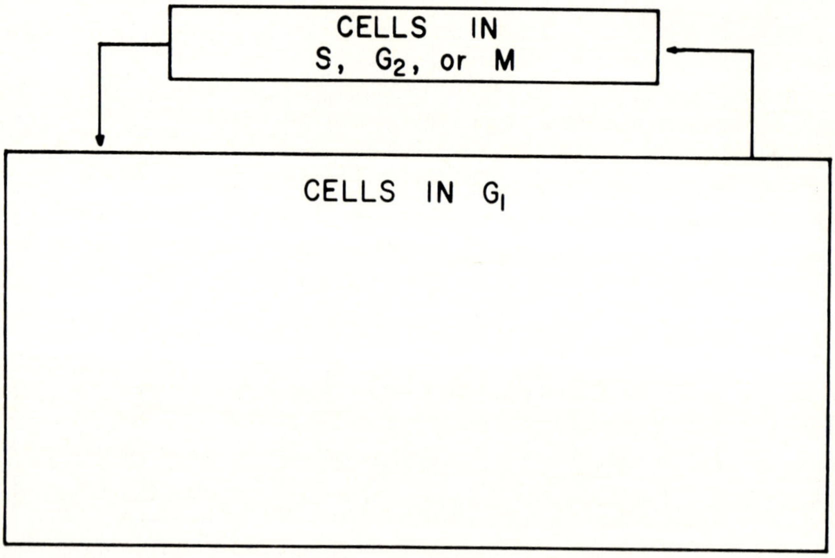

FIG. 3—This illustrates the conceptual distinction to be made in more detail in subsequent figures. All cells are to be considered as members of the same population, however, we wish to distinguish those cells that happen to be in G_1 from those cells that are in the other phases of the generative cycle.

in Figure 4, the distribution function for τ is a gamma function with parameters such that the mean time of S, G_2, and M is about 8–8.5 hours.[14]

The cells in G_1 are assumed to remain in G_1 until a coordinated synthesis of RNA and protein is completed.[15] When this synthesis is completed, the cells soon enter S. The time after birth by mitosis to completion of this synthesis is assumed to be a random variable, t, with a negative exponential distribution function. The parameter in this function is assumed to depend upon the number of cells in the progenitor population. A possible mechanism for this control might be inhibition of the coordinated RNA-protein synthesis [15] by chalone [16] produced by progenitors. Thus, when many progenitors are present, the chalone titer would be high and delay in coordinated synthesis prolonged; G_1 would be long with distribution function of t like that illustrated in Figure 5 by the short dashes (- - - -). When the number of progenitors was small, the chalone titer would be low and there would be no delay in synthesis; G_1 would be short with distribution function of t shown in Figure 5 by the longer dashes. When the population was at some intermediate size, the distribution function would be intermediate as illustrated by the solid line in the upper part of Figure 5.

Now, as is necessary for some other differentiating cell systems,[17] we assume that G_1 must be longer than a certain minimal time, indicated by the first vertical line to the right of the ordinate in the upper part of Figure 5, in order that a cell can be induced to differentiate. In addition, suppose that the ability of the progenitor to interact effectively with ESF begins to decay as G_1 becomes long (indicated by the second vertical line). Thus, there is a "window" in time (during G_1) when a dividing progenitor cell could interact with ESF in a manner that would lead to differentiation. In our model, then, the currently developing concept of "critical mitosis" in differentiating cell systems [18] is the mitosis which immediately succeeds effective interaction with ESF. The interaction takes place during a restricted part of G_1 such that an "effective load" of ESF is carried into S.[6] Thus, only cells that will have a G_1 with length (e.g., 1–3 hours) somewhere in the ESF sensitive "window" will undergo "critical mitosis" and subsequent differentiation.

One can immediately see by comparing the area between the vertical lines under the different distribution functions shown in Figure 5 (upper part) that responsiveness to ESF will be minimal in a rapidly proliferating population of progenitors because only a very small fraction of the total area under the distribution function (long dashes) falls in the "window." This means that the probability is concentrated in the region of small t (short time of G_1), hence very few dividing cells will have G_1 sufficiently long to undergo effective interaction with ESF. Also for a slowly proliferating population, this fraction of the total area under the distribution function (short dashes) is small but for intermediate distributions of t there will be a maximum proportion of probable lengths of G_1 in the window and, therefore, a maximum responsiveness to ESF in the population of progenitors.

We can use these distribution functions to compute the number of cells in a hypothetical population of progenitors that will have G_1 which is: (1) less than 1 hour, (2) in the time window of sensitivity to ESF, or (3) longer than 3 hours (Fig. 6).

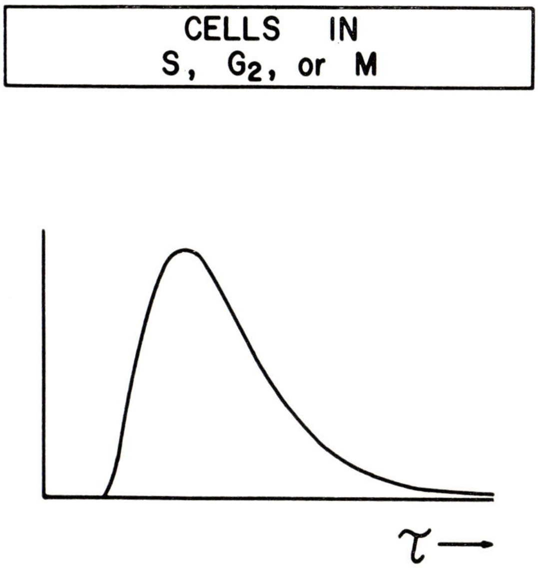

FIG. 4—The length of time (hours) that a cell will spend in S, then G_2, and finally M is defined to be a random variable, denoted by τ, which has a delayed gamma density function illustrated in the lower part of the figure.

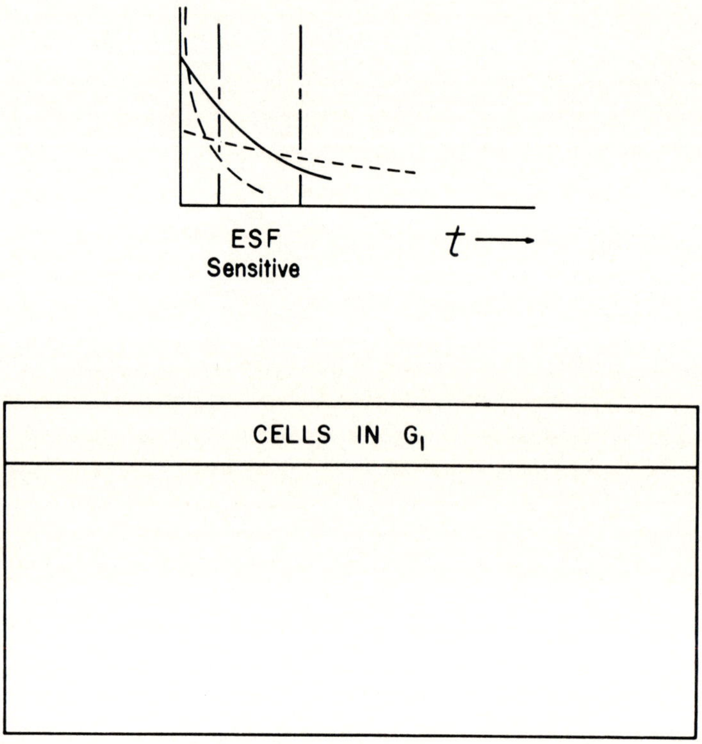

FIG. 5—The length of time (hours) that a cell will spend in G_1 before the coordinated synthesis of RNA and protein [15] necessary for it to proceed through the remainder of its proliferation cycle is defined as a second random variable t. The density function of t is assumed to be a negative exponential whose parameter is a function of the number of cells in the population. See text for an explanation of the different density functions shown (— — — —, ————, - - - - - -).

For the computations shown in Figure 6 we consider that 400 cells/hour are completing mitosis and inquire how many of these would have G_1 of various lengths if the mean generation time in the population were 8.75 hours, 9.0 hours, etc. Of the 400 cells/hour, 396 would have G_1 less than 1 hour long and hence enter cycle in 0–1 hours if the mean generation time were 8.75 hours. Only 4 cells would become sensitive to ESF. With 396 cells entering cycle it is clear that the population would grow rapidly and be unresponsive to ESF. A population of progenitor cells would be most responsive to ESF when the mean generation time was 9.5 hours. This population could still grow even though exposed to high levels of ESF since, even if all 118 sensitive cells were to differentiate, 268 would enter cycle before they could respond to ESF, and only 200 must enter cycle to sustain the 400 cell/hour output. When the mean generation time in the population exceeds about 9.5 hours, a substantial fraction of the population will have G_1 longer than 3 hours. At some particular mean generation time (in the example shown in Figure 6 it is 13.5 hours) exactly 200 cells/hour enter cycle and the output of 400 cells/hour would be sustained and constant. This population could deliver 200 cells/hour for differentiation and still maintain constant size. About 64 per cent of the cells entering cycle would come from the fraction with G_1 longer than 3 hours. One can see from Figure 6 that it would never be possible to differentiate all progenitors no matter how high the dose of ESF.

We can use calculations of the type summarized in Figure 6 to compute the dose-response relation of populations with various specified mean generation times. The dose-response relations shown in Figure 7 were computed by assuming that interaction between cells sensitive to ESF was a random process with Poisson statistics and that for G_1 longer than 3 hours more than one "hit" with ESF would be required for differentiation while cells with G_1 in the ESF sensitive window would respond after only a single "hit."

Consideration of the computations shown in Figure 7 indicates that the dose-response relation changes from a parabolic relation at maximum sensitivity to ESF (mean generation time of 9.5 hours) to one that has diminished slope and has a definite "toe." Referring back to Figure 1, we note that this is the kind of difference between the dose-response relations found when mice 4 days out of hypoxia were compared with mice 7 days posthypoxia.

We suggest, therefore, that the process of inducing polycythemia by hypoxia, since a much elevated rate of erythroid differentiation is required to generate and maintain the increased red cell mass, causes increased proliferation of progenitors. This results in a decrease in mean generation time with increased sensitivity to ESF. When the polycythemic animals are returned to ambient

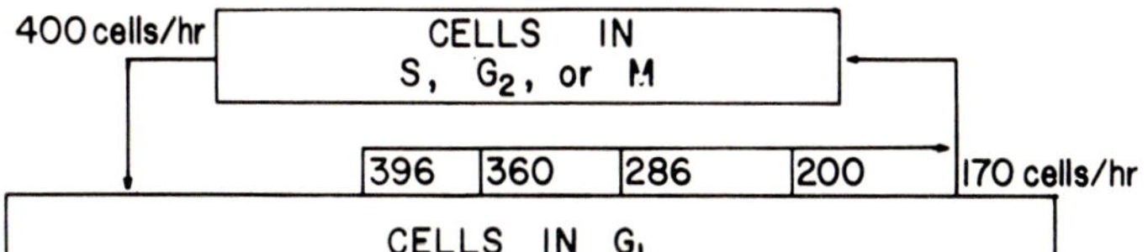

CELLS IN G_1					
Fate	Average Generation Time (hours)				
	8.75	9.0	9.5	13.5	28.5
Enter Cycle In 0-1 Hours	396	360	268	72	20
EP Sensitive	4	39	118	107	37
G_1 Longer Than 3 Hours	0	0	9	951	6009

FIG. 6—Illustrative calculations generated by making the arbitrary assignments for mean generation time shown in the table. The numbers shown in the "Enter cycle in 0–1 Hours" and erythropoietin sensitive ("EP Sensitive") categories refer to the proportion of 400 new cells/hour that would enter these categories. The number shown in "G_1 Longer Than 3 Hours" refers to the number of cells in this category if a population with mean generation times shown were allowed to come to equilibrium. These numbers may be meaningless with regard to response of living animals to erythropoietin and are presented only to bring out the idea that response to erythropoietin may depend upon the rate of proliferation in the population of progenitor cells.

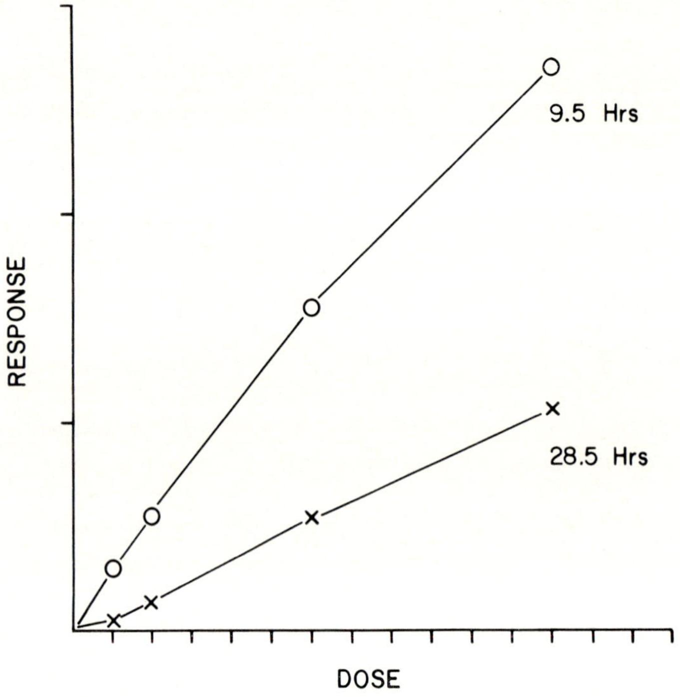

FIG. 7—The dose response relation that could be expected for 2 of the 5 arbitrary assignments of mean generation times shown in Figure 6. The response for a maximum sensitivity population (9.5 hours mean generation time) is compared with the response of a less sensitive population (longer mean generation time). Again the numbers shown may be irrelevant to the response of living animals to erythropoietin. They are presented only to illustrate the point that our model predicts that the dose-response relation may depend upon the rate of proliferation in the population of progenitor cells.

air, red cell differentiation ceases, proliferation in the progenitor population subsides, the mean generation time becomes longer and responsiveness to ESF decreases with a change in the dose-response relation as pictured in Figure 7 and demonstrated in Figure 1.

When a population of progenitors is injured by cytotoxic agents, we suggest that the rapid proliferation among surviving cells that is necessary for recovery leads to a reduction in mean generation time to the minimum possible (8.75 hours in Figure 6), the population is not responsive to ESF and grows rapidly. Note that as the population returns to normal size and proliferation slows down (mean generation time increasing) a very small change in mean generation time results in a very large increase in the number of cells that are sensitive to ESF. Thus, recovery of responsiveness to ESF after injury to progenitors would be sudden and we would expect to find (Fig. 2) among a group of mice recovering from the cytotoxic effects of vinblastine individuals with no response to 0.5 units of ESF, others with normal response, and some with elevated responsiveness depending on the extent of recovery, and the mean generation time in the progenitor population.

SUMMARY

We have presented a model from which predictions can be made about the response to erythropoietin in polycythemic mice. The experimental results presented were consistent with certain predictions that were developed in the text.

REFERENCES

1. Kretchmar, A. L., McDonald, T. P., and Lange, R. D.: Mode of action of erythropoietin in polycythemic mice. I. Response in mice treated with 5-fluorouracil. J. Lab. Clin. Med. In press.

2. McDonald, T. P., Lange, R. D., and Kretchmar, A. L.: Mode of action of erythropoietin in polycythemic mice. II. Response in mice with polycythemia induced by hypoxia or transfusion. In preparation.

3. Bond, V. P., Fliedner, T. M., and Archambeau, J. O.: Mammalian Radiation Lethality. A Disturbance in Cellular Kinetics. New York, Academic Press, 1965, p. 15.

4. Lajtha, L. G., Oliver, R., and Gurney, C. W.: Kinetic model of a bone-marrow stem-cell population. Brit. J. Haemat. 8: 442, 1962.

5. Goldwasser, E.: Biochemical control of erythroid cell development. *In* Moscona, A. A., and Monroy, A. (Eds.): Current Topics in Developmental Biology, Vol. 1. New York, Academic Press, 1966.

6. Kretchmar, A. L.: Erythropoietin: Hypothesis of action tested by analog computer. Science 152:367, 1966.

7. Hodgson, G., and Eskuche, I.: Study of the effects of chemotherapeutic agents on the "early" and "late" responses to erythropoietin. Proc. Soc. Exp. Biol. Med. 127:328, 1968.

8. DeGowin, R. L., Hofstra, D., and Gurney, C. W.: The mouse with hypoxia-induced erythremia, an erythropoietin bioassay animal. J. Lab. Clin. Med. 60:846, 1962.

9. Lange, R. D., Simmons, M. L., and Dibelius, N. R.: Polycythemic mice produced by hypoxia in silicone rubber enclosures: A new technique. Proc. Soc. Exp. Biol. Med. 122:761, 1966.

10. Lange, R. D., McDonald, T. P., and Jordan, Tommye: Antisera to erythropoietin: partial characterization of two different antibodies. J. Lab. Clin. Med. 70:78, 1969.

11. Hodgson, G.: Effect of vinblastine and 4-amino-N^{10} methyl pteroylglutamic acid on the erythropoietin responsive cell. Proc. Soc. Exp. Biol. Med. 125:1206, 1967.

12. Bruchovsky, N., Owen, A. A., Becker, A. J., and Till, J. E.: Effects of vinblastine on the proliferative capacity of L cells and their progress through the division cycle. Cancer Res. 25:1232, 1965.

13. Hoel, D. G., and Mitchell, T. J.: Computer simulation of a cellular proliferation model. Biometrics 25:445, 1969.

14. Morse, B. S., Rencricca, N. J., and Stohlman, F., Jr.: The effect of hydroxyurea on differentiated marrow erythroid precursors. Proc. Soc. Exp. Biol. Med. 130: 986, 1969.

15. Fujiwara, Y.: Role of RNA synthesis in DNA replication of synchronized populations of cultured mammalian cells. J. Cell Physiol. 70:291, 1967.

16. Bullough, W. S.: Mitotic and functional homeostasis: A speculative review. Cancer Res. 25:1683, 1965.

17. Bischoff, R., and Holtzer, H.: Mitosis and the processes of differentiation of myogenic cells in vitro. J. Cell. Biol. 41:188, 1969.

18. Holtzer, R., Bischoff, R., and Chacko, S.: Activities of the cell surface during myogenesis and chondrogenesis. *In* Smith, R. T., and Good, R. A. (Eds.): Cellular Recognition. Appleton-Century-Crofts, New York, 1969.

The Mechanism of Action of Erythropoietin in Relationship to Cell Cycle Kinetics

By Bernard S. Morse, Nicholas J. Rencricca, and Frederick Stohlman, Jr.

The mechanism of action of erythropoietin (EP) and the maintenance of homeostasis within the precursor compartments have dominated the investigative efforts in the field of hemopoietic differentiation for many years. It is clear that the marrow differentiated nucleated erythroid elements are not normally self-sustaining but are continually fed from a more primitive precursor compartment. It is also firmly established that EP transforms precursor cells into recognizable pronormoblasts. The mechanism by which this occurs remains controversial but it might be suggested that the initiation of differentiation is intimately related to the ability to synthesize heme. The observations of Till and McCulloch [1] would indicate that the precursor cell is transplantable and pluripotential. Available evidence suggests the presence of a committed precursor compartment which is intermediate between the pluripotential and the formed hemopoietic elements.[2] In the case of the erythroid series the intermediate compartment appears already committed to erythroid differentiation but requires EP to initiate this process. The intermediate compartment which bears the overwhelming burden for the continual supply of differentiated cells appears to a large extent self-maintaining but when damaged, may be repopulated from the pluripotential compartment.[2] An increased demand for differentiated cells could be in part achieved by a decreased turnover time of the committed compartments, i.e., increasing the proportion of cells in cycle and/or shortening the duration of cycle. Lajtha [3] proposed a kinetic model of the stem cell in which EP differentiates resting or G_0 procursor cells whereas cells in cycle are for the most part unavailable for differentiation. Homeostasis would be maintained by triggering a portion of remaining G_0 cells into the generative cycle culminating in homoioplastic division. Kretchmar [4] in contrast, constructed a model in which precursor cells are in the active generative cycle; however, the time parameters for G_1 are variable. EP must enter the cell during G_1 and remain within the cell for a portion of DNA synthesis (S) so that it may exert its effect during the replication of that part of the genome governing differentiation into a pronormoblast. It is further assumed that the effective intracellular life for EP is shorter than the maximum duration of G_1 and presents a situation in which some of the cells that interact with EP would be differentiated and others would not. Homeostasis would be maintained by shortening the duration of G_1 thereby increasing the probability that cells would enter DNA S and complete the generative cycle.

In an attempt to distinguish between these alternative models the effect

From the Department of Medicine, St. Elizabeth's Hospital, Tufts Medical School, Boston, Massachusetts, and New Jersey College of Medicine and Dentistry, Newark, New Jersey.

Supported in part by HE5600 and HE7542 from the National Heart Institute to St. Elizabeth's Hospital.

of EP was determined in hypertransfused mice treated with hydroxyurea (OHU). OHU has been shown to provide a means by which cell cycle kinetics may be perturbed in a definable and limited fashion in vivo since it exerts a lethal effect on those cells in DNA S.[5] The results to be reported herein appear to favor a variable generation time (t_G) for committed erythroid precursor cells, the consequence of a variable G_1. This compartment when damaged appears to be in part repaired by the influx of cells from the pluripotential precursor compartment.

Materials and Methods

Ten-week-old female CF1 mice were transfused intraperitoneally with washed packed homologous red cells on 2 consecutive days. Six days later they received human urinary EP (3 units) and/or OHU (900 mg./kg.) intravenously via the tail vein. Blood samples were collected from anesthetized animals by cardiac puncture. Reticulocytes were stained with new methylene blue. Animals with hematocrits of less than 59 per cent were excluded from the study. The cellularity of tibial marrow contents was measured by electronic particle counting as described by Fruhman.[6] Bone marrow smears prepared by a brush technic were stained with Wrights and Giemsa. The percentage of nucleated erythroid cells was estimated from differential counts of 500 to 1000 nucleated cells. The percentage of pronormoblasts was based on the further assessment of 500 nucleated erythroid cells. CFU concentration of tibial contents were estimated in heavily irradiated syngeneic recipients as previously described.[7] The spleen colonies were counted independently by 2 observers.

Results

Erythropoiesis was virtually nonexistent in control hypertransfused mice. Six hours after the injection of 3 units of EP there was a significant increase in the numbers of pronormoblasts in the bone marrow ($p < 0.01$) (Table 1). Forty-eight hours after EP, marrow erythropoiesis was maximal and thereafter rapidly declined to transfused control levels by the fourth to fifth day. The rapid decrease of marrow erythropoiesis was associated with the emergence of reticulocytes into the peripheral blood. When EP and OHU were administered

Table 1—*The Appearance of Pronormoblasts After Erythropoietin*

	Pronormoblasts $\times 10^{-6}$/Tibia	Red Cell Precursors $\times 10^{-6}$/Tibia
Control	0.007 ± 0.001	0.05 ± 0.01
EP–6 *	0.030 ± 0.007	0.07 ± 0.01
EP–12	0.066 ± 0.006	0.12 ± 0.01
EP–18	0.076 ± 0.013	0.22 ± 0.03
EP–24	0.070 ± 0.009	0.34 ± 0.04

All animals were transfused 7 days before treatment. Each value represents the mean ± S. E. of at least 7 animals.

* Refers to time of sacrifice in hours after an injection of erythropoietin (EP).

simultaneously the resultant erythroid response was slightly reduced and slightly more prolonged relative to EP treated controls (Fig. 1). When OHU was administered at 4, 7 or 11 hours after EP, there was a progressively greater reduction of erythroid response (Fig. 2). Comparison of areas under the reticulocyte curves to the EP treated controls reveals an exponential decrease of erythroid response (Fig. 3) and indicates a progressively greater kill by OHU with time after EP. When a second dose of OHU was given 7 hours after the simultaneous administration of EP and OHU, the erythroid response was nearly abolished, indicating a synergistic rather than cumulative effect.

After establishing that the response to EP was diminished by OHU, it could be inferred that OHU killed a portion of cells in the committed erythroid precursor compartment, presumably that portion of cells in DNA S. The recovery of this compartment was then studied by administering EP at intervals after OHU. When EP was given 4 hours after OHU, the extent of erythropoiesis was comparable to that observed after the simultaneous administration of EP and OHU. Seven hours after OHU, the response approached that observed

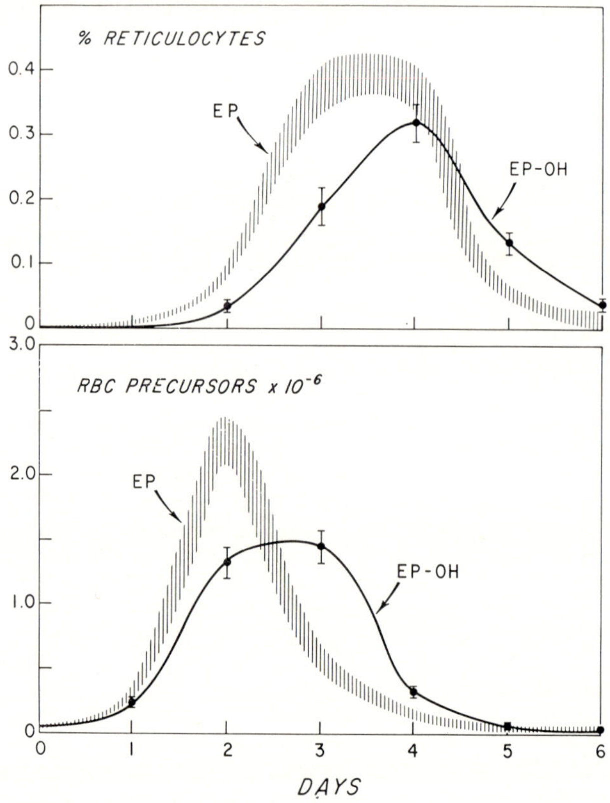

FIG. 1—The shaded area represents the mean ± SE erythroid response of control plethoric mice to 3 units of erythropoietin (EP) administered on day 0. The response to EP and hydroxyurea (OH) when administered simultaneously is depicted by the solid line. Vertical lines through the points indicate ± 1 SEM. (Reprinted courtesy of Blood.[7])

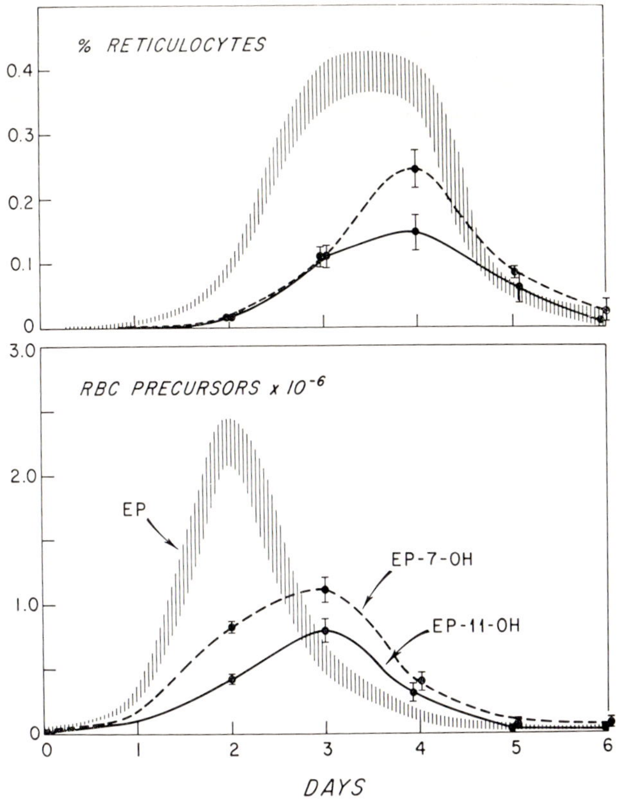

FIG. 2—The erythroid response when hydroxyurea (OH) was administered 7 (solid line) and 11 hours (dotted line) after EP. (Reprinted courtesy of Blood.[7])

in EP treated controls (Fig. 4). Fourteen hours after OHU there was a substantial overshoot in response to EP (Fig. 4).

The effects of OHU on the pluripotential compartment were evaluated by transplantation of bone marrow. There was no appreciable change in the numbers of colony-forming units during the first 4 hours after treatment with OHU. In contrast, at 14 hours there was a significant decrease in colony-forming units ($p = 0.025$) (Fig. 5).

DISCUSSION

The data presented herein bear significantly on the evaluation of several models of erythropoiesis and provide further insight into the maintenance of homeostasis within the precursor compartments. The use of OHU to perturb cell cycle kinetics appears to have several advantages over other cytotoxic agents. It is a simple molecule that rapidly equilibrates in the body water. The duration of action for the dose employed in the present studies is of the order of 2 hours. During this period, cells in DNA S undergo karyorrhexis and karyolysis, whereas cells in other phases of the generative cycle are apparently

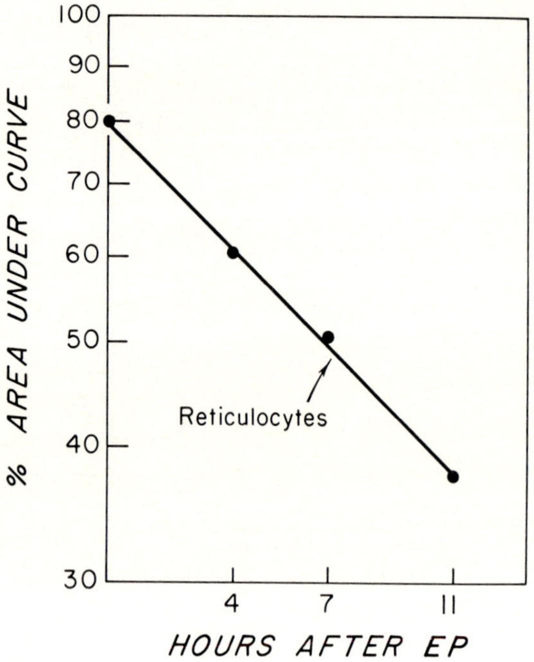

FIG. 3—Areas under the reticulocyte curve are compared to the area obtained in EP-treated control animals. (Reprinted courtesy of Blood.[7])

unaffected as judged by their normal progression through cell cycle.[5] Although OHU exerts a profound inhibitory effect on DNA S, RNA and protein synthesis are unaffected. In the present study, the reduction of erythroid response to the simultaneous administration of EP and OHU appears best explained by the death of a portion of cells capable of responding to EP, presumably that portion in DNA S. Had the compartment been in G_0 as suggested by Lajtha, then no effect from OHU would have been anticipated since none of the cells would have been in DNA S. Relative to EP treated controls, the reduction of erythroid response to the simultaneous administration of EP and OHU was slight which suggests that the fraction of precursor cells in DNA S was necessarily small and by inference the ratio of t_S:t_G was small. This in turn suggests a long t_G which could be accounted for by a long G_1. These data are to be contrasted with the effects of OHU on differentiated marrow nucleated erythroid elements in normal mice.[5] In this situation the ratio of t_S:t_G is large and the reduction of erythropoiesis is quite striking. Studies on the repair of the committed precursor compartment after OHU are also in keeping with the proposed kill of the small portion of cells in DNA S in the hypertransfused mouse. The response to EP administered 7 hours after OHU approached that observed in EP-treated controls and by 14 hours an overshoot was observed. Had the committed precursor compartment been extensively damaged, recovery could have been delayed for several days. These data are to be contrasted with the rates of recovery reported after the administration of other cytotoxic agents and irradiation.

A normal response to EP in the plethoric rodent was observed 8 days after vincristine,[8] 6 days after 150 r,[9] and 6 to 8 days after 5-fluorouracil.[10] Although vincristine manifests the most selective cytotoxic effect in this group, it has also been reported to interfere with nucleic acid and protein synthesis.[11] The duration of action of vincristine is presumably longer than that reported for OHU; dividing cells continue to accumulate in metaphase for at least 8 hours and even at 24 hours, the mitotic index was greater than normal.[2] Irradiation and 5-fluorouracil presumably affect cells in all phases of the generative cycle and thus the rate of recovery is proportional to the extent of injury. Since the rate of recovery after OHU was rapid in comparison to these other agents, it lends further support to the contention that OHU exerts a selective limited effect on the committed precursor cells.

Studies on the pluripotential compartment indicate that these cells are in G_0 or a prolonged G_1 as evidenced by no appreciable change in the numbers of colony-forming units for the first 4 hours after OHU. If a substantial frac-

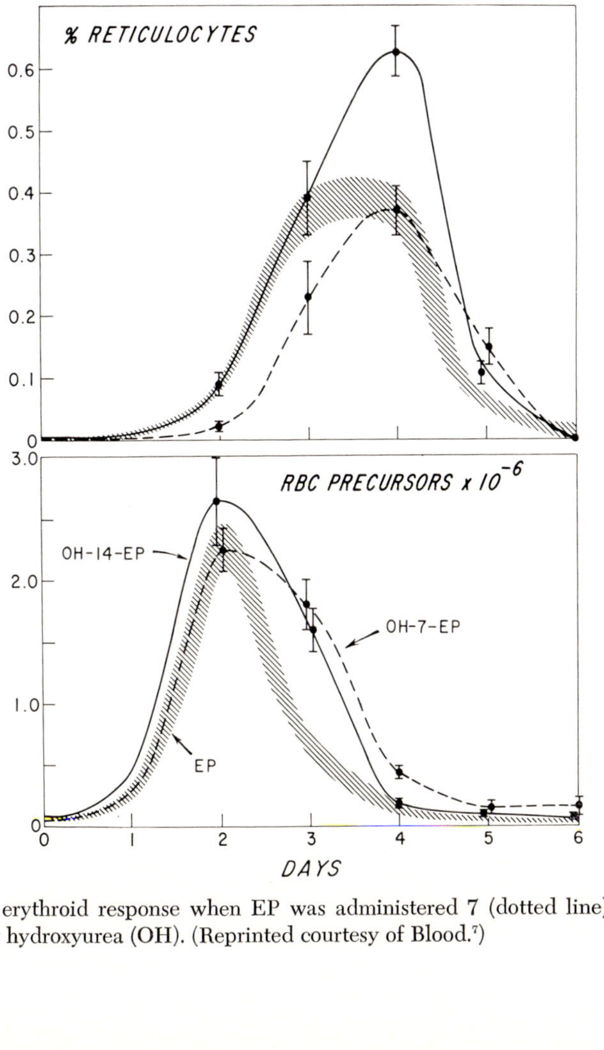

FIG. 4—The erythroid response when EP was administered 7 (dotted line) and 14 hours (solid line) after hydroxyurea (OH). (Reprinted courtesy of Blood.[7])

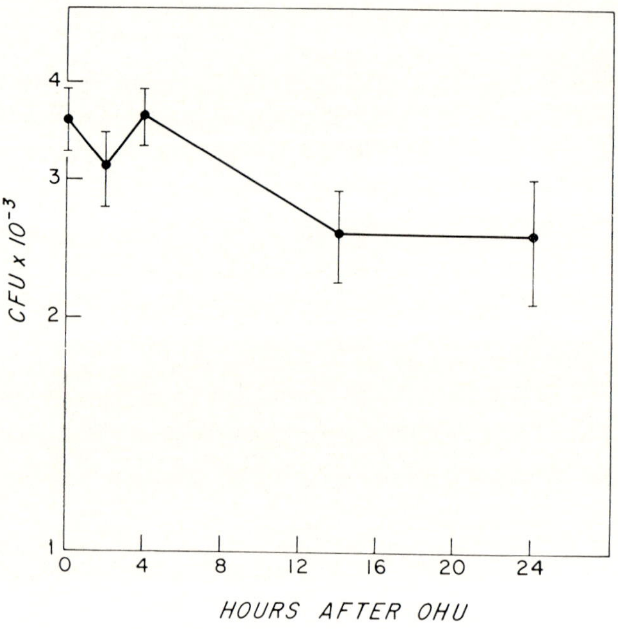

FIG. 5—The changes in colony-forming units of tibial marrow after administration of hydroxyurea (OXU). (Reprinted courtesy of Blood.[7])

tion of these cells were in DNA S, a reduction of colony-forming units would have been anticipated. The data further suggest that the committed precursor compartments were repaired in part by efflux of cells from the pluripotential compartment. Had the committed compartments been repaired solely by an increased turnover of cells within the compartment, then no change in colony-forming units would have been expected. Since colony-forming units were decreased 14 hours after OHU and the committed compartment was repaired as judged by the overshoot in response to EP, it seems reasonable to assume that the damage sustained by the committed compartments was in part repaired by the influx of pluripotential cells, with resultant partial depopulation.

As previously discussed, committed erythroid precursor cells are in cycle in the hypertransfused mouse. After the simultaneous administration of EP and OHU, the relatively small fraction of cells killed indicates that t_S was short compared to t_G and suggests a prolonged G_1. The progressively greater fraction of cells killed as the time interval between EP and OHU was lengthened indicates that the fraction of committed precursor cells in DNA S increased with time after EP. Had the time parameters for the various phases of cell cycle been fixed, then the degree of damage should have been the same irrespective of the time relationships between EP and OHU. From this one may infer either a shortening of t_G or a lengthening of t_S. Since t_S appears to be relatively constant in mammalian hemopoietic tissues and, moreover, a lengthening of t_S would appear to be an inefficient mechanism for response to an increased demand, a shortening of t_G is the more likely explanation for the observed results. The shortening of t_G appears best explained by a variable G_1 and a shortening of G_1 in response to the increased demand imposed by EP (Table 2). A second dose of OHU given 7 hours after the simultaneous injection of EP and OHU nearly abolished the erythroid response again indicating the shortening of G_1 in response to the increased demand. Had the t_G been fixed, then the effects of the second dose should have been additive rather than synergistic. In those animals that received OHU at the later time intervals after EP, the reduction in response by OHU was in part due to the death of some cells which had already differentiated, pronormoblasts having been observed. Schooley [12] reported that the response to EP can be reduced by 85 per cent when anti-EP is given 6 hours after EP. From this it may be inferred that during the initial time intervals investigated in the present studies, the major effect of OHU on the erythroid response was due to its effect on the committed rather than the differentiated compartment.

A shortening of G_1 in response to the increased demand imposed by EP appears consonant with the notion of Kretchmar. Whether or not EP must be present during G_1 and a portion of DNA S cannot be resolved with the present data. Although DNA S would appear to be a prerequisite for the differentiation of some cell systems,[13] the induction of hepatic Δ-aminolevulinic acid(ALA) synthetase is independent of cell cycle phenomena since hepatic parenchymal cells are characterized by their population being normally in G_0. Moreover since the induction of ALA synthetase may well be the first step in the process of erythroid differentiation, it is suggested that EP can effect differentiation throughout most of the cycle with the possible exception of mitosis. A shortening of G_1 would appear to account in significant measure for the homeostatic mechanisms invoked in response to a single injection of 3 units of EP. After multiple injections however, the following have been reported: (1) an accelerated rate of hemoglobin synthesis which is the face of the relatively fixed t_G characteristic of differentiated nucleated erythroid elements results in a skipped terminal division and the production of macrocytes;[14] (2) an enhanced participation of the spleen in the erythroid response; [15] (3) a substantial increase of splenic colony-forming units presumably secondary to migration and/or replication;[15] and (4) recruitment of additional committed erythroid precursor cells.[12] Critical to adequate homeostasis would also be

TABLE 2—*Estimates of Cell Cycle Parameters in Committed Erythroid Precursor Cells*

	t_G*	G_1
Tx control	30	22
Hours after EP		
4	15	7
7	12	4
11	10	2

Cell cycle parameters derived from degree of damage by hydroxyurea with time after erythropoietin assuming fixed periods for DNA synthesis. G_2, and mitosis.

* t_G = Generation time.

control mechanisms to prevent depopulation of the pluripotential and committed compartments. One such control suggested by Kretchmar relates to a very short G_1 and the fraction of the cell population in G_1 too small for sensitivity to EP. Another control is exemplified by the delayed erythroid response to EP in the vincristine treated plethoric rodent.[2] In this situation it would appear that differentiation is delayed until such time as the regeneration of a critical level of committed precursor cells again permits differentiation. These various considerations lead to a tentative model for precursor cell kinetics as shown in Figure 6.

In this model, the 3 proliferative compartments invloved in the formation of the erythrocyte are indicated. Also shown are the interrelationships between cell cycle kinetics and the demand for the delivery of differentiated cells. Many of the relationships remain to a large extent speculative although a number of them are subject to experimental verification. When the demand is nil, e.g., the hypertransfused rodent, the pluripotential compartment is at rest and the cell population of the committed compartment is for the most part in a prolonged G_1. Although erythroid differentiation is not evident, limited injury by OHU presumably results in a shortening of G_1 and an influx of pluripotential cells, both of which would appear to contribute to the overshoot in response when EP was administered 14 hours after OHU. When the precursor compartments are more extensively damaged, i.e., after whole body irradiation, increased responsiveness to EP precedes the recovery of pluripotential cells suggesting that repopulation of the committed compartment has priority over the restoration of the pluripotential compartment.[16] The delayed

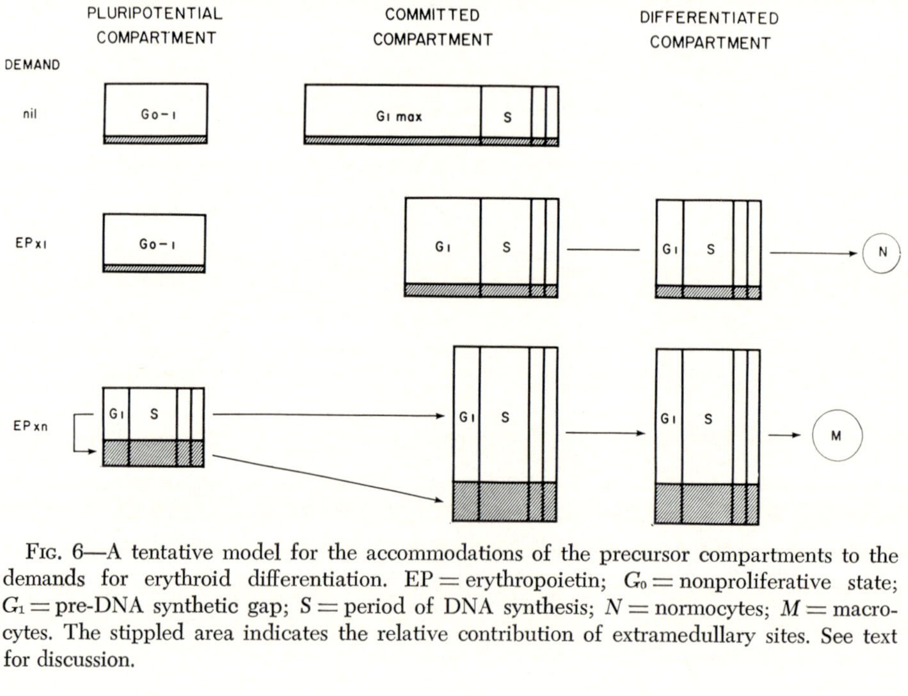

FIG. 6—A tentative model for the accommodations of the precursor compartments to the demands for erythroid differentiation. EP = erythropoietin; G_0 = nonproliferative state; G_1 = pre-DNA synthetic gap; S = period of DNA synthesis; N = normocytes; M = macrocytes. The stippled area indicates the relative contribution of extramedullary sites. See text for discussion.

response to EP following damage by vincristine would appear to constitute a built-in control to prevent depopulation of the committed compartment until regeneration of a critical level of cells is achieved by influx and/or self-repair. In the absence of injury, a single injection of EP differentiates a portion of the committed compartment and increases the turnover of remaining undifferentiated committed cells by shortening the t_G at the expense of G_1. There would appear little need for the active participation of the pluripotential compartment for this limited response. From the data at hand it is difficult to ascertain whether or not the proposed shortening of G_1 attending the administration of EP occurs concurrently with or as a consequence of the loss of cells by differentiation. Although the distinction is moot, it is noteworthy that the accomodation of the committed compartment to the loss of cells occurs rapidly (Table 2). The pluripotential compartment, on the other hand, would appear to respond more sluggishly to cell loss since restoration of compartment size as judged by transplantation was not evident even 24 hours after OHU (Fig. 5). It is recognized however that this does not preclude a change in the rate of proliferation if it is matched by an equal rate of efflux. On the other hand, it seems unlikely that continued efflux of pluripotential cells would be required to sustain the committed compartment in the animal with suppressed erythropoiesis after the limited injury by OHU. When the demands for differentiated cells are increased, i.e., multiple injections of EP, all proliferative compartments participate but the relative contribution of the pluripotential compartment remains to be evaluated. In the committed compartment, G_1 is presumably quite shortened but not quite short enough for a refractory state to EP to develop. Compartment size would be maintained by an increased turnover of committed precursor cells and, in all probability, influx of pluripotential cells. In the differentiated compartment, hemoglobin synthesis is accelerated and as a consequence thereof, macrocytes are released into the peripheral circulation. Studies on the pluripotential compartment in the stimulated animal suggest increased migration of colony-forming units to sites of extramedullary hematopoiesis.[15,17] The rather marked expansion of splenic erythropoiesis in the stimulated animal cannot be explained solely on the basis of an increased turnover of committed precursors. The influx of pluripotential cells is strongly suggested and, moreover, it appears accomplished not only without the depletion of the pluripotential compartment but at a time when splenic colony-forming units are in fact increasing.[15] To insure the numerical integrity of the pluripotential compartment under these conditions, it is necessary to postulate an increased turnover of pluripotential cells. The contribution of recruitment has not been indicated as such in the model since it would appear to combine the effects of skipped terminal division, shortening of G_1, and perhaps the increased migration of pluripotential cells to extramedullary sites.

Summary

In the animal with suppressed erythropoiesis, the committed stem cell is in cycle with a prolonged generation time, the consequence of a variable G_1. After erythropoietin there is a shortening of G_1 and an increase in the rate of turnover of committed stem cells. Damage to the committed compartment appears to be in part repaired by the influx of pluripotential cells. The maintenance of homeostasis by the stem cell compartment is discussed in accord with these findings.

REFERENCES

1. Becker, A. J., McCulloch, E. A., and Till, J. E.: Cytological demonstration of the clonal nature of spleen colonies derived from transplanted mouse marrow cells. Nature (London) 197:452, 1963.

2. Morse, B. S., and Stohlman, F., Jr.: Regulation of erythropoiesis XVIII. The effect of vincristine and erythropoietin on bone marrow. J. Clin. Invest. 45:1241, 1966.

3. Lajtha, L. G.: Recent studies in erythroid differentiation and proliferation. Medicine (Balt.) 43:625, 1964.

4. Kretchmar, A. L.: Erythropoietin: Hypothesis of action tested by analog computer. Science 152:367, 1966.

5. Morse, B. S., Rencricca, N. J., and Stohlman, F., Jr.: The effect of hydroxyurea on differentiated marrow erythroid precursors. Proc. Soc. Exp. Biol. Med. 130:986, 1969.

6. Fruhman, G. J.: Cell counts of mouse bone marrow. Life Sci. 3;737, 1964.

7. Morse, B. S., Rencricca, N. J., and Stohlman, F., Jr.: The relationship of erythropoietin effectiveness to the generative cycle of the erythroid precursor cell. Blood. In press.

8. Morse, B. S., and Stohlman, F., Jr.: Unpublished observations.

9. Gurney, C. W., Lajtha, L. G., and Oliver, R.: A method for investigation of stem-cell kinetics. Brit. J. Haemat. 8:461, 1962.

10. Reissmann, K. R., and Samorapoompichet, S.: Effect of erythropoietin on early recovery of erythropoiesis in mice after sublethal dose of 5-fluorouracil. Proc. Soc. Exp. Biol. Med. 128:898, 1968.

11. Creasy, W. A., and Markiw, M. E.: Biochemical effects of the vinca alkaloids II. A comparison of the effects of colchicine, vinblastine and vincristine on the synthesis of ribonucleic acids of Ehrlich ascites carcinoma cells. Biochem. Biophys. Acta 87: 601, 1964.

12. Schooley, J. C.: Responsiveness of hematopoietic tissue to erythropoietin in relation to the time of administration and duration of action of the hormone. Blood 25:795, 1965.

13. Stockdale, F. E., and Topper, Y. J.: The role of DNA synthesis and mitosis in hormone-dependent differentiation. Proc. Nat. Acad. Sci. U.S.A. 56:1283, 1966.

14. Stohlman, F., Jr., Lucarelli, G., Howard, D., Morse, B., and Leventhal, B.: Regulation of erythropoiesis XVI. Cytokinetic patterns in disorders of erythropoiesis. Medicine 43:651, 1964.

15. Kubanek, B., Tyler, W. S., Ferrari, L., Porcellini, A., Howard, D., and Stohlman, F., Jr.: Regulation of erythropoiesis. XXI. The effect of erythropoietin on the stem cell. Proc. Soc. Exp. Biol. Med. 127: 770, 1968.

16. Schooley, J. C., Carcia, J. F., Cantor, L. N., and Havens, V. W.: A summary of some studies on erythropoiesis using anti-erythropoietin immune serum. Ann. N. Y. Acad. Sci. 149:266, 1968.

17. Hodgson, G., Guzman, E., and Herrera, C.: Characterization of the stem-cell population of phenylhydrazine-treated rodents. *In* Effects of Radiation on Cellular Proliferation and Differentiation. Vienna, International Atomic Energy Agency, 1968.

The Significance of the Spleen in Recruitment of Erythropoietin-Sensitive Cells

By John C. Schooley

Some erythropoietic activity is normally present in the spleen of the adult mouse in contrast to most other laboratory mammals where splenic erythropoiesis is confined to the fetus. However, the spleen of many adult mammals apparently retains throughout life a potential capacity for active erythropoiesis if it is appropriately stimulated. The adult spleen does perform routine functions relative to the maintenance of normal red blood cells other than active erythropoiesis.[1] The present study was undertaken to determine what role, if any, the mouse spleen plays under various conditions of stimulation by the hormone erythropoietin.

Materials and Methods

Female mice weighing about 21–25 Gm. of the C3H/Anf Cum, BC3F1/Cum (C57BL×C3H/Anf), and LAF1/J (C57L/J×A/HeJ) strains were used. The mice were hypertransfused by intraperitoneal injection of 1 ml. of packed isologous saline-washed red blood cells (hematocrit about 80 per cent) on 2 consecutive days. Erythropoietin was first injected in 0.2 ml. of saline, either subcutaneously or intravenously, 5 days after the last intraperitoneal transfusion. Fifty-six hours after the first erythropoietin injection, 0.5 μCi of ^{59}Fe as iron citrate (specific activity approximately 10 μCi/μg.) was injected intravenously, and 72 hours later a sample of blood was taken from the mouse by cardiac puncture. The time of injection of ^{59}Fe and sampling of blood relative to the first erythropoietin injection was always kept constant even if a second dose of erythropoietin or antierythropoietin was injected or a splenectomy performed. The radioactivity in 0.5 ml. of blood was measured, and per cent of injected ^{59}Fe in the total blood volume was calculated. It was assumed that the blood volume of the hypertransfused mouse was 7 per cent of the body weight. Values from any animal which lost weight or whose hematocrit at time of sampling was less than 55 per cent were discarded. Each individual group in each experiment consisted of the values from 6 to 10 mice. The mean of these values and the standard error of the mean are indicated.

Step I sheep plasma erythropoietin (obtained from Connaught Medical Research Laboratories, Toronto, Canada) stated to have a potency of 0.5 IRP (International Reference Preparation) unit/mg. dry weight or human urinary erythropoietin determined to have a potency of 23 IRP unit/mg. dry weight was used.

Splenectomies and sham-splenectomies were performed under methoxy-

From the Lawrence Radiation Laboratory, Donner Laboratory University of California, Berkeley, California.

This work was done under the auspices of the United States Atomic Energy Commission.

flurane anesthesia. In some experiments the operation was performed 2–3 weeks before beginning hypertransfusion, and in other experiments the operation was performed on plethoric mice 24 hours after the first injection of erythropoietin.

Immune serum capable of neutralizing the biologic activity of erythropoietin was obtained from rabbits immunized with human urinary erythropoietin. The schedule of immunizations and some properties of the immune serum have been described previously.[2] One ml. of the serum neutralized about 25 IRP units of human urinary erythropoietin or about 2.5 IRP units of sheep plasma erythropoietin. 0.25 ml. of the serum was injected intravenously in the present experiments.

The erythropoietic stimulating ability of serum obtained from splenectomized or sham-operated C3H mice at various times after exposure to a simulated altitude of 18,000 feet was measured in plethoric C3H mice using the schedule of ^{59}Fe injection and sampling described above. One ml. of the serum was injected intraperitoneally into the assay mice.

Results

The effect of hypoxic exposure on the erythropoietin content of serum and the hematocrit of splenectomized and sham-operated C3H mice are shown in Figure 1. Serum taken from both groups of mice before hypoxic exposure did not stimulate erythropoiesis in assay mice. Within one day of hypoxic exposure a significant elevation in erythropoietin content of the serum occurred, which was much higher in the splenectomized mice than in the controls. The splenectomized mouse serum had a higher titer than the control until the sixth day of hypoxia when the titers of both groups were no longer different. The hematocrits of the splenectomized mice were slightly lower than the controls at the beginning of hypoxic exposure and remained so during the period of observation. The hematocrits of the splenectomized mice after 6 days of hypoxia were significantly higher than at the beginning of the experiment, but significantly less than the controls ($P < 0.001$). Thus, in spite of intense stimulation by high levels of erythropoietin, the splenectomized mice responded less than control mice to hypoxic exposure, yet they did respond slightly.

The response of splenectomized and sham-operated C3H mice to various doses of sheep plasma erythropoietin administered either intravenously or subcutaneously is shown in Figure 2. It is evident that splenectomized mice made plethoric by transfusion 2–3 weeks after operation respond much less efficiently to larger doses of erythropoietin than their appropriate controls. When erythropoietin is injected intravenously, the observed erythropoietic response is much less than when the same dose is injected subcutaneously. Following intravenous doses of erythropoietin, significant differences in the erythropoietic response between splenectomized and sham-operated mice are not observed until about 4.0 units of erythropoietin are given. However, when erythropoietin is given subcutaneously, significant differences between

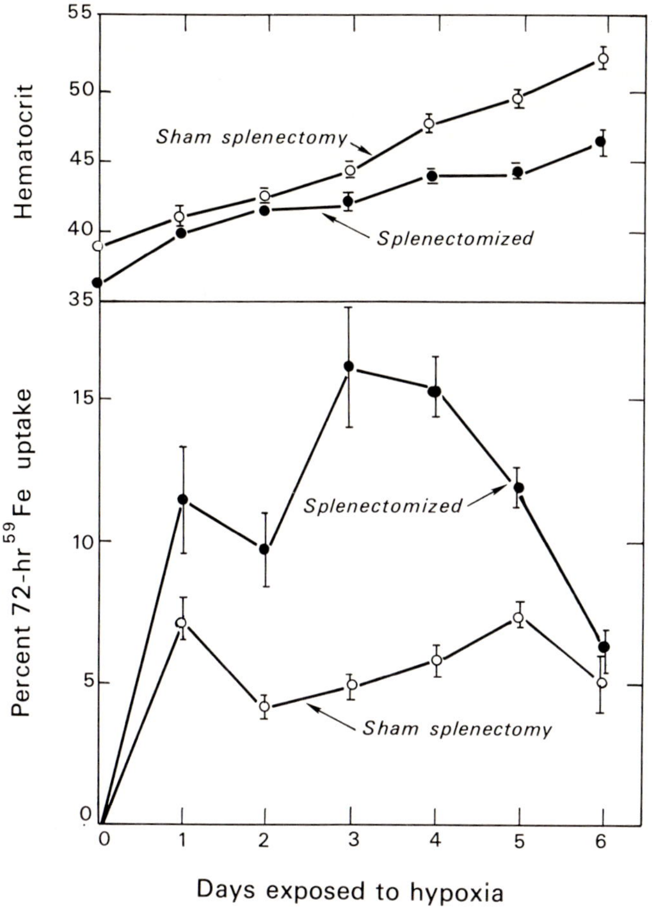

FIG. 1—Erythropoietic activity of the serum and hematocrit of splenectomized and sham-operated female C3H mice during exposure to a simulated altitude of 18,000 ft. Operation performed 2–3 weeks before experiment.

the splenectomized mice and their controls are observed after only 1 unit of erythropoietin.

The effect of varying the interval between the intravenous injection of 2 identical 0.5 unit IRP doses of erythropoietin on the erythropoietic response of splenectomized and sham-operated plethoric C3H mice is shown in Figure 3. The ^{59}Fe uptakes are plotted as a function of the interval between the first and second erythropoietin doses. The values for the ^{59}Fe uptake are given relative to the ^{59}Fe uptake observed when no interval existed between the first and second erythropoietin injection, i.e., a total of 1.0 IRP unit of human urinary erythropoietin was given in one single injection. It is evident that an increased erythropoietic response occurs in control mice even when the interval between the 2 injections is only 6 hours. The erythropoietic response observed when the interval between the first and second injection

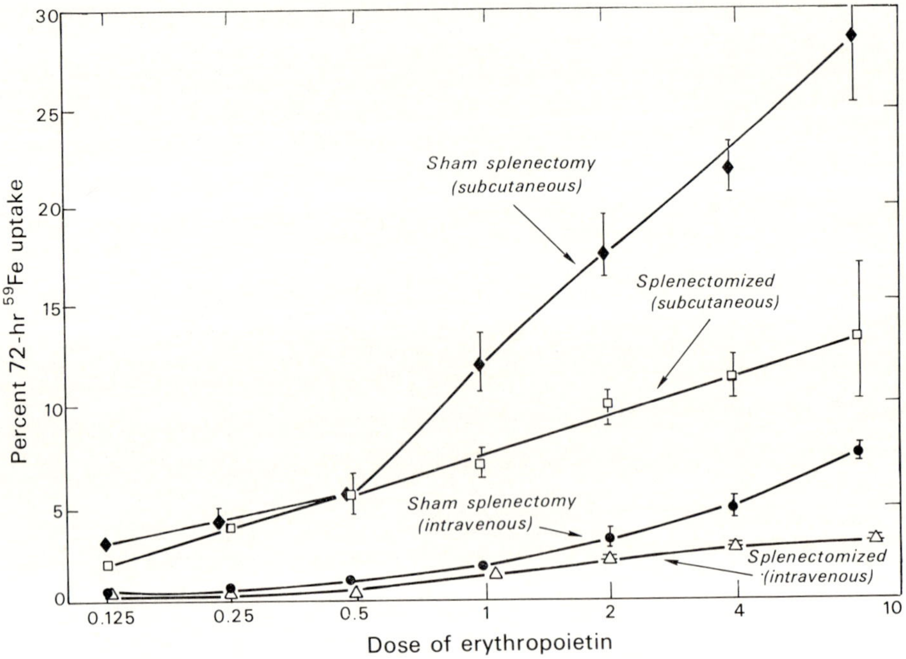

FIG. 2—Response of sham-operated and splenectomized female C3H mice to various doses of sheep plasma erythropoietin. Assay performed 2–3 weeks postoperatively.

of erythropoietin is 24 hours is almost 3 times that observed when the same total dose is administered at one time. Interestingly, very little increase in the erythropoietic response occurred when plethoric splenectomized mice received fractionated doses of erythropoietin.

In Figure 4 the results of a series of fractionation experiments using the BC3F1 mouse are given. The interval between the first and second doses of 0.5 IRP unit of human urinary erythropoietin was 24 hours. It is evident that the response of plethoric BC3F1 mice is similar to C3H mice in that the erythropoietic response in splenectomized and sham-operated mice to 1.0 IRP unit of erythropoietin is not significantly different. Fogh (unpublished results) in this laboratory has more recently demonstrated similar results in the female CF1 mouse. If this same total dose of erythropoietin is given in fractionated doses, the splenectomized mice give a lower response than do control mice. However, the response of the splenectomized mouse to the fractionated dose is not significantly different from the response of control mice receiving 1.0 IRP unit without fractionation. If the availability of the second dose of erythropoietin is limited to an 8-hour period in the assay mice (by the injection of enough antierythropoietin to neutralize the biologic activity of all erythropoietin present in the mouse), a measurable erythropoietic response occurs in both the splenectomized and control mice. However, in the control mice the response is significantly higher than in splenectomized mice. In fact, the response of the control mice is almost as great as that observed when anti-

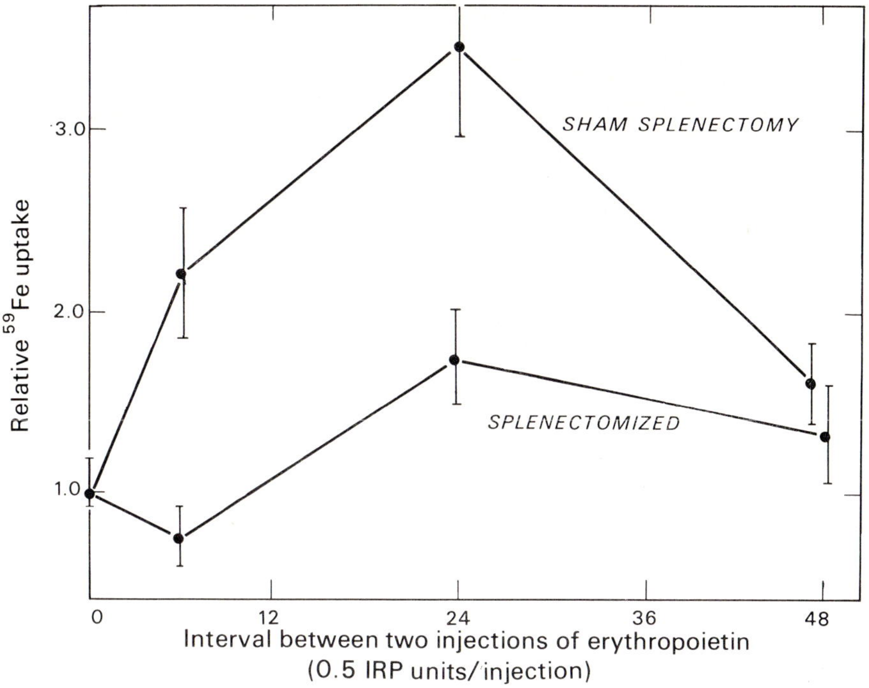

FIG. 3—The effect of divided intravenous dose of erythropoietin relative to the effect of the same amount of erythropoietin given as a single dose. Unit response is defined as that observed following a single intravenous injection of 1.0 IRP unit of erythropoietin. The curve represents the response observed, relative to this unit response, when the dose was divided into two 0.5 IRP intravenous doses given at times separated by the intervals in hours indicated on the abscissa. Operations performed 2–3 weeks before experiment.

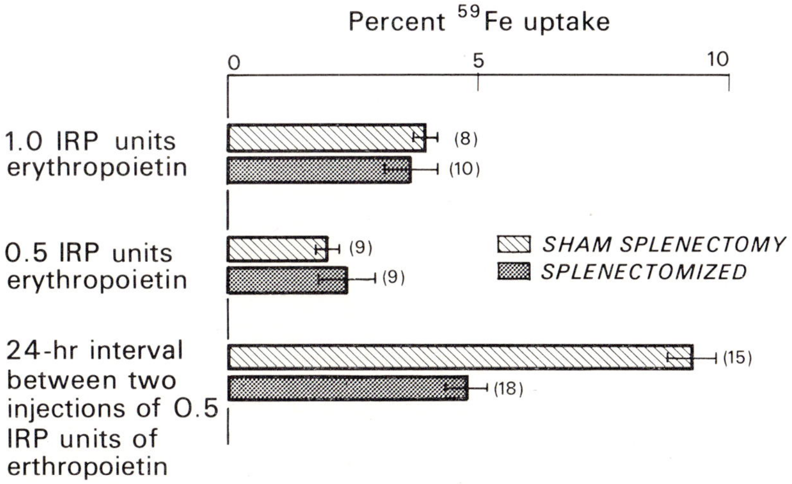

FIG. 4—Erythropoietic response of plethoric splenectomized and sham-operated female C3H mice to human urinary erythropoietin. Operations performed 24 hours after the first but before the second injection of erythropoietin (number of mice in each experiment is indicated by figures in parentheses).

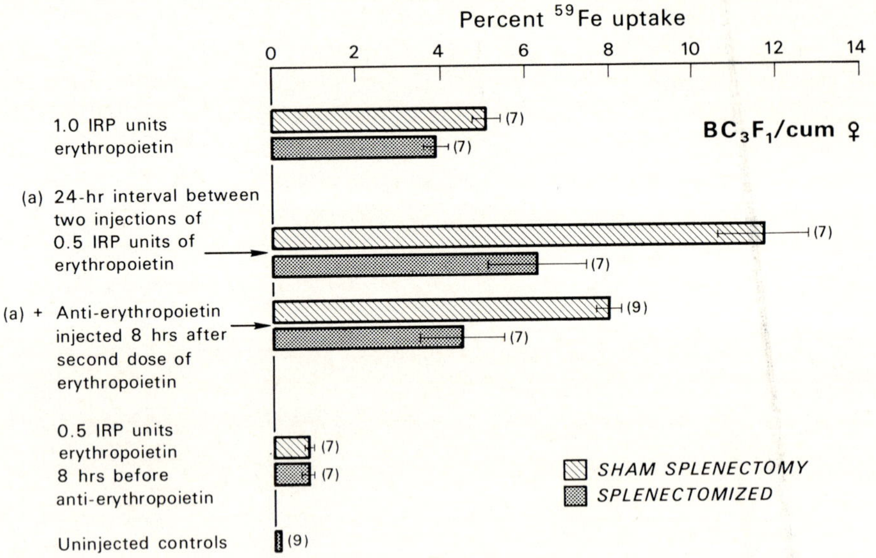

FIG. 5—Erythropoietic response of plethoric splenectomized and sham-operated female BC3F1 mice to human urinary erythropoietin. Operations performed 2–3 weeks before the experiment (number of mice in each experiment is indicated by figures in parentheses).

erythropoietin is not injected. Moreover, if the second 0.5 IRP dose is injected without the previous 0.5 IRP dose of erythropoietin and then antierythropoietin is injected 8 hours later, a measurable erythropoietic response occurs, but there is no difference between splenectomized and control mice. Thus, greater numbers of cells sensitive to erythropoietin are found in the spleen in these experiments only if a "priming" dose of erythropoietin is given.

In all of the previous experiments mice were made plethoric 2 or 3 weeks after splenectomy and then stimulated with erythropoietin. In Figure 4 the results are shown of an experiment where splenectomy was not performed until 24 hours after the intravenous injection of erythropoietin into plethoric C3H mice. It is evident that a significant difference between the response of splenectomized and sham-operated mice to 1.0 or 0.5 IRP units of human urinary erythropoietin did not occur. However, if 1.0 IRP unit of erythropoietin was given in 2 equal doses of 0.5 unit with 24 hours between each injection, the response in splenectomized mice was significantly less than the controls and equal to the response of 1.0 IRP unit given without fractionation.

In contrast to the above experiment using C3H mice, the response of LAF1 mice to fractionated doses of erythropoietin is not depressed by splenectomy. Thus, the ^{59}Fe uptakes after injection of 1.0 IRP unit of human urinary erythropoietin were respectively 9.60 ± 0.91 and 9.00 ± 0.56 per cent in 10 sham-operated and 10 splenectomized mice. With fractionation the ^{59}Fe uptakes were 16.4 ± 1.1 in 10 control mice and 13.3 ± 0.99 in 12 splenectomized mice. In addition to this difference between the two strains, the magnitude of the erythropoietic response after fractionation is much less in LAF1 mice than

C3H mice. The reason for these strain differences is unknown, but they may be related to the observations of Bernstein et al.[3] that the LAF1 mouse is more sensitive to erythropoietin than the C3H mouse.

Discussion

It appears likely that the spleen colony-forming cell or unit (CFU) of Till and McCulloch[4] corresponds to the hematopoietic "stem cell": it is capable of extensive proliferation, it is capable of self-renewal, and it is capable of giving rise to differentiated cells such as erythroid cells. It is the most primitive precursor of the hematopoietic system whose behavior can be measured. The production of erythroid cells within spleen colonies depends upon the presence of erythropoietin, yet the available evidence suggests that the hormone does not act directly on the CFU, but upon a population of cells derived from the CFU.[5] These progeny of the CFU may be termed erythropoietin-sensitive cells (ESC). The evidence that ESC themselves are capable of division, and that their proliferation, as well as their formation from colony-forming cells, is regulated, has been reviewed recently.[6]

The spleen of the C3H mouse contains slightly more CFU than are found in the femoral marrow;[6] nevertheless, in the normal mouse much more active erythropoiesis is found in the marrow than in the spleen. After 5 days of hypoxic exposure, however, the total cellularity of the spleen more than doubles, and most of the cells produced belong to the erythroid series.[7] Obviously, the production of nucleated erythroid cells in the spleen is different than in the bone marrow. Splenic erythropoiesis appears to be of physiologic importance only during periods of intense erythropoietic stimulation. Thus, only a slight increase in hematocrit occurred in splenectomized mice exposed to hypoxia compared to control mice even though marked increases in the blood erythropoietin levels occurred. Boggs et al.[8] observed that the hematocrits of splenectomized mice recovered more slowly than those of control mice after hemorrhage.

The spleen, therefore, appears to have a reservoir of potential ESC or CFU which can give rise to ESC upon intense erythropoietic stimulation. Even in plethoric mice rather large doses of erythropoietin must be given before a difference between the erythropoietic response in splenectomized and control mice is observed. The differences in the erythropoietic response observed in plethoric splenectomized and control mice when the same erythropoietin dose is given intravenously or subcutaneously suggests that it is not necessarily the local concentration of erythropoietin presented to the spleen at any one time which influences the responsiveness of the spleen, but the pattern of changing concentrations of erythropoietin presented to the spleen over a period of time. Two units of erythropoietin would have a high blood concentration immediately after intravenous injection, which would then decrease, whereas the same dose of erythropoietin after subcutaneous injection would never achieve a concentration in the blood equal to that found after intravenous injection because of the time required for absorption from the subcutaneous site. Moreover, erythropoietin levels in the blood would steadily increase until complete absorption from the injection site occurred, and then the level would decrease. This pattern of increasing concentration of erythropoietin in the blood is much more analogous to the changing concentrations of serum erythropoietin observed after hypoxic exposure.

Gurney, Wackman, and Filmanowicz [9] demonstrated that a smaller erythropoietic response occurs after a single submaximal dose of erythropoietin than when the same total dose is given in divided doses. This problem was further studied by Schooley [10] using antierythropoietin to limit the time of action of 2 doses of erythropoietin. It was suggested that cells are sensitive to erythropoietin only at some limited time of their individual life-cycle, and this time probably corresponds to the G_1 phase of interphase. It was concluded that following a single injection of erythropoietin, a recruitment of more cells into a state receptive to the differentiative action of subsequent erythropoietin injections occurs. This recruitment may be due to both increased divisions of ESC and more rapid movement of CFU into ESC. Fogh [11] has recently found that this recruitment of ESC in plethoric mice, by a priming dose of erythropoietin, occurs even if differentiation into the erythron is prevented by the simultaneous injection of actinomycin D.

The results of the present experiments indicate that the spleen is a major site for the recruitment of ESC after the injection of fractionated doses of erythropoietin. It is of interest that in these experiments splenic erythropoiesis was initiated by a small total dose of erythropoietin. This same total dose of erythropoietin, if injected at one time, stimulated erythropoiesis similarly in mice whose spleens were removed either before or after administration of erythropoietin. This agrees with the data of Gurney and Rosett [12] who observed that splenectomy did not significantly decrease the response of plethoric CF1 mice to erythropoietin.

The results of the fractionation experiments add support to the view that it is not just the total dose of erythropoietin which is important in initiating splenic erythropoiesis but also the manner in which the hormone is presented to the spleen. A high local concentration of erythropoietin would initially occur in the spleen as a result of the first erythropoietin injection which would progessively decrease, but the second erythropoietin injection would again produce a high concentration of erythropoietin in the spleen. Thus, a single erythropoietin injection may, because of its high splenic concentration, recruit ESC in the spleen, but the probability of these cells subsequently becoming triggered into recognizable erythroid cells is low because the concentration of circulating erythropoietin is continually decreasing with time. Experiments utilizing infusions of increasing and decreasing concentrations may clarify this point.

Summary

Plethoric splenectomized mice respond poorly to large doses of erythropoietin after both intravenous and subcutaneous injections, yet the erythropoietic response to small doses of erythropoietin is not measurably different from sham-operated control mice unless the small dose of erythropoietin is fractionated into 2 doses. It is suggested that the spleen is a major site for the recruitment of erythropoietin-sensitive cells, and that the magnitude of splenic recruitment depends upon increasing levels of erythropoietin in the blood circulating through the spleen.

ACKNOWLEDGMENTS

The technical assistance of Mrs. Virginia W. Havens and Miss Lynn J. Mahlmann and the secretarial assistance of Mrs. Grace Walpole are greatly appreciated.

REFERENCES

1. Crosby, W. H.: Normal functions of the spleen relative to red blood cells: A review. Blood 14:399, 1959.

2. Schooley, J. C., and Garcia, J. F.: Some properties of serum obtained from rabbits immunized with human urinary erythropoietin. Blood 25:204, 1965.

3. Bernstein, S. E., Keighley, G., and Keighley, M. G.: Strain differences in erythropoietin sensitivity. Abstracts, XII International Society of Hematology. New York, 1968, p. 151.

4. Till, J. E., and McCulloch, E. A.: A direct measurement of the radiation sensitivity of normal mouse bone marrow cells. Radiat. Res. 14:213, 1961.

5. Schooley, J. C.: The effect of erythropoietin on the growth and development of spleen colony-forming cells. J. Cell. Physiol. 68:249, 1966.

6. Schooley, J. C.: Studies of stem cell kinetics utilizing the spleen colony technique and the erythropoietin sensitivity test. *In* V. P. Bond, and T. Sugahara (Eds.): Comparative Cellular and Species Radiosensitivity. Tokyo, Japan, Igaku Shoin, 1969, p. 125.

7. Schooley, J. C., Garcia, J. F., Cantor, L. N., and Havens, V. W.: A summary of some studies of erythropoiesis using anti-erythropoietin immune serum. Ann. N. Y. Acad. Sci. 149:266, 1968.

8. Boggs, D. R., Geist, A., and Chervenick, P. A.: Contribution of the mouse spleen to post-hemorrhagic erythropoiesis. Life Sci. 8:587, 1969.

9. Gurney, C. W., Wackman, N., and Filmanowicz, E.: Studies on erythropoiesis. XVII. Some quantitative aspects of the erythropoietic response to erythropoietin. Blood 17:531, 1961.

10. Schooley, J. C.: Responsiveness of hematopoietic tissue to erythropoietin in relation to the time of administration and duration of action of the hormone. Blood 25: 795, 1965.

11. Fogh, J.: Studies on the mechanism of the increased dose-response of erythropoietin after stimulation with erythropoietin. Blood. In press.

12. Gurney, C. W., and Rosett, W.: Splenic erythropoiesis in the mouse. Exp. Hemat. 15:94, 1968.

Ineffective Erythropoiesis and the Erythropoietic Component of Early-Labeled Bilirubin

By STEPHEN H. ROBINSON

ALTHOUGH MOST of the bile pigment formed under normal conditions is derived from the hemoglobin of senescent red blood cells, a small but significant fraction arises from other sources. Erythrocytes containing labeled hemoglobin heme are produced within a few days after the administration of ^{15}N- or ^{14}C-labeled glycine. At the end of the physiologic life-span of these cells, the fall in specific activity of hemoglobin heme is associated with a large "late peak" of labeled bile pigment formation.[1–3] In addition, a smaller "early-labeled peak" of bile pigment is formed during the first few days, before there is significant label in circulating red cells.[1,2] In the past few years work from several laboratories has helped to clarify the pathophysiology of early-labeled bilirubin production.[3–18] A summary of the findings from this laboratory [3,8,9] is presented in Figure 1. In the rat, and also in man,[11,16] the early-labeled peak is heterogeneous in origin and consists of an early sharp component followed by a more prolonged plateau phase. The initial peak appears to originate exclusively from nonerythroid sources, primarily from the turnover of nonhemoglobin hemes in the liver.[9,10,15] The plateau phase also is derived largely from nonhemoglobin sources in rats under normal conditions, but contains a small erythropoietic component which becomes prominent when red cell production is either accelerated or abnormal.[3]

The early-labeled peak is enlarged in pathologic conditions associated with ineffective erythropoiesis, such as pernicious anemia,[19] thalassemia,[20,21] and refractory normoblastic anemia.[5,14] On the basis of this indirect evidence, it seems probable that the erythropoietic bilirubin fraction in these disorders is derived from the hemolysis of immature erythroid cells. The origin of this component in subjects with physiologically regulated erythropoiesis has been less clear. Experiments were therefore performed to determine whether or not destruction of young erythroid cells (most readily studied as reticulocytes in the peripheral blood) contributes to the enlarged erythropoietic bilirubin component both in normal rats with stimulated erythropoiesis and animals with abnormal red cell production associated with iron deficiency anemia.[22]

An enlarged erythropoietic component was generated in normal rats by subjecting them to two major hemorrhages. In some animals the early-labeled peak was measured directly with glycine-2-^{14}C[3] (Fig. 2), while in others reticulocytes containing ^{14}C-labeled hemoglobin heme were removed 24 hours after the administration of glycine-^{14}C and then transfused into unlabeled recipients with external bile drainage. As shown in Figure 3, a mean of 6.5

From the Departments of Medicine, Beth Israel Hospital and Harvard Medical School, Boston, Massachusetts.

This work was supported in part by Grant AM 09843 and Career Development Award 19600 from the U. S. Public Health Service.

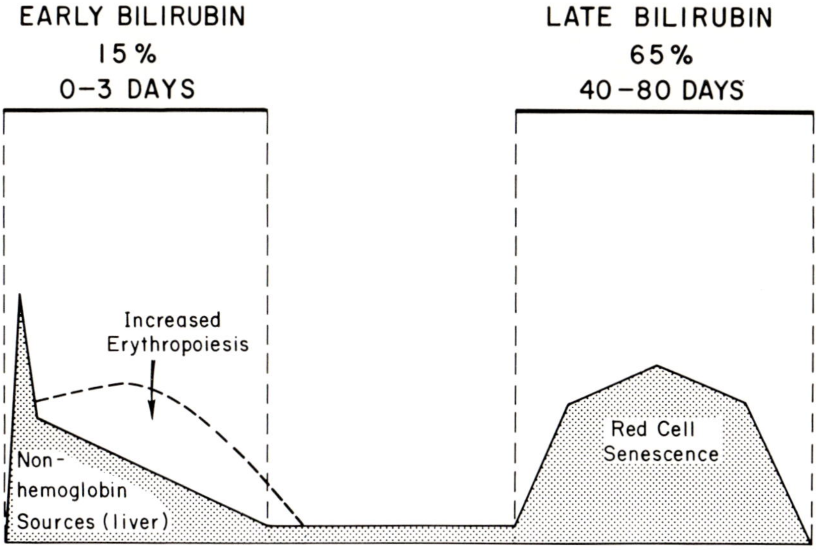

FIG. 1—Sources of bile pigment in the rat, based on studies with glycine-2-^{14}C[3,8,9].

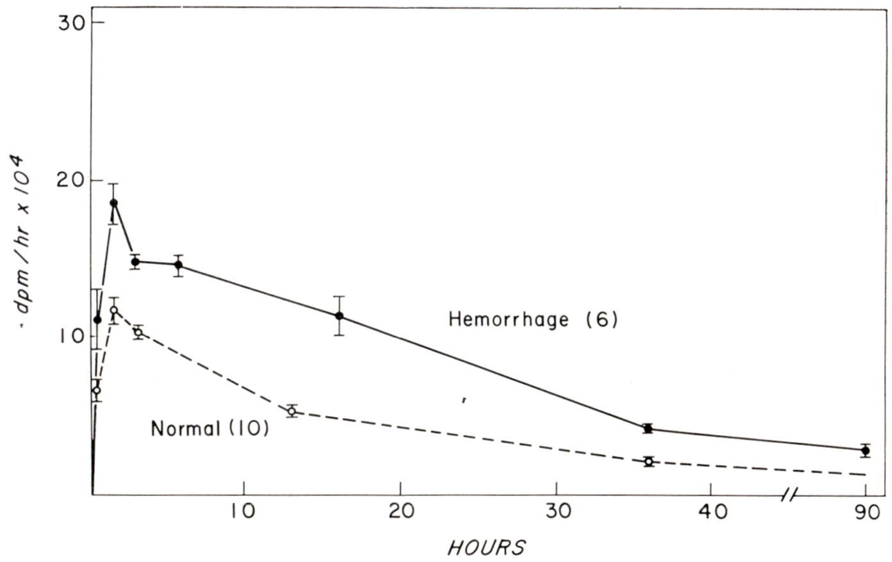

FIG. 2—Production of early-labeled bilirubin in rats with external bile drainage. Hemorrhaged rats were bled 35 per cent of blood volume 3 and 1 days before the administration of glycine-2-^{14}C. Means ± SE are plotted at the median times of each period of bile collection. The scale is adjusted to corespond to a dose of 1 mc. glycine-^{14}C. Glycine-^{14}C incorporation into hemoglobin heme was 0.86 per cent (SE 0.07) in hemorrhaged rats and 0.38 per cent (SE 0.12) in controls at the end of the experiment at 96 hours.

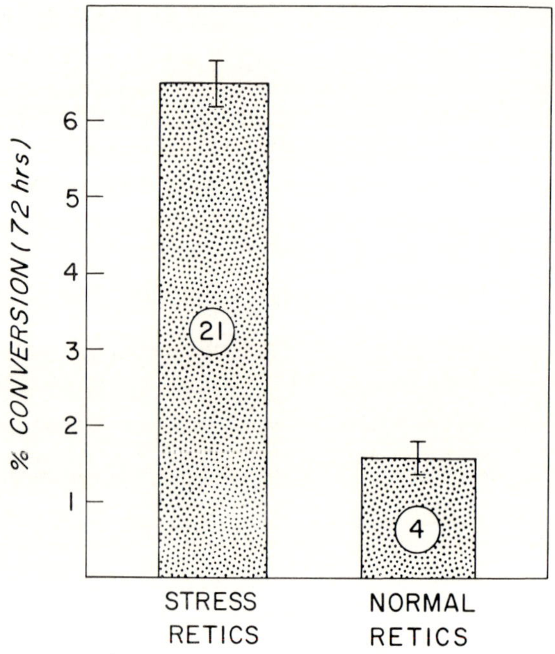

FIG. 3—Per cent conversion of hemoglobin heme-^{14}C in transfused reticulocytes to bilirubin-^{14}C excreted in the bile of recipient rats over 3 days. Labeled stress reticuloytes were obtained 24 hours after glycine-^{14}C administration from rats bled according to the same schedule described in Figure 2. Normal reticulocytes were derived from unstimulated rats given glycine-^{14}C 24 hours earlier. Values from splenectomized and sham-operated donors and recipients were similar and have therefore been pooled. Circles refer to the number of experiments. Means $\pm$ SE are plotted.

per cent of transfused hemoglobin heme in "stress" reticulocytes from hemorrhaged donors was converted to bilirubin-^{14}C over the 3 days of observation. By contrast, only 1.6 per cent conversion was found with normal reticulocytes from rats not subjected to hemorrhage. Parallel results were observed when the disappearance of ^{59}Fe-labeled reticulocytes, rather than the excretion of ^{14}C-labeled bilirubin, was measured. In none of these experiments did splenectomy of donor and/or recipient rats alter the findings.

To study the relationship of reticulocyte hemolysis to the phase of the reticulocyte response, transfusion experiments were performed at different intervals after erythroid stimulation. In general, per cent conversion of hemoglobin heme to bilirubin varied with the height of the reticulocytosis. The maximum occurred when bleedings were performed 4 and 2 days prior to transfusion, when donors had a mean reticulocyte count of 22 per cent. As the interval between hemorrhage and transfusion was widened, reticulocytosis was associated with progressively lower rates of bilirubin production. When bleedings were performed 7 and 5 days earlier, reticulocytes were still 14 per cent, but bilirubin production had fallen to only 1.9 per cent as compared to 1.6 per cent with cells from normal donor animals. This finding is in agreement with previous studies,[23,24] and suggests that reticulocytes produced after the peak response to an erythroid stress have progressively more normal features.

The fourfold difference in per cent conversion to bilirubin of hemoglobin-^{14}C in stress as compared to normal reticulocytes (Fig. 3) is an index only of the hemolytic susceptibility of each cell type, and fails to take into account the much greater number of reticulocytes produced by hemorrhaged donors. In order to account for both of these factors, the observed rates of bilirubin-^{14}C production were adjusted to correspond to transfusions of the entire red cell mass of the average donor used in each experiment. Assessed in this manner, the total production of bilirubin-^{14}C from hemolysis of stress reticulocytes was almost twentyfold higher than that observed with normal reticulocytes (Fig. 4).

One may now estimate the contribution of reticulocyte hemolysis to the increase in erythropoietic bilirubin production from all potential sources in intact rats. In Figure 5 the upper panel shows the increment in early-labeled bilirubin production stimulated by hemorrhage in intact animals (from Fig.2), while the lower panel illustrates total bilirubin-^{14}C production from degradation of hemoglobin in stress as compared to normal reticulocytes (from Fig. 4). It should be noted that the transfusion experiments were performed 24 hours after the administration of glycine-^{14}C to donor rats, and the first 3 days of these studies therefore conform to the interval from 24 to 96 hours after glycine

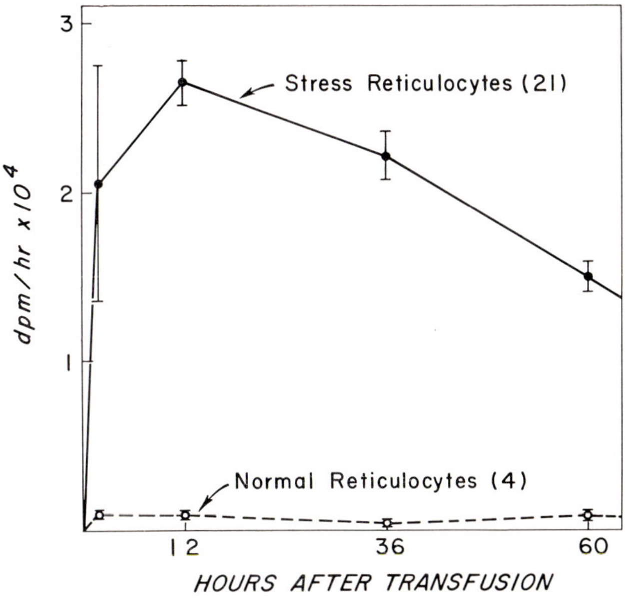

FIG. 4—Total bilirubin-^{14}C production in recipients of ^{14}C-labeled red cells. The observed rates of bile bilirubin-^{14}C excretion have been adjusted to correspond to transfusions of an entire donor red cell mass. Scale has been adjusted to a dose of 1 mc. glycine-^{14}C in donor rats. Means ± SE are plotted.

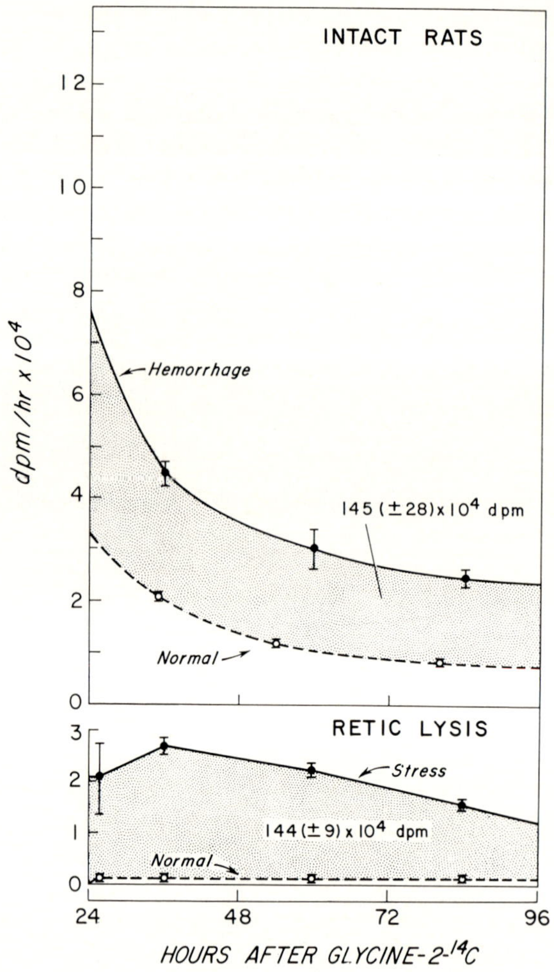

FIG. 5—Comparison of the increment in early-labeled bilirubin formation in intact hemorrhaged rats (from Fig. 2) with the rise in bilirubin-^{14}C production after transfusion of labeled stress as compared to normal reticulocytes (from Fig. 4).

administration in intact rats. During this period early-labeled pigment production rose the equivalent of 145×10^4 dpm in bilirubin-^{14}C. Virtually the same value was obtained for labeled pigment production from hemolysis of stress reticulocytes, 144×10^4 dpm. In transfusion experiments corresponding to the interval from 12 to 24 hours, however, reticulocyte hemolysis accounted for only about ⅓ of the total rise in early-labeled bilirubin formation. Although it is somewhat imprecise to compare results obtained by two different technics, it is clear that degradation of hemoglobin from stress reticulocytes is a major mechanism of erythropoietic bilirubin formation, a possibility suggested by a few earlier investigators.[4,25] Since this mechanism is most important during the later phases of early pigment production, it appears likely that this is part of a spectrum of ineffective erythropoiesis, and that hemolysis of nucleated erythroid cells in the marrow accounts for much of the earlier rise.

These findings are consistent with the concept that erythroid stimulation leads to the release of macrocytic reticulocytes with a shortened life-span.[23,24,26–28] However, they do not exclude the possibility [29] that these large cells survive normally and give rise to early-labeled bilirubin through the loss of fragments of hemoglobinized cytoplasm. The former concept seems more attractive in many ways, and is supported by radioautographic studies.[24]

It is of interest to compare these findings in rats undergoing a physiologic response to erythroid stress with those in an experimental model of pathologic ineffective erythropoiesis, i.e., iron deficiency anemia in the rat.[22] In both of these conditions there is enlargement of the plateau phase of early bilirubin production (Figs. 2 and 6). However, contrary to the findings in the hemorrhaged animals, the production of red cell hemoglobin heme is depressed in rats with iron deficiency (Fig. 6). This disparity between the formation of early-

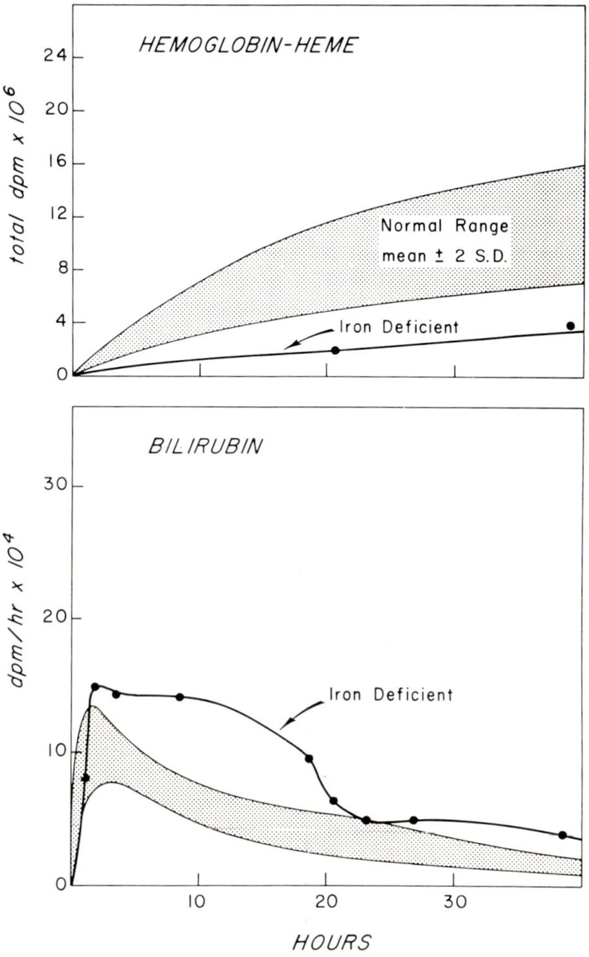

FIG. 6—Production of hemoglobin heme and early-labeled bilirubin in a rat with iron deficiency anemia induced by special diet.[22] The stippled areas refer to means ± 2SD in normal rats. Scale has been adjusted to correspond to a dose of 1 mc. glycine-2-^{14}C. (Reproduced from Robinson [22] with permission of the publisher.)

labeled bilirubin and hemoglobin heme is quite characteristic of pathologic conditions associated with ineffective erythropoiesis.[3,19–21] With the physiologically regulated erythroid response, on the other hand, there is a more or less proportionate rise in the labeling of both of these entities.[3]

Rats with iron deficiency anemia have moderate reticulocytosis, averaging 8–12 per cent, in contrast to the usual findings in humans with this disease. In preliminary studies, reticulocyte hemolysis was investigated by transfusion of labeled reticulocytes from iron deficient donors into normal recipients with external bile drainage. A mean of 37 per cent of transfused hemoglobin heme-^{14}C in iron deficient reticulocytes was converted to bilirubin-^{14}C in 3 days (Fig. 7). This is in contrast to the experiments with stress reticulocytes, in which reticulocyte counts averaged 22 per cent while mean conversion to labeled bilirubin was 6.5 per cent. Thus, in rats with ineffective erythropoiesis related to iron deficiency anemia, reticulocyte hemolysis assumes quite large

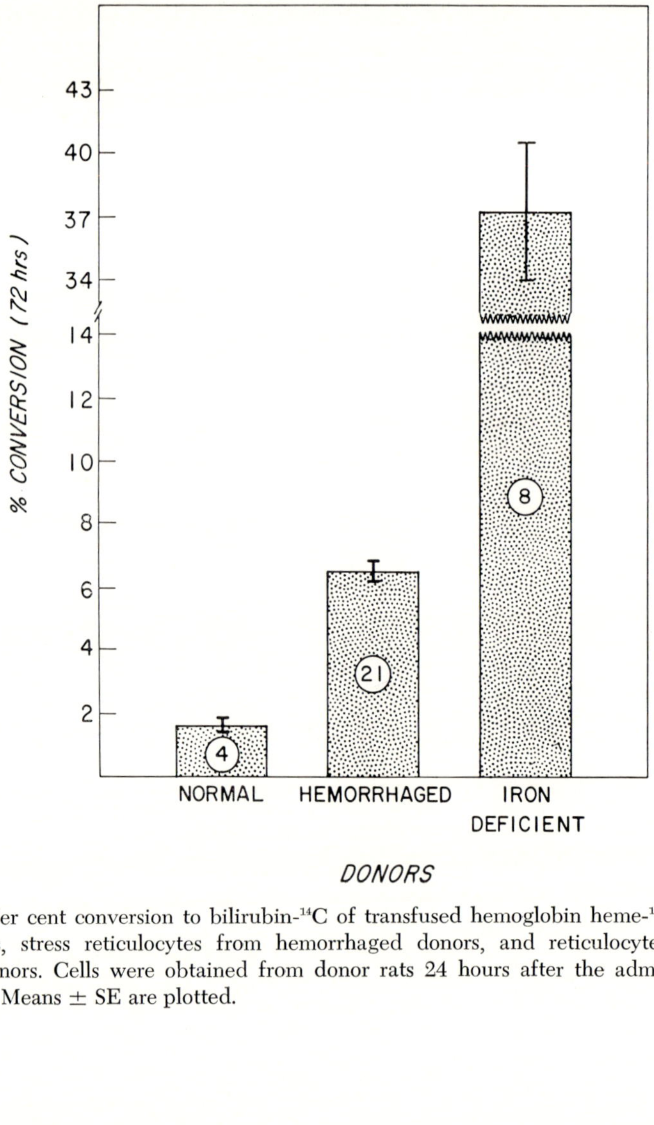

FIG. 7—Per cent conversion to bilirubin-^{14}C of transfused hemoglobin heme-^{14}C in normal reticulocytes, stress reticulocytes from hemorrhaged donors, and reticulocytes from iron deficient donors. Cells were obtained from donor rats 24 hours after the administration of glycine-^{14}C. Means ± SE are plotted.

proportions, despite a comparatively modest number of reticulocytes in the peripheral blood.

It may be concluded that reticulocyte hemolysis, and probably ineffective erythropoiesis in a more general sense, are important mechanisms of erythropoietic bilirubin production under both physiologic and pathologic conditions. The characteristic disproportionality between the formation of early-labeled bilirubin and hemoglobin heme in the disease states may be largely a quantitative phenomenon. Destruction of a large fraction of the abnormal erythroid precursors in iron deficiency anemia would lead to both increased formation of erythropoietic bilirubin and depressed production of circulating erythrocytes. In hemorrhaged rats, comparatively less hemolysis of the more normal red cell precursors would also cause an increase in early pigment formation, but would permit a net increase in hemoglobin synthesis.

REFERENCES

1. London, I. M., West, R., Shemin, R., and Rittenberg, D.: On the origin of bile pigment in normal man. J. Biol. Chem. 184: 351, 1950.

2. Gray, C. H., Neuberger, A., and Sneath, P. H. A.: Studies in congenital porphyria. 2. Incorporation of ^{15}N in the stercobilin in the normal and in the porphyric. Biochem. J. 47:87, 1950.

3. Robinson, S. H., Tsong, M., Brown, B. W., and Schmid, R.: The sources of bile pigment in the rat: Studies of the "early-labeled" fraction. J. Clin. Invest. 45:1569, 1966.

4. Gray, C. H., and Scott, J. J.: The effect of haemorrhage on the incorporation of α-^{14}C glycine into stercobilin. Biochem. J. 71:38, 1959.

5. James, G. W., III, and Abbott, L. D.: Stercobilin N^{15} excretion in refractory anemia. Trans. Amer. Clin. Climat. Ass. 73: 110, 1961.

6. Israels, L. F., Skanderbeg, J., Guyda, H., Zingg, W., and Zipursky, A.: A study of the early-labelled fraction of bile pigment: The effect of altering erythropoiesis on the incorporation of 2-^{14}C glycine into haem and bilirubin. Brit. J. Haemat. 9:50, 1963.

7. Giblett, E. R., Coleman, D. H., Pirzio-Biroli, G., Donhue, D. M., Motulsky, and Finch, C. A.: Erythrokinetics: Quantitative measurement of red cell production and destruction in normal subjects and patients with anemia. Blood 11:291, 1956.

8. Robinson, S. H., and Schmid, R.: The relation of erythropoiesis to bile pigment formation. Medicine (Balt.) 43:667, 1964.

9. Robinson, S. H., Owen, C. A., Jr., Flock, E. V., and Schmid, R.: Bilirubin formation in the liver from nonhemoglobin sources. Experiments with isolated, perfused rat liver. Blood 26:823, 1965.

10. Schwartz, S., and Cardinal, R.: Incorporation of ALA-C^{14} into dog tissue hemes. Fed. Proc. 24:485, 1965.

11. Yamamoto, T., Skanderbeg, J., Zipursky, A., Israels, L. G.: Early appearing bilirubin: Evidence for two components. J. Clin. Invest. 44:31, 1965.

12. Ibrahim, G. W., Schwartz, S., and Watson, C. J.: Early labeling of bilirubin from glycine and delta-aminolevulinic acid in bile fistula dogs with special reference to stimulated versus suppressed erythropoiesis. Metabolism 15:1129, 1966.

13. White, P., Coburn, R. F., Williams, W. J., Goldwein, M. I., Rother, M. L., and Shafer, B. C.: Carbon monoxide production associated with ineffective erythropoiesis. J. Clin. Invest. 46:1986, 1967.

14. Barrett, P. V. D., Cline, M. J., and Berlin, N. L.: Association of urobilin "early peak" and erythropoiesis in man. J. Clin. Invest. 45:1657, 1966.

15. Schmid, R., Marver, H. S., and Hammaker, L.: Enhanced formation of rapidly labeled bilirubin by phenobarbital: Hepatic microsomal cytochromes as possible source. Biochem. Biophys. Res. Commun. 24:319, 1966.

16. Robinson, S. H., Lester, R., Crigler, J. F., Jr., and Tsong, M.: The early-labeled peak of bile pigment in man: Studies with glycine-C^{14} and delta-aminolevulinic acid-H^3. New Eng. J. Med. 277:1323, 1967.

17. Levitt, M., Schacter, B. A., Zipursky, A., and Israels, L. G.: The nonerythropoietic component of early bilirubin. J. Clin. Invest. 47:1281, 1968.

18. Robinson, S. H.: Increased bilirubin formation from nonhemoglobin sources in rats with disorders of the liver. J. Lab. Clin. Med. 73:668, 1969.

19. London, I. M., and West, R.: The formation of bile pigment in pernicious anemia. J. Biol. Chem. 184:359, 1950.

20. Grinstein, M., Bannerman, R. M., Vavra, J. D., and Moore, C. V.: Hemoglobin metabolism in thalassemia: In vivo studies. Amer. J. Med. 29:18, 1950.

21. Robinson, S. H., Vanier, T., Desforges, J. F., and Schmid, R.: Jaundice in thalassemia minor. A consequence of "ineffective erythropoiesis." New Eng. J. Med. 267:523, 1962.

22. Robinson, S. H.: Increased formation of early-labeled bilirubin in rats with iron deficiency anemia: Evidence for ineffective erythropoiesis. Blood 33:909, 1969.

23. Brecher, G., and Stohlman, F., Jr.: Reticulocyte size and erythropoietic stimulation. Proc. Soc. Exp. Biol. Med. 107:887, 1961.

24. Stryckmans, P. A., Cronkite, E. P., Biocomelli, G., Schiffer, L. M., and Schnappauf, H. P.: The maturation and fate of reticulocytes after in vitro labeling with tritiated amino acids. Blood 31:33, 1968.

25. Berendes, M.: The proportion of reticulocytes in the erythrocytes of the spleen as compared with those of circulating blood, with special reference to hemolytic states. Blood 14:558, 1959.

26. Neuberger, A., and Niven, J. S. F.: Haemaglobin formation in rabbits. J. Physiol. (London) 112:292, 1951.

27. Berlin, N. I., and Lotz, C.: Life span of the red blood cell of the rat following acute hemorrhage. Proc. Soc. Exp. Biol. Med. 78:788, 1951.

28. Card, R. T., and Valberg, L. S.: Characteristics of shortened survival of stress reticulocytes in the rabbit. Amer. J. Physiol. 213:566, 1967.

29. Ganzoni, A., Hillman, R. S., and Finch, C. A.: Maturation of the macroreticulocyte. Brit. J. Haemat. 16:119, 1969.

ACKNOWLEDGMENTS

The invaluable assistance of Miss Linda Lavidor and Miss Maria Tsong is gratefully acknowledged.

Radiosensitivity of the Hematopoietic Colony-Forming Cells of the Mouse

By J. F. Duplan

Since Till and Mc Culloch[1] described the spleen colony assay, the radiosensitivity of the hemopoietic colony-forming units (CFU) has been the subject of extensive investigation. From the various values found in the literature it can be stated that the D_0 varies between 65 and 100 r for adult bone marrow CFU and between 80 and 140 r for fetal liver CFU. These wide variations appear to be mainly related to strain differences. In all cases, however, within the same strain the radiosensitivity of bone marrow CFU is 25 to 40 per cent greater than that of the fetal liver CFU. It has recently been demonstrated[2] that less than 10 per cent of the CFU present in a suspension of normal bone marrow are susceptible in vitro to the lethal effect of tritiated thymidine (T^3H) but when the same procedure is applied to a suspension of fetal liver cells it kills 30–50 per cent of the CFU. This technic is thought to selectively kill cells in DNA synthesis. One can assume, therefore, that bone marrow CFU are predominatly in a resting state (G_0) but that the majority of the fetal liver CFU are in active cycle. This would be consistent with steady state kinetics in the adult marrow, whereas the hemopoietic stem cell pool of the fetus is in active cycle to support a rapidly growing population.

I have investigated the changes of radiosensitivity that result when (a) the cells in S were killed by a suicidal dose of tritiated thymidine, and (b) a significant number of bone marrow cells were triggered into cycle.

Materials and Methods

The experimental animals were male and female mice of the pure bred line XVII. Recipients of hemopoietic cells were 80 to 100 days old when irradiated, donors of bone marrow were 55 to 65 days old and fetal liver cell suspensions were obtained from 15–17-day-old fetuses. Several mice were bled from a retro-orbital vein, the amount of blood removed varying between .35 to .45 ml. After 20–24 hours microhematocrit determinations were performed and the animals were sacrificed as donors of bone marrow cells.

Bone marrow of several femora or livers from several fetuses were suspended in Tyrode's solution in order to produce suspensions of approximately 2×10^7 nucleated cells per ml. for bone marrow or 1×10^8 for fetal liver. This suspension was divided into two aliquots. To the first was added an equal volume of Tyrode containing 6–8 μg. of thymidine (TdR) per ml.; the second was diluted with an equal volume of Tyrode containing 700–800 μCi of T^3H (specific activity: 12–16 Ci/mM) per ml. Both suspensions were incubated for 25 minutes at 37 C. and then diluted with 10–12 times their volume of Tyrode containing 3μg. of TdR per ml. This technic is similar to that previously described by

From the Institut du Radium, Laboratoire Pasteur and Fondation Curie, Paris, France.
This work was supported by a grant from DRME #487–68.

Becker et al.[2] The nucleated cells were counted in an hemocytometer, both suspensions were adjusted to the desired concentrations and injected into irradiated recipients.

The recipients were divided into 5 groups of 12 to 16 mice which had been exposed to 800, 700, 650, 600, and 500 r of x-radiation delivered under the following conditions: 225 kv., 11mA, 0.3 cu. mm., and 2 mm. Al added filtration. Immediately after this initial irradiation each group was divided into 2 lots of 6–8 animals; one received an intravenous injection of the hemopoietic cells incubated with TdR and the other the suspension which had been incubated with T^3H. Except for the groups which had received 800 r, all individual groups were exposed to a second irradiation of 100, 150, 200, and 300 r respectively. In this fashion the cells were irradiated with 0 to 300 r while all the recipients received 800 r.

The recipients were killed 8–9 days later, their spleens removed and fixed in Bouin's fluid for 4–6 hours. The spleen colonies were enumerated under a $2.5\times$ magnifying lens. The mean values for D_0 and extrapolation number are given together with the corresponding 95 per cent confidence limits.

Results

Hemopoietic Cells of the Fetal Liver

Nine experiments were performed to study the radiosensitivity of the fetal liver CFU. The fetuses were obtained from mice in their fifteenth to seventeenth days of pregnancy and 2–3 litters were pooled in order to reach the required cellular concentration. In each individual experiment the following parameters were measured: (1) the T^3H suicide effect expressed as a percentage of CFU killed by incubating the fetal liver suspension with T^3H; (2) the D_0 and the extrapolation number related to either the over-all CFU population or to the fraction of cells surviving the T^3H suicide.

When the total CFU population was tested, the variations between individual experiments were limited. The mean value for D_0 was 83 ± 8 r and the extrapolation number was 1.6 ± 0.4. In the suspensions which had been incubated with T^3H the surviving CFU had a D_0 ranging from 76 to 138 r. But as shown in Figure 1 a positive correlation existed between the percentage of CFU killed by T^3H and the corresponding D_0 values. It was thus possible to divide the 9 experiments into 2 groups according to the percentage of CFU incurring T^3H suicide. There were 3 experiments in which the percentage of suicide was less than 20 (average 17 per cent); the pooled D_0 from these experiments fell within the normal range, being equal to 82 r. In the 6 other experiments forming the second group the percentage of CFU killed by T^3H was over 30 except for one case (25 per cent) and the mean D_0 value was 121 ± 20 r with an extrapolation number of 1.8 ± 0.3.

Figure 2 shows the 2 dose-effect curves which represent the radiosensitivity of the over-all CFU population of fetal liver and of this same population from which cells in S phase have been removed by incubation with a suicidal dose of T^3H.

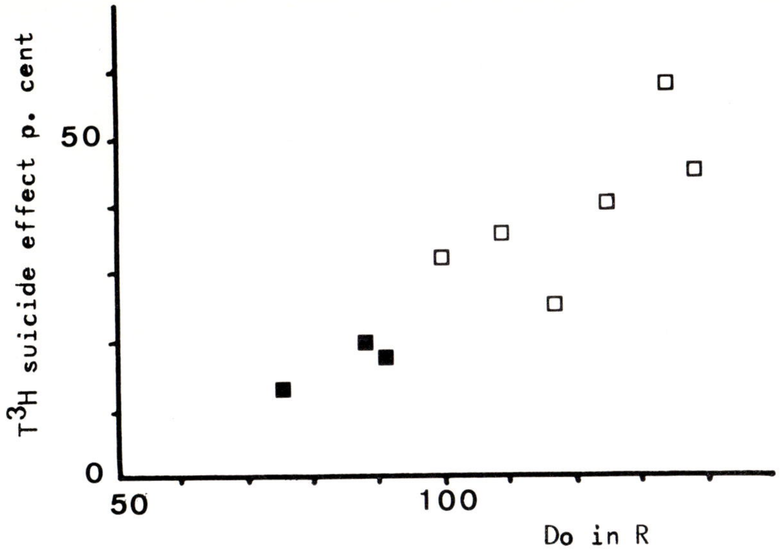

FIG. 1—CFU of the fetal liver—Correlation between the suicide effect by T^3H and the radiosensitivity (D_o) of the surviving CFU population: ■ Experiments in which the suicide effect was less than or equal to 20 per cent; □ Experiments in which the suicide effect was more than 20 per cent.

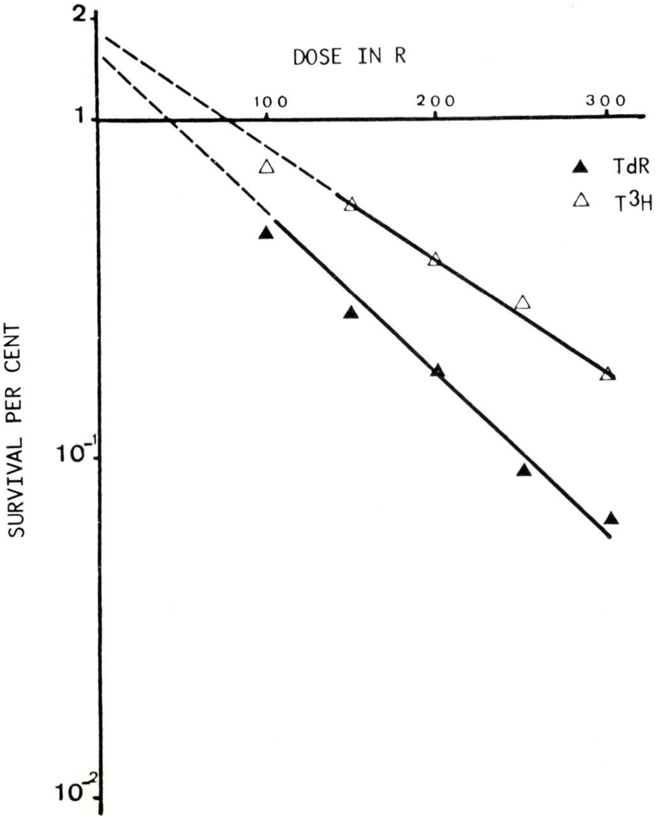

FIG. 2—Dose survival curves of CFU from fetal liver showing a suicide effect higher than 20 per cent incubated with TdR (▲) or with T^3H (△) prior to irradiation.

Hemopoietic Cells of the Adult Bone Marrow

As previously stated very few adult bone marrow CFU are killed by incubation with T^3H. Therefore the previously reported experiment could not be done on the bone marrow population without first triggering the CFU into cycle, and increasing the number of cells in S. This was done by bleeding the donor mice, a method expected to be harmless for the cells. Preliminary experiments were performed for the purpose of assessing the hematocrit value and the optimum time interval between bleeding and sacrifice which would give the greatest suicide effect. It was found that the hematocrits had to be 28–32 per cent and that the time between bleeding and sacrifice had to be in the vicinity of 24 hours. Fourteen experiments were done fulfilling these two requirements.

As far as the over-all CFU population was concerned the characteristics of radiosensitivity for bled mice were found to fall within the same range as for normal mice of the XVII strain with a mean D_0 value of 65 ± 6 r and an extrapolation number of 1.9 ± 0.4.

In contrast with this result the D_0 of the CFU previously incubated with T^3H showed wide variations, but each individual value correlated with the percentage of cells killed by T^3H (Fig. 3). Consequently the 11 experimental groups were divided into 3 subgroups according to the magnitude of the suicide effect.

In the first group (3 experiments) no more than 20 per cent of the CFU were killed by T^3H and the D_0 was 62 r, i.e., a normal value. The percentage of

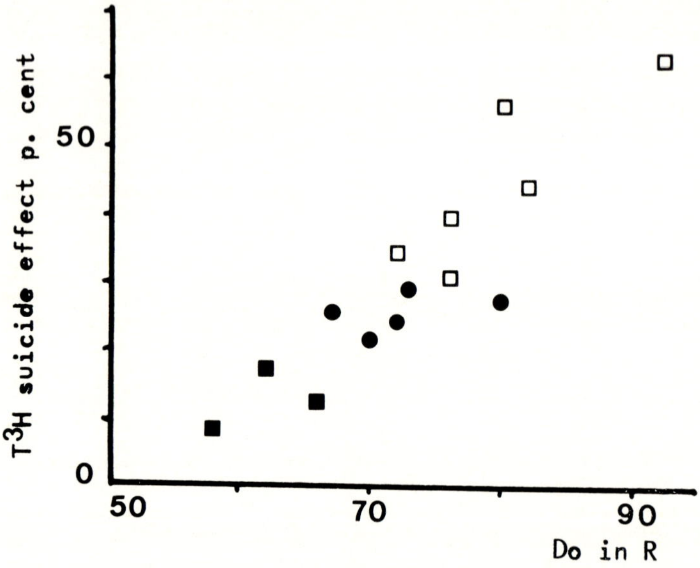

FIG 3—CFU of the adult bone marrow after bleeding—Correlation between the suicide effect by T^3H and the radiosensitivity (D_0) of the surviving CFU population: ■ Experiments in which the suicide effect was less than 20 per cent; ● Experiments in which the suicide effect ranged from 20 to 29 per cent; □ Experiments in which the suicide effect was more than 29 per cent.

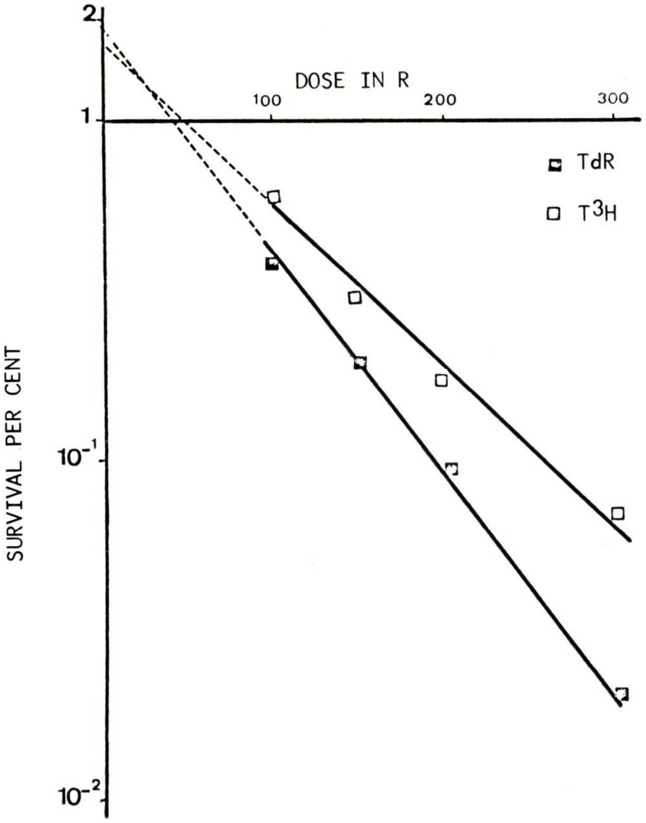

FIG. 4—Dose survival curves of CFU from the bone marrow of bled mice showing a suicide effect higher than 29 per cent. Cell suspensions incubated with TdR (■) or with T^3H (□) prior to irradiation.

suicide ranged from 21 to 29 for the 5 experiments included in the second group, and the corresponding D_0 was 72 r, a value which can be accepted as greater than normal. For the 6 experiments in the last group the suicide effect was greater than 30 per cent and the mean value for D_0 was 80 ± 9 r with an extrapolation number of 1.7 ± 0.3. The curves for the latter group are shown in Figure 4. Table 1 summarizes the results obtained for the fetal liver and the adult bone marrow from normal or bled animals. From this it is clear that for cells of both origins the CFU population which did not incorporate T^3H had a lower radiosensitivity than the entire CFU population.

DISCUSSION

The radiosensitivity of fetal liver or bone marrow CFU incubated with 3–4 μg. of TdR is identical to the radiosensitivity of nonincubated cells of the same origin. Furthermore this treatment does not induce any significant change in the cloning efficiency. Accordingly the reported findings cannot be ascribed to an incidental selection of a cellular population or to a change in sensitivity

TABLE 1—*Radiosensitivity of the CFU According to Their Origin and Their Position in the Cell Cycle*

Origin of CFU	Incubation	Suicide per cent	D_0 † in *R*	E.N.‡
	no	—	82 ± 7	1.4 ± 0.2
Fetal liver	TdR	—	83 ± 8	1.6 ± 0.4
	T³H	39	121 ± 20	1.8 ± 0.3
	no	—	63 ± 5	1.7 ± 0.3
Normal bone marrow	TdR	—	62 ± 5	1.5 ± 0.4
	T³H	10	60 ± 6	1.6 ± 0.3
	no *	—	66	1.8
Bled bone marrow	TdR	—	65 ± 6	1.9 ± 0.4
	T³H	43	80 ± 10	1.7 ± 0.3

* Mean of 4 experiments.
† Mean value ± 95 per cent confidence limits.
‡ Extrapolation number.

related to the incubation with TdR.[3] Our data confirm a previously reported experiment[2] which demonstrated that an average of 40–45 per cent of fetal liver CFU are in S phase and less than 10 per cent of the corresponding bone marrow cells are able to incorporate T³H.

From the above it may be assumed that the total CFU population of the fetal liver is in cycle and the D_0 of 83 r is actually the mean of the various D_0 which characterize the different phases of the cycle. The removal of the cells which were synthetizing DNA led to an increase of the D_0. This means that the average D_0 value for the cells in S phase must be lower than 83 r. The remaining population contained cells in G_1, G_2, and mitosis. Since it is well documented that mitosis is generally radiosensitive, it may be concluded that the D_0 for cells in G_1 and G_2 is higher than 121 r.

These same considerations apply to the bone marrow CFU. In the normal steady state, the D_0 of 65 r corresponds to the radiosensitivity of CFU in G_0. The 15–20 per cent of the cells in cycle in the normal steady state apparently do not affect the D_0 as evidenced by the fact that when almost all CFU were in cycle in the bled animals, the D_0 was still 62 r, a phlebotomy of about one-third of the circulating blood resulted not only in anemia but also in leukopenia. It is likely that both participate in the compensatory reaction through which the stem cells including the CFU are triggered into cycle. When more than 30 per cent of the CFU were in S phase nearly the whole CFU population was in cycle, i.e., no or very few cells were still in G_0. This is easily demonstrated because when 45–60 per cent of the CFU are in S there are only two possibilities: (1) the remaining CFU are all in cycle, or (2) a significant fraction stays in G_0 and the cycling CFU population is synchronized. This second hypothesis is obviously wrong because removing the cells in S did not lead to a D_0 of 65 r which corresponds to that for G_0 cells but of 80 r which can only apply to cells which are neither in G_0 nor in S. It could in fact be foreseen that bleeding was unlikely to synchronize the stem cell pool. Therefore it can be assumed that the

D_0 of 62 r was actually the mean of different individual D_0 each corresponding to fractions of the CFU population which were in various stages of the cycle. Consequently the average D_0 value of the cells in S phase must be lower than 62 r and the minimum value for the cells in G_1 and G_2 must be close to 80 r. Since the population is not synchronized and the quantitative distribution of the CFU throughout the phases of the cycle is unknown it does not seem possible to obtain a more accurate evaluation of the D_0 for various phases of the cycle.

Data obtained in vitro with synchronized cell populations point to the complexity of the relationships between radiosensitivity and cell cycle. For Chinese hamster cells [4] and to some extent for some human HeLa cell lines [5] the most sensitive period is in G_1 and mitosis, but for several lines of mouse L cells [6–7] part of the S phase is as sensitive as mitosis. These results have two features in common: one is the high sensitivity of the mitotic stage, and the other is the absence of correlation between the changes in radiosensitivity and the cell transit through the phases of the cycle other than mitosis. This is especially clear for S which can be accurately followed by the synthesis of DNA and is actually not associated with any specific change in radiosensitivity. Therefore, it can be concluded: (1) that several biochemical processes linked to the cycle but different from DNA synthesis may influence to a large extent the cellular radiosensitivity [7]; (2) that since the cycle is presently divided according to DNA synthesis, the D_0 of a given stage may be the mean of widely different values.

Differences have been found between fetal liver and adult bone marrow CFU from many standpoints, such as capacities for differentiation [8–9] and for self-replication as well as therapeutic and restorative efficiencies.[10–11] The differences in radiosensitivity could be ascribed to the different status of the cell population, i.e., steady state for the bone marrow and a rapidly increasing population for the fetal liver. In view of the present findings the latter hypothesis does not hold. Since fetal G_1 and G_2 cells had a D_0 of 121 r and adult marrow CFU in the same phase of cycle had a D_0 of 80 r, it seems reasonable to assume that the fetal liver stem cells constitute a cellular population different from the adult bone marrow stem cells pool although the former may provide the precursor elements for the latter.

REFERENCES

1. Till, J. E., and McCulloch, E. A.: A direct measurement of the radiation sensitivity of normal mouse bone marrow cells. Radiat. Res. 14:96, 1961.

2. Becker, A. J., McCulloch, E. A., Siminovitch, L., and Till, J. E.: The effect of differing demands for blood cell production on DNA synthesis by hemopoietic colony-forming cells of mice. Blood 26:296, 1965.

3. Djordjevic, B., and Kim, J. H.: Modification of radiation response in synchronized HeLa cells by metabolic inhibitors: Effects of inhibitors of DNA and protein synthesis. Radiat. Res. 37:435, 1969.

4. Sinclair, W. K., and Morton, R. A.: X-ray sensitivity during the cell generation cycle of cultured Chinese hamster cells. Radiat. Res. 29:450, 1966.

5. Terasima, T., and Tolmach, L. J.: X-ray sensitivity and DNA synthesis in synchronous populations. Science 140:490, 1963.

6. Whitmore, G. F., Till, J. E., and Gulyas, S.: Radiation induced mitotic delay in L cells. Radiat. Res. 30:155, 1967.

7. Sinclair, W. K.: Cyclic X-ray responses in mammalian cells *in vitro*. Radiat. Res. 33: 620, 1968.

8. Silini, G.: Control of haemopoiesis and the action of radiation on the blood forming organs. *In* Effects of Ionizing Radiations on the Haematopoietic Tissue. Vienna, International Atomic Energy Agency, Panel Proceedings Series, 1967, p. 52.

9. Duplan, J. F.: Étude comparative des cellules souches hématopoïétiques de la moelle osseuse et du foie foetal. Nouv. Rev. Franç. Hemat. 8:445, 1968.

10. Duplan, J. F., Efficacités thérapeutiques comparées des injections de foie foetal ou de moelle osseuse isogéniques chez les souris irradiées. C. R. Acad. Sci. [D] (Paris) 267:227, 1968.

11. Duplan, J. F., and Estienne, E.: Relation entre la mortalité des souris irradiées et les délais écoulés entre l'irradiation X et l'injection des cellules restauratrices. C. R. Soc. Biol. (Paris) 158:228, 1964.

Humoral Regulation of Fetal and Neonatal Erythropoiesis

By G. Lucarelli, V. Rizzoli, R. Delsignore, N. Rencricca, A. Porcellini, L. Ferrari, C. Carnevali, D. Howard, and F. Stohlman Jr.

During the past several years we have been studying control mechanisms and morphologic characteristics of fetal and neonatal erythropoiesis.[1] From these studies, as well as others, certain facts have evolved which suggest that fetal and neonatal erythropoiesis differ in part from that of the adult. These are summarized below.

In the rodent on the sixteenth day of fetal life the liver is the primary erythropoietic organ. All stages of normoblastic erythropoiesis are present and in addition there is a cell which is frequently seen to occur in syncytia and which may serve as the precursor cell for the erythroid cells. At birth the femur is devoid of erythroid cells but there is an early onset of erythropoiesis and an explosive increase in erythroid cellularity so that 72 hours after birth up to 60 per cent of the nucleated bone marrow cells are erythroid. During this phase of development a cell similar to the syncytial cell is seen in the bone marrow. Gradually, however, these cells disappear and by 4 weeks of life the bone marrow shows characteristics similar to that of the adult.

Direct observations on the control of red cell production in the fetus have not been possible. Jacobson et al., as well as ourselves, observed that polycythemia in the mother did not affect red cell production in the fetus.[2] When Schooley gave antierythropoietin to pregnant rats it did not affect fetal erythropoiesis.[3] In part this may have been due to failure of the protein to cross the placental barrier. We further observed that although starvation depressed erythropoiesis in the mother there was no effect on red cell production in the fetus.[1]

Transplantation of fetal liver into heavily irradiated animals resulted in a different pattern of erythroid colonizationt than was seen with adult marrow.[4] There was significant repopulation of the bone marrow of recipients of fetal liver, whereas after transplantation of adult marrow erythroid colonization was primarily splenic. Duplan has discussed earlier at this conference that there is a significant difference in radiosensitivity between the fetal and adult stem cell.[5] Thus there is indirect evidence that fetal erythropoiesis differs from that of adults both in its morphologic characteristics, i.e., from the standpoint of the syncytial cell and also in some of its physiologic characteristics.

We have also studied in the newborn the effect of classic technics used to suppress erythropoiesis in the adult animal, i.e., nephrectomy, hypertransfusion, and starvation.[1-6] As judged by bone marrow differentials, splenic histology and peripheral blood reticulocytes, nephrectomy, and starvation

From the Instituto di Semeiotica Medica, Universita' di Parma, Italy, and St. Elizabeth's Hospital, Tufts University, Boston, Massachusetts.

This investigation was supported by Consiglio Nazionale delle Ricerche, Roma, Italy and by HE 07542 and HE 5600 National Heart Institute, Bethesda, Maryland.

did not importantly alter red cell production during the first 2 weeks of life. Hypertransfusion produced some suppression of erythropoiesis but the extent of suppression is nowhere nearly as great as observed in the adult animal. These observations may be interpreted as indicating that red cell production is independent of renally produced erythropoietin, that there is a difference in sensitivity of the cell to erythropoietin or that because of differences in metabolic rate in the rapidly growing newborn animals starvation does not change the oxygen supply-demand ratio to the extent seen in the adult. Alternatively it might be suggested that fetal and neonatal erythropoiesis are independent of erythropoietin. Pertinent to these considerations are the observations of Carmena et al. who not only demonstrated erythropoietin in the plasma of newborn rats exposed to hypoxia but also observed that unlike adults renoprival newborn animals produced erythropoietin.[7] Schooley also reported erythropoietin production in newborn rats and observed that antierythropoietin reduced red cell production.[3] These observations attest to the importance of erythropoietin in the neonatal rat.

Our recent studies have been directed at further evaluation of red cell production in the fasted and nephrectomized animal. In a significant number of newborn animals fasted prior to the fifteenth day of life an increase in mitotic index was observed due in large part to an accumulation of prophases and metaphases. The reticulocyte count in the peripheral blood in the fasted newborn animals was similar to that observed in control animals. The reticulocytes, however, appeared to be more mature and there was a shift to the right in the Heilmeyer index, as shown in Table 1. These observations suggested a prolonged transit time and delayed release.

To further document this, iron incorporation studies were carried out in vivo and in vitro on animals which were fasted from the seventh to the tenth day of life. Pertinent red cell parameters are shown in Tables 2–3 and in Fig. 1. In one group of animals iron was injected intraperitoneally and the red cell iron incorporation measured 18 hours later. The iron uptake in RBC in the normal was 33.0 per cent, while in the starved the iron uptake was 12.2 per cent. Administration of 2 units of erythropoietin daily during the period of starvation increased the red cell iron incorporations to a value of 17.5 per cent.

TABLE 1—*Maturation Stage of Reticulocytes*

	% Retics.	Stage I	Stage II	Stage III
Normal control	36.9 (3.9)	44.5 (1.5)	28.0 (1.0)	27.5 (0.5)
Fasted *	30.7 (1.1)	20.0 (1.0)	29.5 (0.5)	50.5 (0.5)
Fasted + EP †	31.0 (2.8)	28.5 (2.5)	30.5 (3.5)	41.0 (6.0)

The values are from the peripheral blood of 10-day-old rats. The stage of maturation was determined by the amount of reticulum present in the cell. Values in parentheses are standard errors of the mean.

* Animals were fasted for 72 hours prior to sacrifice.

† Erythropoietin was given in the amount of 2 units daily during the 72 hour period of fasting.

TABLE 2—*Effect of Erythropoietin on Erythropoiesis in Starved 10-Day-Old Rats*

	No. Animals	Weight (GM.)	HCT. %	Retics. %	% $^{59}Fe/10^9$ Retic.
Control	53	21.8 (1.0)	25 (0.36)	29.7 (3.72)	49.24 (1.6)
Fasted	50	13.0 (1.0)	33 (0.73)	27.2 (4.26)	16.04 (2.95)
Fasted + Erythropoietin	52	13.3 (1.5)	32 (2.36)	28.3 (4.12)	30.40 (0.61)

The animals were fasted for 72 hours prior to sacrifice and in the case of erythropoietin treatment received 2 units daily throughout the period of starvation. Per cent ^{59}Fe inc./10^9 retic. = 2 hour in vitro iron incorporation per 10^9 reticulocytes. Values in parentheses are standard errors of the mean.

When peripheral blood reticulocytes were incubated with radioiron for a period of 2 hours there was a 48 per cent uptake of radioiron. In contrast in animals fasted for 72 hours there was only 16 per cent incorporation of iron by the reticulocytes. In those animals receiving erythropoietin throughout the period of starvation the in vitro iron uptake increased to 30 per cent.

Red cell iron uptake in bone marrow and spleen incubated for 2 hours showed no substantial difference between that of control and fasted animals,

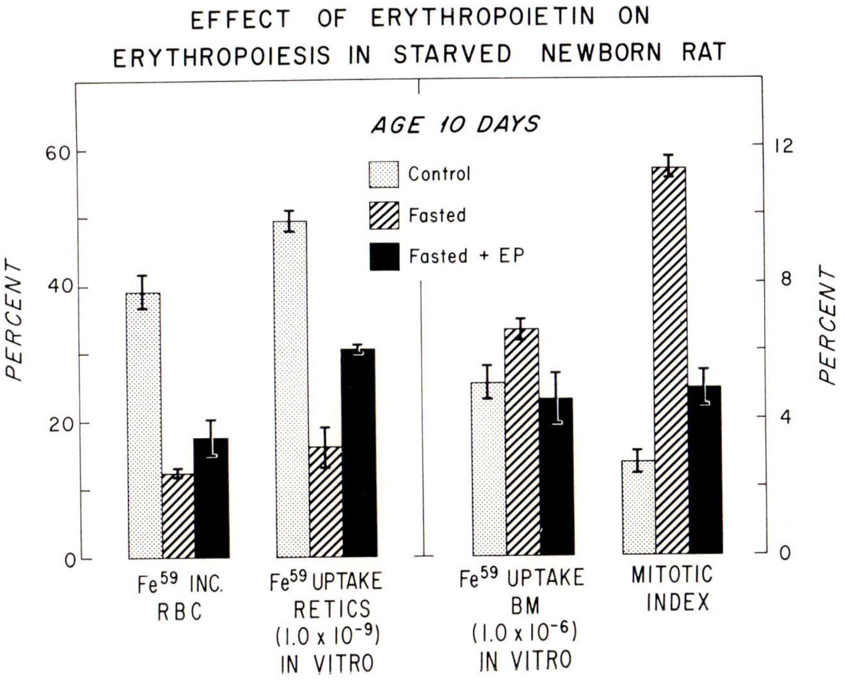

FIG. 1—Age refers to the day of sacrifice. Animals were fasted for 72 hours. Erythropoietin was given 2 units daily for 3 days. RBC ^{59}Fe Inc.: per cent iron uptake in peripheral red cells 24 hours after intraperitoneal administration of radioiron. Retics.: in vitro uptake of ^{59}Fe by 1×10^{-9} reticulocytes after 2 hours of incubation with radioiron. BM: in vitro uptake of ^{59}Fe by 1×10^{-6} red cell precursors after 2 hours of incubation with radioiron. Mitotic index: based on per cent of red cell precursors in mitoses. (Bars refer to SE of the mean.)

Table 3—*Effect of Erythropoietin on Erythropoiesis in Starved 10-Day-Old Rats*

	No. of Animals	% Erythroid*	Pronormoblast	Basophilic Normoblast	Polychromatophilic Normoblast	M I†
Control	60	43.7 (1.65)	4.6 (0.27)	8.7 (0.76)	30.4 (0.54)	2.7 (0.30)
Fasted	70	54.0 (1.89)	3.7 (0.16)	7.1 (0.53)	43.2 (1.20)	11.3 (0.30)
Fasted + Erythropoietin	75	60.3 (1.35)	5.8 (0.18)	9.8 (0.30)	44.7 (1.20)	4.9 (0.51)

The animals were fasted for 72 hours prior to sacrifice and in the case of erythropoietin treatment received 2 units daily throughout the period of starvation.

* The percentage of total femoral nucleated cells which were erythroid. These are further subclassified as per cent pronormoblast, basophilic normoblast, and polychromatophilic normoblast.

† Mitotic Index of erythroid cells.

Values in parentheses are standard errors of the mean.

TABLE 4—*In Vitro % ^{59}Fe Incorporation in Bone Marrow Cell Suspensions From 10-Day-Old Rats*

	Bone Marrow	Spleen
Control	5.08 (0.47)	3.99 (0.30)
Fasted	6.60 (0.34)	4.28 (0.48)
Fasted + Erythropoietin	4.61 (0.75)	3.42 (0.50)

Animals were fasted for 72 hours prior to sacrifice. Erythropoietin was given 2 units daily for 3 days. Cell suspensions of femoral bone marrow and spleen were prepared in TC 199 medium and incubated 2 hours with radioiron. Cells were then washed 3 times in EDTA and saline, and ^{59}Fe uptake expressed per 1×10^{-6} erythroid cells. Average of 8 to 11 determinations. Values are given as mean ± S.E.

if anything there was a slightly greater uptake of iron in the fasted animals (Table 4).

Of particular interest to us was the rather striking increase in mitotic index in the fasted animals as compared with the control animals (Table 3, Fig. 1). Values of 10–12 per cent were seen in fasted animals as compared with 3 to 4 per cent in the control animals. Administration of erythropoietin during the period of starvation reduced the mitotic index and there was a significant decrease in the numbers of prophases and metaphases observed in the bone marrow. The mitotic block was present in 67 per cent of fasted 10 day old rats but only in 19 per cent of fasted animals treated with erythropoietin.

In a second set of experiments rats were bilaterally nephrectomized on the eighth day of life and sacrificed on the tenth day of life. Since it was not possible to routinely maintain bilaterally nephrectomized animals for a period of more than 48 hours, these were compared with animals fasted for 48 hours as well as with normal controls. As may be seen in Tables 5 and 6, and Fig. 2, there was no significant change in either the bone marrow or peripheral blood parameters or in the iron incorporation between nephrectomized and control animals. In the fasted animals, however, there was a reduction in the iron incorporation of fasted animals after 48 hours.

TABLE 5—*Effect of Nephrectomy and Starvation on Erythropoiesis in the 10-Day-Old Rat*

	No. of Animals	Weight (GM.)	HCT.	Retics. %	% ^{59}Fe In Vitro Uptake 1×10^{-9} Retics.
Control	9	22 (1.1)	28 (1.7)	27.5 (2.5)	42.0 (3.8)
Nephrectomized	7	20 (1.3)	27 (2.9)	28.7 (2.1)	43.6 (4.1)
Fasted	10	15 (0.8)	30 (1.8)	30.2 (3.0)	16.0 (2.9)

The animals were sacrificed on the tenth day of life. Nephrectomy or fasting was initiated 48 hours before sacrifice. Iron incorporation is expressed as per cent uptake/10^9 reticulocytes after 2 hours in vitro incubation. Values in parentheses are standard errors of the mean.

TABLE 6—*Effect of Nephrectomy and Starvation on Erythropoiesis in the 10-Day-Old Rat*

	No. of Animals	Erythroid	Pronormoblast	Basophilic Normoblast	Polychromatophilic Normoblast	Mitotic Index
Control	10	40.0 (3.6)	4.4 (0.9)	6.7 (0.7)	28.8 (3.1)	3.7 (0.8)
Fasted	7	41.2 (1.2)	5.5 (0.9)	9.9 (1.8)	24.3 (2.1)	3.3 (0.5)
Fasted 48 hours	10	47.8 (2.9)	4.0 (0.7)	11.2 (1.2)	32.6 (2.1)	3.4 (0.9)

The animals were sacrificed on the tenth day of life.

The animals were fasted for 72 hours prior to sacrifice and in the case of erythropoietin treatment received 2 units daily throughout the period of starvation.

The percentage of total femoral nucleated cells which were erythroid. These are further subclassified as per cent pronormoblast, basophilic normoblast, and polychromatophilic normoblast.

Mitotic index of erythroid cells.

Values in parentheses are standard errors of the mean.

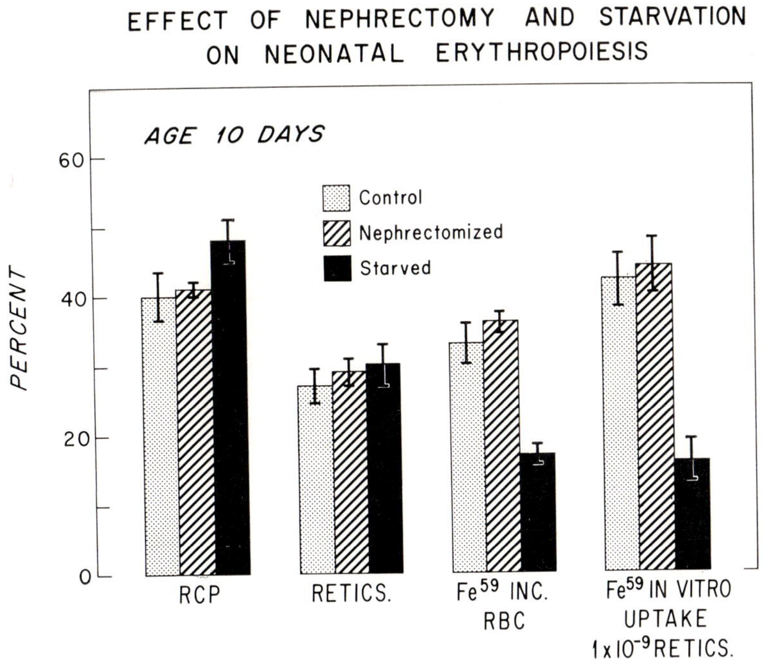

FIG. 2—Age refers to the day of sacrifice. Animals were fasted or nephrectomized 48 hours before sacrifice. RCP: red cell precursors counted on smears of femoral bone marrow. Retics.: per cent in peripheral blood. ^{59}Fe Inc.: iron incorporation into red blood cells 24 hours after intraperitoneal administration of radioiron. ^{59}Fe in vitro uptake: iron incorporation by 1×10^{-9} reticulocyte after 2 hours incubation with radioiron. (Bars refer to SE of the mean.)

DISCUSSION

Although the numbers of bone marrow precursors and peripheral reticulocytes were similar in fasted and control newborn animals there was a difference in the rate of red cell production. Thus there was a decrease in 18 hour peripheral iron incorporation, decrease in the rate of plasma iron clearance,[8] a shift to the right in the peripheral reticulocytes and a decrease in iron uptake in vitro. These observations appear to be best explained by a prolongation of the marrow transit time and the release of more mature reticulocytes into the peripheral blood. In part these events were reversed by erythropoietin and it is indeed possible that more frequent administration of perhaps larger doses would have completely reversed these changes. In the nephrectomized animal, where erythropoietin production persisted, these changes did not occur. Thus the increased maturation time appears to be due, in significant measure, to a decreased influx of cells from the precursor compartment into the differentiated erythroid compartment. The bone marrow differential was similar to that seen in control animals which together with an increase in transit time must have reflected a decrease in influx.

Of particular interest to us were the mitotic changes and the presence of a metaphase block in the fasted animal. Six per cent of 70 control animals exhibited this metaphase block and the mitotic index was 11 per cent. Erythropoietin reduced the frequency of the metaphase block in fasted animals so that only 19 per cent of 75 animals observed had such a block. A decrease in the mitotic index was associated with this. The change in the mitotic index is best explained by prolongation of the mitotic time. Whether or not these cells eventually escape from the block or die remains to be established. The most likely explanation for prolongation of the mitotic time would appear to be a decrease in metabolic rate associated with high energy requirements for mitosis. One would not anticipate, however, that such a mechanism should be affected by erythropoietin and further study will be required to determine the mechanism by which erythropoietin affects mitosis in the newborn.

In summary then there are certain differences between fetal and neonatal erythropoiesis, a difference in transplantation characteristics, in radiosensitivity, the production of erythropoietin by extrarenal sites, and the failure of starvation and hypertransfusion to produce erythroid aplasia. It is clear, however, that red cell production in the newborn animal is at least in part influenced by erythropoietin. Thus, in the starved animal while aplasia is not produced there is a prolongation of the maturation interval and an increase in the mitotic index which is thought to reflect an increase in mitotic time, and a decrease in the rate of influx of precursor cells into the differentiated erythroid compartment.

REFERENCES

1. Lucarelli, G., Porcellini, A., Carnevali, C., Carmena, A., and Stohlman, F.: Fetal and neonatal erythropoiesis. Ann. N.Y. Acad. Sci. 14:9,544, 1968.

2. Jacobson, L. O., Marks, E. K., and Gaston, E. O.: Studies on erythropoiesis. XII. The effect of transfusion-induced polycythemia in the mother on the fetus. Blood 14:644, 1959.

3. Schooley, J. C., Garcia, J. F., Cantor, L. N., and Havens, V. W.: A summary of some studies on erythropoiesis using anti-erythropoietin immune serum. Ann. N.Y. Acad. Sci. 14:9, 266, 1968.

4. Kubanek, B., Rencricca, N., Porcellini, A., Howard, D., and Stohlman, F.: The pattern of recovery of erythropoiesis in heavily irradiated mice receiving transplants of fetal liver. Radiat. Res. 30:2, 1969.

5. Duplan, J. F.: Radiosensitivity of the hematopoietic colony forming cells of the mouse. In this volume.

6. Lucarelli, G., Howard, D., and Stohlman, F.: Regulation of erythropoiesis. XV. Neonatal erythropoiesis and the effect of nephrectomy. J. Clin. Invest. 43:2, 1964.

7. Carmena, A. O., Howard, D., and Stohlman, F.: Regulation of erythropoiesis XXII. Erythropoietin production in the newborn animal. Blood 32:376, 1968.

8. Rizzoli, V., Rencricca, N., Monette, F., and Howard D.: Personal communication, 1969.

Growth Patterns of Hemopoiesis Derived from Fetal Liver Transplanted into Heavily Irradiated Mice

By Bernhard Kubanek, Nicholas Rencricca, Adolfo Porcellini, and Frederick Stohlman, Jr.

Previous observations have suggested that the regulation of erythropoiesis in the newborn rodent may differ from that of the adult.[1] In part, however, these differences may be related to extrarenal erythropoietin production [2] together with differences in the metabolic rate and hence oxygen supply-demand relationship, or even to a different sensitivity of the erythropoietic cells to erythropoietin. If in fact a difference exists between fetal and adult erythropoiesis it might be attributed to a different regulatory mechanism or a different generation time of the hemopoietic stem cells. In an effort to study some of these problems we turned to transplantation of fetal hemopoietic tissue into heavily irradiated hypertransfused and normal mice.[3] In contrast to Feldman [4] a complete suppression of red cell production and erythroid colony formation was achieved in the plethoric heavily irradiated recipients of 16½-day-old fetal liver cells. Recently Bleiberg and Feldman [5] reported that there was a significantly greater suppression of erythroid colony formation in female plethoric mice receiving fetal liver than in male recipients. He relates this to a higher endogenous erythropoietin production in the male animal. In our studies it was possible to initiate red cell production in 6- and 9-day-old fetal transplants by treatment of the recipients with exogenous erythropoietin for 2 days prior to sacrifice. Noteworthy, however, was the greater response to erythropoietin of hypertransfused recipients transplated with fetal cells compared to recipients of adult bone marrow cells, which suggests that a larger number of erythropoietin sensitive cells are available in the fetal transplants. Accordingly, recovery patterns of erythropoiesis and proliferation of colony-forming units (CFU) were compared in heavily irradiated mice after transplantation of either fetal liver cells or adult bone marrow cells.

The numbers of cells transplanted were adjusted so that the number of CFUs present in fetal liver suspensions was similar to those in adult bone marrow.

Materials and Methods

Virgin mice of thc CF1 strain 10–12 weeks of age received 800 r whole body radiation (dose rate 48 r/min.; 250 kv.; 15 mA with 1.0 mm. copper and 1.0 mm. aluminium filtration). Immediately after irradiation they were injected with either 12×10^6 nucleated fetal liver cells or 1.8×10^6 nucleated

---

From the Department of Medicine, St. Elizabeth's Hospital, and Tufts Medical School, Boston, Massachusetts.

Supported by USPHS Grants HE 7542 and HE 5600 National Heart Institute and a travel grant of the "Deutsche Forschungsgemeinschaft."

bone marrow cells from adult female mice of the same strain. Liver cell suspensions were prepared from 16½-day-old fetuses as previously described.[6] Adult bone marrow suspensions were made from the femora of 6–8 female mice 10–12 weeks of age. The plating efficiency (f-value) and the growth curve of colony-forming units in the spleen were measured by the technic of Siminovitch et al.[7] To measure the CFU present in the suspensions, 6×10^4 marrow cells or 4×10^5 fetal liver cells were given to 10-week-old CF1 mice immediately after irradiation (800 r). The animals were sacrificed on the ninth day, the spleens fixed in Bouins and the colonies counted independently by two observers. The number of nucleated cells in the spleen and a 3 hour ^{59}Fe uptake in bone marrow and spleen were measured at intervals after transplantation in the same animals as previously described.[8]

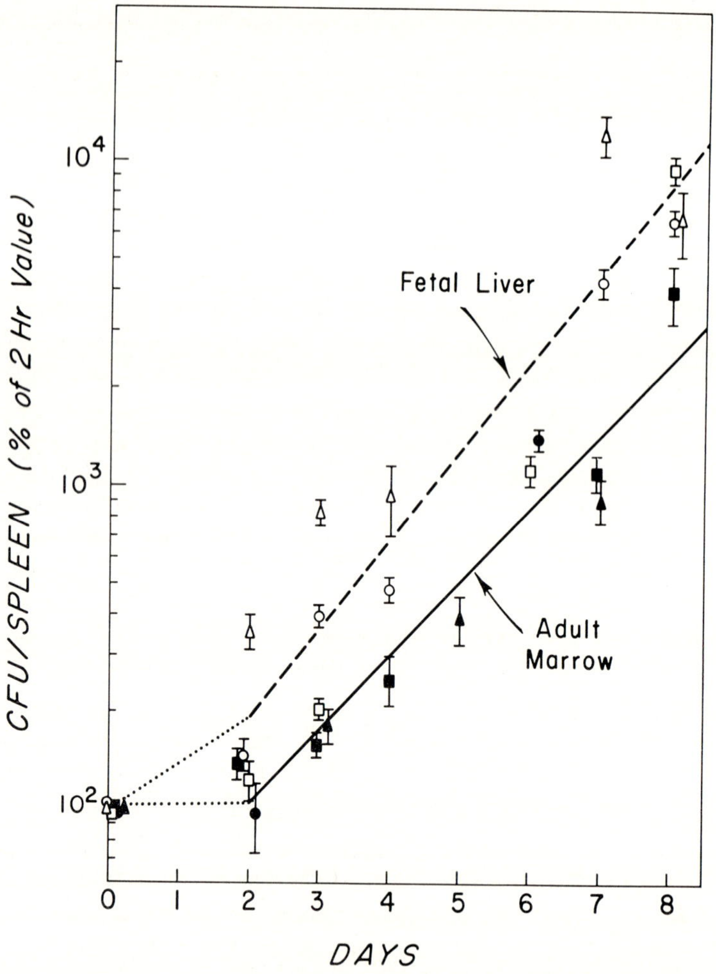

FIG. 1—The growth curve of splenic CFU in heavily irradiated animals receiving 16½-day-old fetal liver or adult bone marrow cells. Each point represents the value ± 1 SE from an individual experiment and is the mean number of colonies from 8–10 recipients. The curves were fitted by the methods of the least squares.

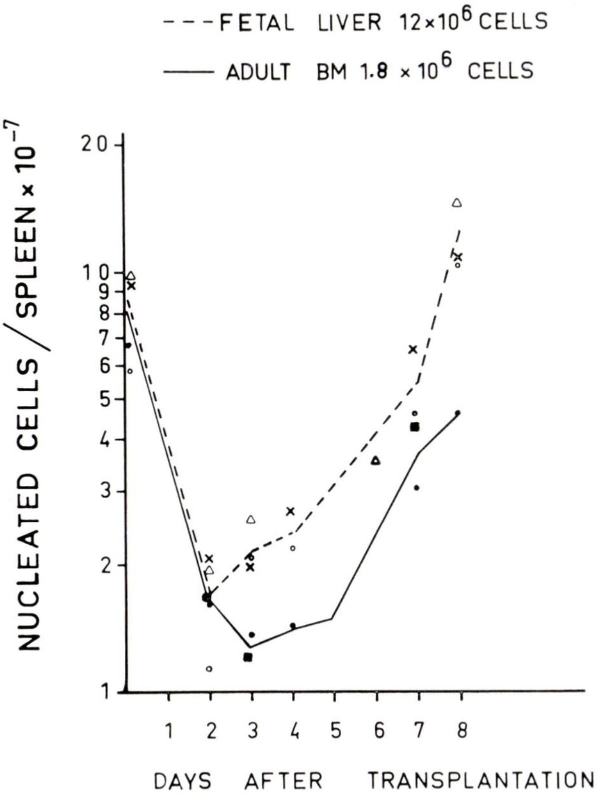

FIG. 2—The numbers of nucleated cells present in the spleen of heavily irradiated animals at intervals after transplantation with 16½-day-old fetal liver or adult bone marrow. Each point is the value from a pooled suspension of 5 spleens. The various symbols represent different experiments. The solid circle and square are adult bone marrow.

RESULTS

The mean number of CFUs in 12×10^6 nucleated liver cells was 517 with a range of 364–779; in 1.8×10^6 bone marow cells there were 504 CFU with a range of 409–641. As shown in Figure 1 the slopes of the growth curves of splenic CFU derived from fetal liver cells or for the adult bone marrow cells differ. The doubling time for fetal cells calculated from the slope of the growth curve between the second and eight day is about 24 hours as compared with 32 hours for the adult cells. The plating efficiency of the fetal cells was somewhat lower (9.6 per cent) than that of the adult bone marrow (14 per cent). It is of interest, however, that the fetal growth curve indicates an increase of CFUs over the 2 hour value as early as on the second day, whereas with the adult bone marrow, CFUs assayed at 2 hours and on the second day are the same. Absolute nucleated cell counts in the spleen on various days after transplantation of adult bone marrow cells and fetal liver cells are shown in Figure 2. Repopulation appears to occur earlier with fetal liver cells after a similar decline in cellularity in the first days after irradiation and transplantation. The cell counts of the fetal transplants rise above normal values on the eighth day whereas the absolute cell counts in the mice transplanted with adult bone marrow were about 50 per cent of the normal controls at this time. Changes in splenic erythropoiesis as measured by 3 hour radioiron incorporation at intervals after transplantation are shown in Figure 3. Again in the recipients of hemopoietic cells of fetal origin there is a striking difference in iron uptake on the third and fourth day as compared with the irradiated recipients given adult bone marrow. Figure 4 indicates that the recovery pattern of erythropoiesis in the femur estimated by a 3 hour radioiron incorporation differs from that seen in the spleen. In recipients, given adult or fetal tissue, femoral iron uptake decreases through the second post-transplantation day and on the third day there is a slight but equal increase in both groups. Thereafter there is a significant increase of femoral erythropoiesis in recipients of fetal liver through the sixth post-transplantation day when it reaches about 50 per cent of that seen in normal control mice of the same

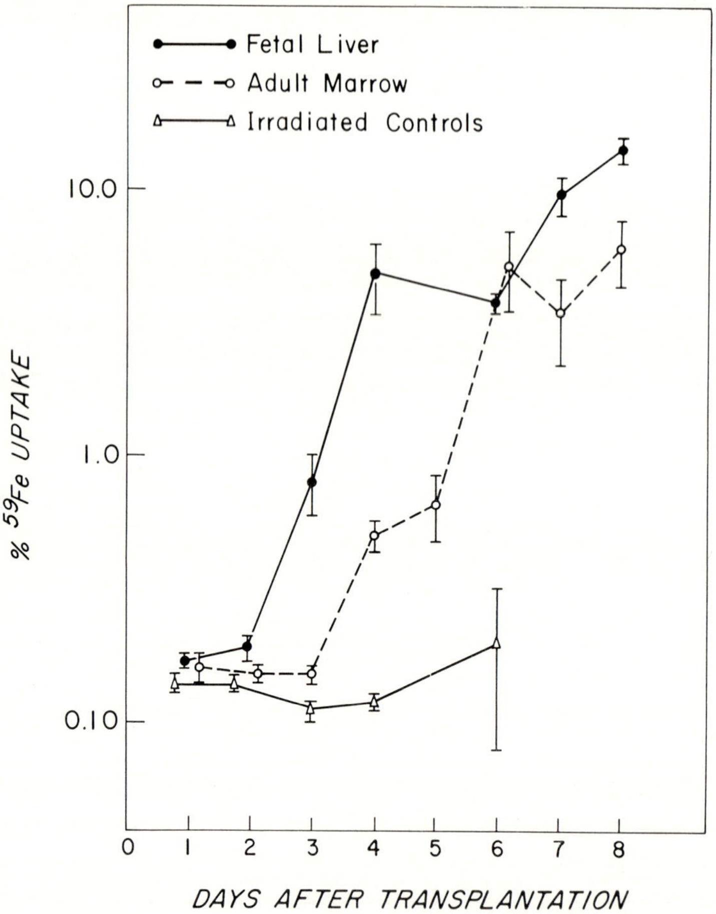

Fig. 3—3-hour splenic ^{59}Fe uptake in lethally irradiated mice receiving fetal liver or adult bone marrow: each point represents the mean ± 1 SE of 5–24 animals.

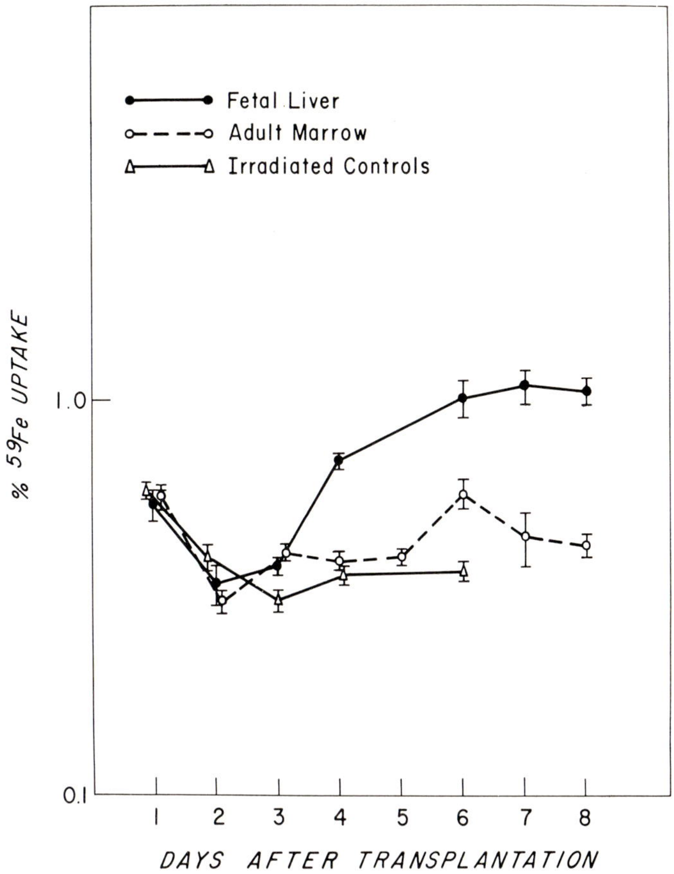

FIG. 4—3-hour femoral ^{59}Fe uptake in lethally irradiated mice receiving fetal liver or adult bone marrow: each point represents the mean ± 1 SE of 5–25 animals.

strain. In contrast, mice receiving adult bone marrow had only a slight increase in ^{59}Fe uptake above irradiated controls by the sixth day.

Discussion

The somewhat lower f-value for the 16½ day fetal liver in our experiment was comparable to that reported by Silini.[9] In recipients receiving fetal hemopoietic tissue the CFU content of the spleens increased somewhat earlier after transplantation than in those receiving adult bone marrow. An explanation for this earlier rise of CFU could be that CFU derived from fetal liver are actively in cell cycle when transplanted, whereas the stem cells of adult bone marrow are mainly a resting population. The growth rate for the CFU of fetal origin was greater than that for the CFU derived from adult cells. The measured growth rate of CFU is a resultant of the true growth rate and the loss of CFU by differentiation out of the stem cell compartment, which are therefore not included in the measurement. Differentiation occurs earlier in recipients of fetal liver tissue than in recipients of adult bone marrow, as is indicated by the nucleated cell counts as well as the more rapid recovery of erythropoiesis. From this it can be appreciated that the doubling time recorded herein minimizes the differences in generation time between transplanted fetal and adult tissue. The early increase of CFU in the spleen of recipients of fetal liver and the shorter doubling time may be responsible, in part, for the difference in the pattern of recovery of erythropoiesis as judged from the splenic iron incorporation. But the apparently greater colonziation of the femoral marrow in recipients of fetal liver cannot be explained on the basis of generation time alone and must reflect a more fundamental difference between the fetal and adult hemopoietic stem cell. The change in slope of the splenic radioiron uptake curves between the fourth and the sixth post-transplantation day for fetal liver and between the sixth and seventh day for adult bone marrow could be explained by a diversion of stem cells from erythroid to myeloid differentiation.

Thus there are significant differences in the pattern of recovery of nucleated cells, erythropoiesis and splenic CFU in heavily irradiated animals receiving fetal liver as compared with those receiving adult bone marrow. Earlier at this conference Dr. Duplan discussed the significant difference in radiosensitivity between the fetal and adult stem cell. Dr. Lucarelli described a syncytical precursor cell, which is seen in fetal liver and bone marrow of newborn rats, but not in adult bone marrow. In conclusion it appears that there are differences in transplantation characteristics, in radiosensitivity, and in cell morphology between the adult and fetal hemopoietic precursor cells.

REFERENCES

1. Lucarelli, G., Porcellini, A., Carnevali, C., Carmena, A., and Stohlman, F., Jr.: Fetal and neonatal erythropoiesis. Ann. N.Y. Acad. Sci. 149:544, 1968.

2. Carmena, A., Howard, D., and Stohlman, F., Jr.: Regulation of erythropoiesis. XXII. Erythropoietin production in the newborn animal. Blood 32:376, 1968.

3. Kubanek, B., Rencricca, N., Porcellini, A., Howard, D., and Stohlman, F., Jr.: The pattern of recovery of erythropoiesis in heavily irradiated mice receiving transplants of fetal liver. Radiat. Res. 39:105, 1969.

4. Feldman, M., and Bleiberg, I.: Studies on the feedback regulation of haemopoiesis. *In* Ciba Foundation Symposium, "Cell Differentiation." London, Churchill, 1968, pp. 79–89.

5. Bleiberg, I., and Feldman, M.: On the regulation of hemopoietic spleen colonies produced by embryonic and adult cells. Develop. Biol. 19:566, 1969.

6. Kubanek, B., Rencricca, N., Porcellini, A., Howard, D., and Stohlman, F., Jr.: The patterns of stem cell repopulation in heavily irradiated mice receiving transplants of fetal liver. Blood. 35:64, 1970.

7. Siminovitch, L., McCulloch, E. A., and Till, J. E.: The distribution of colony forming cells among spleen colonies. J. Cell. Physiol. 62:327, 1963.

8. Kubanek, B., Tyler, W. S., Ferrari, L., Porcellini, A., Howard, D., and Stohlman, F., Jr.: Regulation of erythropoiesis. XXI. The effect of erythropoietin on the stem cell. Proc. Soc. Exp. Biol. Med. 127:770, 1968.

9. Silini, G., Pozzi, L. V., and Pons, S.: Studies on the haemopoietic stem cells of mouse fetal liver. J. Embryol. Exp. Morph. 17:303, 1967.

10. Becker, A. J., McCulloch, E. A., Siminovitch, L., and Till, J. E.: The effect of differing demands for blood cell production on DNA synthesis by hemopoietic colony forming cells of mice. Blood 26: 296, 1965.

Granulocytopoiesis

By Eugene P. Cronkite and Paul C. Vincent

We have recently completed a review on granulocytopoiesis [1] in which we analyzed the contributions made toward an understanding of granulopoiesis by the proportional distribution of cells in the bone marrow, the mitotic index, labeling of DNA in the proliferative compartment of the bone marrow with tritiated thymidine, and the results of labeling peripheral blood cells with diisopropylflurophosphate ($DF^{32}P$) followed by autotransfusion. This analysis will be a condensation of the earlier review. A good starting point is a discussion of peripheral blood granulocytes. Figure 1 shows the relationship of the bone marrow to the circulating granulocyte pool (CGP), marginal granulocyte pool (MGP), and tissues from Cartwright et al.[2] The CGP is derived from blood volume and direct blood counts. The MGP is derived from the dilution of autotransfused $DF^{32}P$ labeled granulocytes. The average ratio of the CGP to the total blood granulocyte pool (TBGP) is 0.44 with a range of 0.16 to 0.99. The sum of CGP and MGP equals the TBGP. The model shown in Figure 1 implies equilibrium between the CGP and MGP which is attained rapidly after introduction of labeled granulocytes into the circulation. The disappearance of autotransfused $DF^{32}P$ labeled cells is random with a half-time of 6.7 hours.[2] The notion of the random loss of granulocytes from the blood received direct support from the observation that tritiated thymidine labeled granulocytes appear in the blood and extravascular sites (saliva) almost simultaneously.[3] It has also been shown that the granulocytes die by senescence since labeled pykotic forms are seen in the blood commencing about 30 hours after labeled granulocytes first appear in the peripheral blood.[3] Thus, the random loss of granulocytes from the blood is truncated by senescent loss about 30 hours after the entrance of cells into the blood.

The fractional turnover rate per hour (λ) for a compartment from which there is random loss is $\log_e \div T\,½$. The reciprocal of λ is the average time spent in the compartment. The granulocyte turnover rate (GTR) is, therefore,

$$GTR = TBGP \times \lambda,$$

where GTR equals the number of granulocytes lost from the blood per hour.

$$\lambda = 0.103 \text{ per hour}$$

$$\frac{1}{\lambda} = 9.7 \text{ hours}$$

and for the subjects reported by Cartwright et al.[2] in whom TBGP and $T\,½$ were both measured the GTR was 6.79×10^7 per hour or 163×10^7/Kg./day.

Until recently all investigators have considered that blood granulocyte production follows a relatively strict steady state system in which TBGP

From the Medical Research Center, Brookhaven National Laboratory, Upton, New York.
Research supported by the U. S. Atomic Energy Commission.

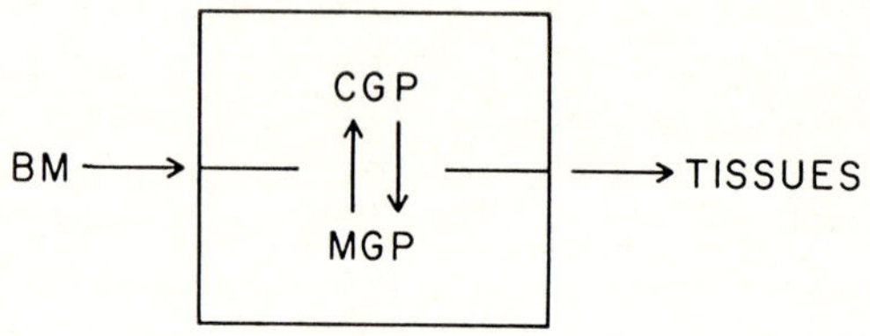

FIG. 1—Model for the total blood granulocyte pool (TBGP) after Cartwright et al.[2] The circulating granulocyte pool (CGP) and the marginal granulocyte pool (MGP) are shown in free exchange, although the equal probability of loss into the tissues implied in this figure may not necessarily be correct. (Reproduced by courtesy of the authors and publishers of Blood, Grune & Stratton.)

is relatively constant and the marrow release rate (MRR) is equal to the GTR in the blood at any one time. To accomplish this then the marrow production rate (MPR) must equal the MRR if there is no loss in the production line. The notion of a steady state where the instantaneous GTR and MPR are equal must be questioned in view of the long term oscillation of the granulocyte count described by Morley.[4] He found a cycle length varying from 14 to 23 days in males, thus disassociating it from the menstrual cycle. The cycling has been interpreted by King-Smith and Morley[5] as being compatible with a system that has a negative feedback circuit acting on the marrow stem cell (self-perpetuating level) with a time delay corresponding to the time from induction of the stem cell into granulopoiesis until it appears in the blood as a granulocyte. In Figure 2, blood granulocyte concentration is shown schematically as a sine wave. The time delay approximates half the cycle time.

Normal cycling of granulocyte concentration in the blood would be expected from what is known about the structure of granulocytopoiesis, although Morley's description has yet to be confirmed. Many tissues are in a state of flux, the size varying with changes in rates of production on one hand or in the rates of disappearance on the other hand. Therefore, steady state calculations of granulocyte kinetics to get at the question of myelocyte sinks or losses in some part of the production pathway should ideally be based on data averaged over a time equal to the cycle length for both MPR and GTR. For single comparative studies it would be necessary to compare the MPR at one time to the GTR at an appropriate later time dependent upon the time delay (stem cell to blood granulocyte transit time) for the given subject. So far estimates of the time spent in the various stages of the maturation, proliferation, and storage compartments of the marrow are based on single or a very few observations. Hence there is inadequate knowledge of the variance involved.

In Figure 2 the granulocyte concentration, the solid line, is shown as a sine wave. The reciprocal of this curve, the dotted line, represents the strength of the signal that drives stem cells down the granulopoietic pathway. This signal presumably initiates a complex sequence of biochemical events that constitute the process of differentiation into a cell type characterized by numerous specific proteins, enzymes, granules, and organelles. In keeping with contemporary knowledge on transfer of biochemical codes, the signal is almost certainly chemical and initiates derepression of transcription sites for mRNA coded for the enzymes and proteins that are characteristic of the cell line. Since many specific proteins and enzymes are going to be synthesized, it is conceivable that the chemical signal derepresses the transcription sites for a single protein and that the subsequent cascade of derepression of transcription sites for different proteins is related to derepression of other sites by the successive protein products. These in turn may suppress their own transcription sites after attaining a specific concentration by "product" inhibition.

Although there is some knowledge about the time parameters of proliferation of granulopoietic cells in man, data are inadequate to determine the mean values and their variance. It has been shown by Stryckmans et al.[6] that the average DNA synthesis time is about 12 hours. From the data summarized in reference [1] it has been deduced that the generative cycle of myelocytes is roughly 52 hours. Labeled myelocytes replaced metamyelocytes at a rate of about 3.3 per cent per hour in one patient.[7] The minimum transit times through the nonproliferating marrow compartment are, 3 hours for the myelocytes to metamyelocytes, 36 hours to the band neutrophil, and 48 hours to the segmented neutrophil. These are minimum transit times determined by

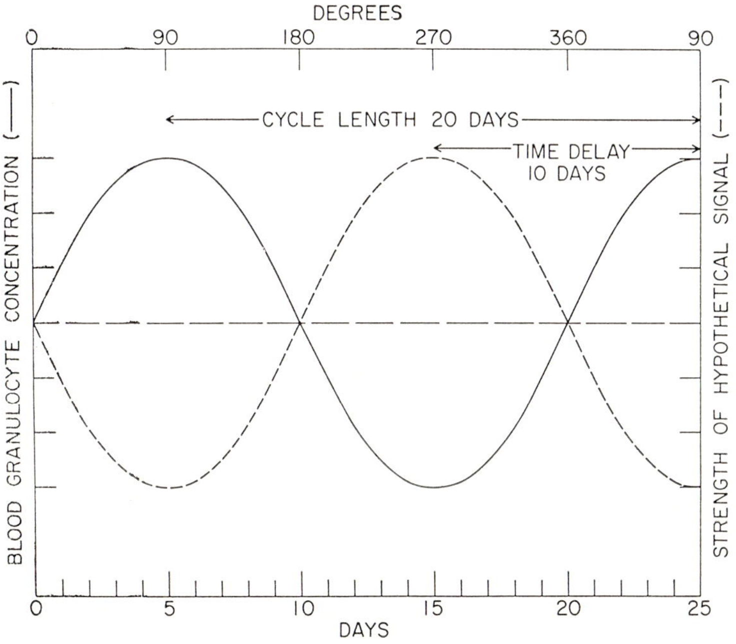

FIG. 2—Diagram of the concept of cyclic variations in the blood granulocyte concentration, based on the observations of Morley.[4] Granulocyte concentration is shown as a solid line and the strength of a hypothetical signal stimulating granulopoiesis is shown as a broken line. Sine waves have been used in this model, although in fact the shape of the curves is probably more complex than this. The cycle length is 20 days and the time delay from maximum signal to maximum response is 10 days.

the movement of the leading edge of a wave of labeled cells progressing from the proliferating compartment through the nonproliferating compartment.[8] The emergence time (ET) of granulocytes into the peripheral blood has been defined by Fliedner et al.[9] as the time from labeling the myelocyte until the first granulocytes appear in the peripheral blood. In a series of 15 patients, the closest approximation to the normal is an emergence time of 96–144 hours. In severe infection this was reduced to 48 hours.

Several factors influence the TBGP, CGP, MGP, and GTR. Adrenal corticosteroids increase the total granulocyte count apparently by a combination of an increase in marrow release rate (MRR) and a decreased efflux from the blood as indicated by a lesser rate of loss of granulocytes through small areas of inflammation in subjects given corticosteroids.[10] Bacterial endotoxin produces a granulocytosis preceded by a fleeting granulocytopenia with a shift in cells from the CGP to the MGP. This is followed by an increase in TBGP with a decrease in the granulocytic specific activity of $DF^{32}P$ labeled cells indicating an increased MRR of unlabeled cells.[11] Etiocholanolone releases granulocytes from the bone marrow.[12] It is only effective in human beings and has been used to measure the marrow granulocyte reserve.

The TBGP is increased in infection.[13,14] Surprisingly, the half-time in the blood is normal or increased. The GTR is almost always increased. The increased GTR demands an increase in MRR and MPR. The increased half-time in infection is rather surprising. It appears inescapable that infection alters the generalized random loss presumably by decreasing loss through capillary beds not associated with the infectious process. The mechanism of action is obscure. Such a mechanism of course must be nonoperative at the site of infection or it would be self-defeating.

The only reserve granulocytes available to be drawn on in need are the mature granulocytes present in the marrow. This pool is about 4 to 6 times the number present in the peripheral blood. The total nonproliferating pool of marrow cells (metamyelocyte, band, and segmented neutrophils) is about 12.5 times the normal number present in the blood. Cells leave the marrow more or less by a pipeline process with the older cells being released first.[15] If granulocytes were to be lost randomly from the marrow with a half-time of 6.7 hours vast numbers would have to be produced to outlast the estimated range in a marrow granulocyte transit time of 37–94.7 hours. The transit time of mature segmented granulocytes in the marrow is 1–3 times the senescence time in the peripheral blood. Accordingly, it is highly unlikely that senescence of cells occurs in the bone marrow. One might postulate that senescence occurs at some set time after the last myelocyte division. This would be approximated by the sum of the range of the ET from the marrow to the blood plus the senescence time in the blood giving a rough range of 126–174 hours after the last myelocyte division. Therefore, if granulocytes were released prematurely from the marrow they may have a longer potential life span in the blood. This may in part explain the longer time in blood during infection.

There is considerable information now on the nature of substances which cause the release of granulocytes from the bone marrow.[16] The most interesting substance is the leukocytosis-inducing factor (LIF) that causes the release of granulocytes from the marrow and shortens the emergence time of cells from the myelocyte stage in the marrow to the blood. LIF is increased when granulocytopenia is induced by apparently any cause.[16–18] LIF presumably is the chemical mediator of a freedback loop from the plasma to the marrow granulocyte reserve, the concentration of which increases as the granulocyte count diminishes and thus increases the MRR.

Next, a general consideration of the marrow blood relationships is pertinent. The MPR could be equated with the GTR if an instantaneous steady state exists.

$$\frac{N_S}{T_S}=\frac{N_m}{T_m}=\text{MPR}=\text{GTR}=\frac{\text{Granulocytes per Kg. in blood}}{\text{average life span}} \quad (1)$$

N_S = total marrow cellularity per Kg. $\times f_g \times f_S$
N_m = total marrow cellularity per Kg. $\times f_g \times f_m$
T_S = average DNA synthesis time
T_M = average mitotic time
f_g = fraction of marrow in granulopoiesis
f_S = fraction of granulopoietic cells in DNA synthesis
f_m = fraction of granulopoietic cells in mitosis

The preceding calculation assumes a continuous steady state relationship between MPR and GTR. However, as mentioned earlier, Morley [4] has produced good evidence that a periodic state exists. Accordingly, to compare validly MPR to GTR requires measurement of GTR at the appropriate time delay after measurement of the MPR or to integrate both over a complete cycle.

In Figure 3 a scheme of hemopoiesis is shown in which the input from the stem cell is listed as K_{in} followed by a series of mitoses to amplify the stem cell input. The number of mitoses is n. K_S is rate of flow through S and equals K_{in}. The K_{out} of the proliferating pool into the nonproliferating pool is equal to

$$K_{\text{out}}=\frac{\sum N_S}{T_S}+K_{\text{in}} \quad (2)$$

$$=K_B+K_{\text{in}} \quad (3)$$

This will be considered in more detail later.

We are interested not only in MPR and GTR under conditions of health but also by what mechanisms can granulocyte production be increased and what is a plausible limit to the maximum granulocyte production.

The size of any compartment is a function of the flux into the compartment, the birth rate in the compartment, the time in the compartment less the flux out. The bone marrow is a relatively rigid compartment. An additional mitosis would double the output and provide cells within a time that is equal to the sum of the time to provide another mitosis and the minimum transit time through the nonproliferating pool. There is evidence that the latter may be decreased to as short as 48 hours.

There is some evidence [1] suggesting that generation times (T_C) are of the order of 18 to 24 hours for myeloblast and promyelocyte and about 52 hours for the myelocyte.[1] Thus, the transit time from the stem cell to the metamyelo-

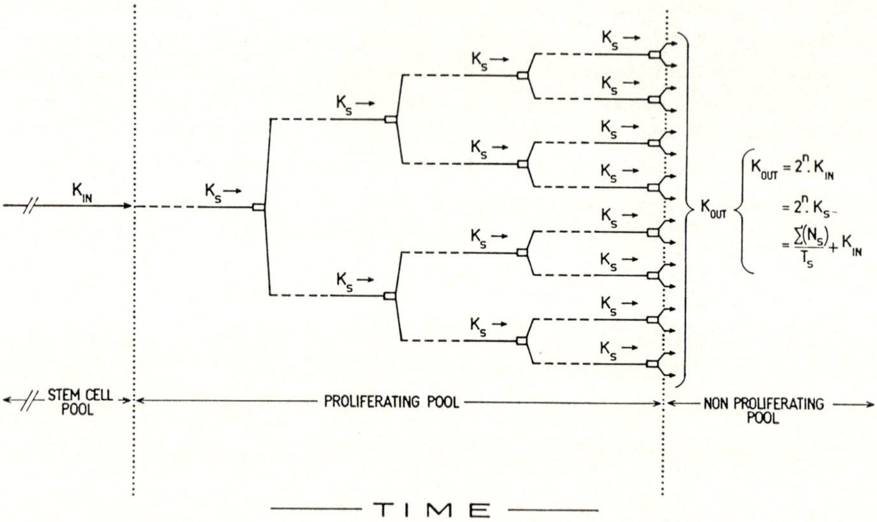

FIG. 3—Model of cell flow through a proliferating pool. The pool boundaries are defined by vertical dotted lines; cells enter from a stem cell pool and leave by reaching a stage at which they can no longer divide.

Each horizontal line represents one generative cycle, with cells moving through G_1 (– – –), S (——), G_2 (=), and M (<) in sequence. The rate at which cells move through each phase must be a constant in order to satisfy steady-state conditions, so that

$$K_{G1} = K_S = K_{G2} = K_M,$$

of which K_S only is shown for simplicity in the diagram. In the case of the first generative cycle, $K_{in} = K_S$ (and K_{G1}, etc.).

The first mitosis gives rise to 2 generative cycles, the second mitosis to 4, and so on. The flow rate through each of these cycles is equal to K_S, so that the total flow rate at any point is equal to $2^n \cdot K_S$ (or $2^n \cdot K_{in}$) where n is the number of preceding sequential mitoses. The final output of the system, K_{out}, is thus equal to $2^n \cdot K_S$ (or $2^n \cdot K_{in}$) where n is the number of sequential mitoses in the proliferating pool.

This model is valid regardless of where or whether or not compartmental boundaries are set within the proliferating pool.

cyte is in the vicinity of 140 hours under normal conditions. Since ET lies between 96–144 hours a minimum estimate of stem cell to blood granulocyte is 236–284 hours. Thus, if the only means of increasing granulocyte production were to be an increased stem cell input without changing transit time, this would not be reflected in the blood until at least 236–284 hours (9.8–11.7 days) after recruitment of stem cells. Doubling stem cell input under these conditions would, after 236–284, hours double the volume occupied by granulopoiesis and clearly increase the myeloid-erythroid ratio, but would not deliver cells soon enough.

In making calculations on the maximum possible production rate of granulopoiesis it will be assumed that the volume of marrow occupied by erythropoiesis and cell production other than for granulocytes remains constant. If stem cell to blood transit time remains constant, added volume required can be realized by expanding into intrasinusoidal blood volume and more slowly by expanding into the fatty bone marrow. If one assumes the average size of a granulopoietic cell in the marrow is 400 cubic microns then normal granulopoiesis requires a volume of 4.56 ml. per Kg. Standard man (70 Kg.) is considered by Lisco [19] to have 1500 grams of red marrow and 1500 grams of yellow marrow or 21 grams per kilogram of each type. Replacement of all yellow marrow by granulopoietic marrow would allow space for a total of 25.6 ml./Kg. or 5.6 times normal.

This would allow a production rate 5.6 times normal without changing any time parameters in granulopoiesis and the same amplification from stem cell to metamyelocyte of 1 to 16 that is postulated in Figure 3. This increase in production rate is adequate to maintain a granulocytosis of 33,600 per $mm.^3$ with an unchanged granulocyte half-time but such a sustained granulocytosis could not be realized for at least 236 hours unless transit time is decreased. Transit through the nonproliferating pool has been observed to decrease to 48 hours. This would allow a sustained granulocytosis to appear in 7.8 days rather than 9.8 days. Since sustained granulocytoses are observed more quickly, one must seek a solution other than those postulated so far.

The only alternative appears to be tacking another mitosis onto the amplification system. This will double the volume occupied by the proliferative compartment of granulopoiesis and will still only be realized in the peripheral blood after the sum of the minimum transit time in the nonproliferating marrow compartment (48 hours) and the myelocyte generation time (52 hours) or 100 hours. Sustained granulocytoses are seen sooner. This is not an adequate model.

Further consideration is required to understand the flow of cells through a nonproliferating compartment. The compartment transit time (CTT) equals

$$\mathrm{CTT}=\frac{N}{K_{\mathrm{in}}} \tag{4}$$

where N equals compartment size. Hence,

$$K_{\mathrm{out}}=K_{\mathrm{in}}=\lambda N=\mathrm{MRR}=\mathrm{GTR} \tag{5}$$

for the entire nonproliferating compartment (metamyelocytes, bands, and segmented neutrophils) when expressed as cells per Kg. and the respective λ. An orderly flow of cells as in a pipeline is assumed. It is self-evident that time in a compartment determines its size. The output is, with constant fractional turnover (λ), proportionate to N. When more cells are needed it is desirable to produce them in a shorter time, but there must be an irreducible minimum time necessary for cells to mature. We assume, from data derived in the presence of severe infection,[9] that this is of the order of 48 hours.

From the preceding equations the transit times for metamyelocytes, bands, and marrow neutrophils can be estimated. The fractional turnover (λ) of metamyelocytes of 3.3 per cent/hour was determined in a patient by flow of labeled cells from the myelocyte stage to metamyelocyte stage.[8] However, one can also estimate λ from absolute cellularity of the marrow metamyelocyte (N) of 270×10^7/Kg. and GTR (6.79×10^7/Kg./hr).

$$\lambda=\frac{\mathrm{GTR}}{N}=\frac{6.79}{270}=0.255 \tag{6}$$

We suspect 0.225 to be an approximation of the average for fractional turn-

TABLE 1—*Relative and Absolute Sizes of Marrow Compartments with Estimates of Transit Time*

		Proportional‡ Distribution	CTT Hours	Number per‖ Kg. × 10^7	CTT Hours
	Stem cell				
PP *	Myeloblast (M_1)	100	18	5.5	18
	Promyelocyte (M_2)	283	24	15.6	24
	Myelocyte M_{3+4}	4720	104	260	104
	Metamyelocyte M_5	3784	40 ¶	270	40
NPP †	Band M_6	6227	65.8	360	53
	Segmented M_7	8964	94.7	250	37
B **	CGP	0.45		31	
L	MGP	0.55		39	
O	TBGP	1		70	
O	Random loss	500–2000	9.7		9.7
D	Senescent loss	1	30.0		30.0
	Stem cell to blood		346		276
	Emergence time		200		130

* PP represents proliferating pool.
† NPP represents nonproliferating pool.
‡ Based on average of 5 normal males.
¶ Replacement rate of 2.5 per cent/hour determined from granulocyte turnover rate and absolute metamyelocyte number.
‖ From Donohue et al.[20]
** From Cartwright et al.[2]

over rates of the metamyelocytes since it is based on an average GTR and an average absolute cellularity based on many patients. The 0.033 determined in a single patient may be an upper limit. In Table 1, the relative proportion of cell types and the absolute marrow cellularity with their respective transit times are tabulated. The large discrepancy in transit through the marrow segmented compartment depending on whether or not one uses relative or absolute marrow cellularity is not understood. The shorter times obtained with absolute cellularity is more compatible with the range in emergence time of 96–144 hours.

We have assumed that there is an obligatory minimum time required for a cell to mature from terminal myelocyte mitosis to a marrow granulocyte. Evidence suggests this may be 48 hours. We assume that under stress of great need, segmented granulocytes immediately enter the blood. Under these circumstances the maximum flux that can be maintained through a compartment with a shorter transit time is the normal flux times the ratio of T to T_1, where T is the normal transit time of M_5 through M_6 and T_1 is the shortest possible transit time or 48 hours. Using the values in Table 1, based on absolute cellularity, a flux $93 \div 48$ or a flux 1.94 times normal is the maximum flux that can be carried without expanding the metamyelocyte and band compartments.

If all marrow granulocytes (about 4 times blood granulocytes) were dumped into the blood, one would have an immediate granulocytosis of nearly 4 times normal or 20,000/mm.[3] This level would fall to 1250 per mm.[3] within 26 hours if there were no further release of granulocytes from the marrow, and the half-time in the blood remained normal. Even if granulocyte release continued at a normal rate, the count would still fall to 6100 per mm.[3] at 26 hours. More rapid maturation of bands and metamyelocytes could increase the granulocyte release rate and help sustain the TBGP, but this would be achieved at the expense of refilling the depleted granulocyte marrow reserves. However, if the total pool of nonproliferating granulocytes were to enter the blood at a constant MRR of 18×10^7/Kg./hr. this would expend all stored cells in nonproliferating compartment within 48 hours. This would allow an increase in GTR to 18×10^7/Kg./hr. which, with a normal half-time in the blood, would sustain an increase in TBGP to about 2.5 times normal. A longer half-time in the blood, often seen in infection,[13,14] would allow a greater increase in the TBGP, but the GTR would be unchanged. Granulocyte utilization in excess of the GTR immediately from the marrow stores will result in granulocytopenia in serious infection, and one can set up conditions of MRR and GTR where severe granulocytopenia will be observed within a few hours. The crucial issue is delivery of granulocytes to the sites of infection and this depends on an increased delivery of granulocytes to the blood. We think it is essential that increased granulocyte production must involve, in addition to the probable increased stem cell input, additional mitoses with a shortening of the generation time. Otherwise a granulocytopenia would be a more frequent observation in course of infections.

Next we'll try to establish a generalized model for granulopoiesis. A periodic state is believed to exist with the rate of entry of stem cells into granulopoiesis varying with the intensity of the stimulus received from the periphery. As the granulocyte count falls in the blood, the intensity of the stimulus increases. The rate of entry of stem cells into granulopoiesis increases. This will be reflected by an increased MRR into the peripheral blood after the stem cell to blood granulocyte transit time of 11.5–15 days. The cycle time of maximum to maximum has been reported to be 14–23 days.[4] The time delay should be half of the cycle time or 7–11.5 days and should in principle be an independent estimate of stem cell to blood transit time. Experimental estimates of stem cell to blood granulocyte transit times are in the same ball park. There are not yet enough observations to refine either the statistics of the transit time of stem cell to blood granulocyte or the variance in the cycle times.

In the ensuing discussion of our proposed model for granulopoiesis, we assume under normal conditions that there is very little variance in transit time through the marrow. However, the cellular flux and population size of the compartments must vary as the stem cell entry into granulopoiesis fluctuates. It follows, therefore, that the proportional distribution of cells in granulopoiesis will change relatively little during a peripheral blood cycle but the myeloid:erythroid ratio will fluctuate. For mathematic simplicity we will consider at any moment in the marrow that a steady relationship exists between stages in the cell cycle and the fluxes through succeeding cellular compartments although this will not be strictly true.

By convention in Figure 3, the model is set up to show a series of parallel arms which double in number with each mitosis. Each of these arms represents one generative cycle. In view of the earlier assumptions, the following are assumed to be true: (1) cells must move at the same rate through each arm of the system or cells will be piling up at particular stages; (2) cells move through phases of the cell cycle in any and all of the arms at a constant rate; (3) it can be deduced that the fractional rate of cell movement through any phase of cell cycle in all cytologic compartments must also be a constant. That is, they move through at the same rate in the myeloblast and the myelocytic compartments; (4) the rate at which cells enter from the stem cell pool also equals the rate of cell movement through each arm.

If the influx from the stem cell pool is called K_{in}, the production rate (K_{out}) of the proliferating pool will be $2^n \cdot K_{in}$, where n=the number of sequential mitoses in the system.

K_{in} is also equal to the rate at which cells enter DNA synthesis (K_S) in one arm of the system. Thus,

$$K_{\text{out}} = 2^n \cdot K_S = 2^n \cdot K_{\text{in}}$$
$$= [\Sigma\, 2^{n-1} + 2^{n-2} \ldots + 2^0]\, K_S + K_{\text{in}} \tag{7}$$

The first term equals

$$\frac{\Sigma N_s}{T_s}$$

which is equal to K_B in the usual equation

$$K_{\text{out}} = K_B + K_{\text{in}} \tag{8}$$

This formula allows the calculation of the production rate (K_{out}) from the proportional distribution of cells and labeling indexes without any knowledge of the number of mitoses. Stem cell input K_{in} will be discussed later.

In Figure 4 our provisional scheme for granulopoiesis is shown with relative sizes of compartments, fluxes into, through, and out of the cellular compartments. The numbers used are obtained from a study in which serial autoradiographic studies were performed on a single patient after in vivo administration of ^{3}HTdR.[7,8] From this study, the cells in DNA synthesis are distributed between myeloblasts, promyelocytes, large and small myelocytes in the ratio of 85:190:770:330. For reasons to be discussed later we have pooled large and small myelocytes into one compartment giving a ratio of labeled myeloblasts to promyelocytes to myelocytes of 85:190:1100 or 1:2.1:13.0. This ratio appears best fitted by assigning one mitosis to myeloblast and promyelocyte and giving the myelocyte 2 sequential mitoses as shown in Figure 4.

The myelocyte compartment can be constructed with 12 packets of cells in S, 4 before the first mitosis and 8 before the second mitosis. When 1100 myelocytes in DNA synthesis are divided into 12 equal packets each will consist of nearly 92 cells. With a DNA synthesis time of 12 hours, the number of cells flowing through S per hour (K_s) is $92 \div 12$ or K_s is 7.6 cells per 100 myeloblasts per hour in each arm, the production rate or the K_{out} of the proliferating pool is

$$K_{\text{out}} = 16 \times 7.6 = 122 \text{ cells/100 myeloblasts/hour}$$

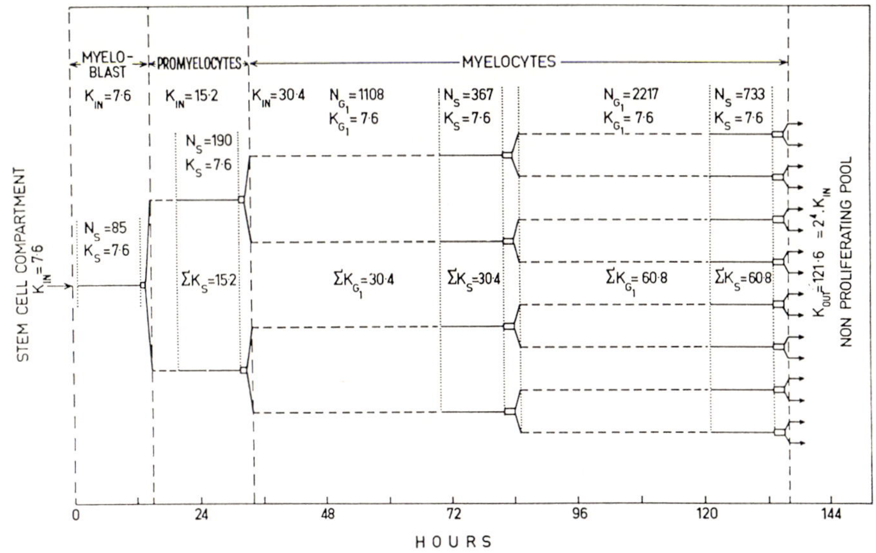

FIG. 4—Application of the model of Figure 3 to the granulocytopoietic proliferating pool. The numerical data are derived from Table 4B (Ref. 1) and are all relative to 100 myeloblasts. The unit of rate is cells/100 myeloblasts/hour.

Compartmental boundaries are shown as vertical dashed lines, and boundaries between the phases of the cell cycle G_1, S and $(G_2 + M)$ as vertical dotted lines. The horizontal lines represent successive generative cycles, subdivided into G_1 (– – –), S (——), G_2 (=), and M (<). The duration of each cycle, and its phases, can be read off the time scale.

Cells enter from a stem cell compartment at a rate (K_{in}) equal to $K_S \cdot$(7.6 cells/100 myeloblasts/hour), as shown in Figure 3. After division, twice as many promyelocytes move into the next generative cycle, so that the total flow rate is equal to $2 \times K_S$ or 15.2.

Following promyelocytic division, cells enter the myelocyte compartment. The data are best satisfied by postulating 2 successive mitoses within this compartment, of which the first is homomorphogenic (giving rise to myelocytes) and the second heteromorphogenic (giving rise to metamyelocytes). The final

$$K_{out} = 16 \times K_S,$$

or 121.6 cells/100 myeloblasts/hour.

Myelocytes labeled with ^{3}HTdR in the final S phase (121–133 hours) will lead, after G_2 and M, to the appearance of labeled metamyelocytes at a rate of 121.6 cells/100 myeloblasts/hour. This is very close to the observed rate of 147.1.

G_1 is quite long (36.3 hours) in myelocytes, and behaves as an "on-line" nondividing compartment. Cells labeled with ^{3}HTdR either as promyelocytes, or in the first myelocyte cycle (70–82 hours) will lead, again after G_2 and M, to the appearance of labeled G_1 myelocytes at a rate of 2.74 per cent per hour. This is similar to the rate at which labeled small myelocytes appear, and for this reason it is probable that most of the small myelocytes are G_1 cells.

If all cells were to enter the metamyelocyte compartment of 4460 per 100 myeloblasts, the fractional turnover of the metamyelocytes would be 0.0273 per hour. In this patient there was an observed replacement of about 0.033 per hour.[1]

The myelocyte compartment is 4700/100 myeloblasts.[1] The ratio of K_{out} estimated above divided by the size of the myelocyte compartment yields a fractional production rate for myelocytes of about 2.6 per cent per hour. One can also calculate the fractional production rate independently using absolute marrow cellularity and the peripheral blood GTR. Thus, from a marrow myelocyte mass[20] of 260×10^7/Kg. and GTR of 6.79×10^7 granulocytes/Kg. from Cartwright et al.[2]

$$\lambda = \frac{6.79}{260} \times 100 = 2.6 \text{ per cent per hour.}$$

In the model shown in Figure 4 and from the restrictions imposed by us, the stem cell input in absolute values must be around $\frac{1}{16}$ of the granulocyte turnover rate or 0.42×10^7 cells/Kg./hour.

In this analysis we choose not to discuss the literature on the identity of the stem cell. This has been reviewed.[1] However, in terms of this we will predict where it might be found. It is apparent that the input of the stem cell compartment into granulocytopoiesis is considerable in both relative and absolute terms. When cells are removed from the stem cell compartment by differentiation there must be a mechanism to replenish the stem cell compartment by mitosis at the stem cell level. In Figure 4 stem cells must, under the conditions imposed earlier, be moving through S at an average rate of 7.6 cells per 100 myeloblasts per hour. Let's assume that T_S for the stem cell compartment is also 12 hours. Then the number of labeled cells (N_s) is

$$N_s = K_{\text{in}} \cdot T_s = 7.6 \times 12 = 91$$

Since all relative values are expressed per 100 myeloblasts there should be 91 labeled stem cells for every 100 myeloblasts, 85 labeled myeloblasts or 4700 myelocytes or 1100 labeled myelocytes. This means there should be one labeled stem cell per 52 myelocytes. The ratio of all stem cells to all myelocytes depends on the fraction of stem cells labeled. If 50 per cent of stem cells were labeled there would be 4 stem cells per 100 myelocytes. If only 10 per cent of stem cells were labeled there would be 20 stem cells per 100 myelocytes.

The absolute stem cell production rate of 0.42×10^7 cells per Kg. per hour which was calculated earlier requires a minimum of 5×10^7 stem cells in DNA synthesis per Kg. If one assumes that only 10 per cent of the stem cells are in DNA synthesis there must be a minimum of 50×10^7 stem cells per Kg. that is feeding only granulocytopoiesis. This is about $\frac{1}{5}$ of the absolute number of myelocytes (260×10^7 per Kg.). Since stem cells also feed erythropoiesis and megakaryocytopoiesis, we suggest that with the exception of the small lymphocyte-like cell in the bone marrow there are no morphologically distinct cells present in sufficient numbers in the marrow to account for values of this magnitude.

Another approach to the magnitude of the stem cell input is based on the following equation:

$$K_{\text{out}} = K_{\text{in}} \cdot 2^{t/T_c} = \text{CTR} \tag{9}$$

K_{in} would be the stem cell input and t is the time from differentiation of the stem cell until it matures to the metamyelocyte. The t/T_c equals the average number of mitoses. An approach to estimation of t follows.

The peripheral blood granulocyte cycle time is 14–23 days [4] and in the Morley concept the time delay is one-half the cycle time and this should represent the time from the stem cell to the blood granulocyte time. Let's assume that the midpoint of the cycle time observed is the average or 18.5 days. Then 9.25 days is average stem cell to blood granulocyte time. Subtracting the midpoint of the observed ET (96–144 hours) or 120 hours, one estimates average time from stem cell to metamyelocyte to be 102 hours. By substituting the values in Equation (9) one can calculate K_{in} for various T_c using GTR of 6.79×10^7 per Kg. per hour and t of 102 hours

For T_c of 26 hours—$K_{in} = 0.449 \times 10^7$ per Kg. per hour

For T_c of 52 hours—$K_{in} = 1.746 \times 10^7$ per Kg. per hour

One is inclined to reject K_{in} of 1.746×10^7 per Kg. per hour because this would make the size of the stem pool so large in reference to other cells that one could not be in doubt about its identity. A K_{in} of 0.449×10^7 per Kg. per hour is compatible with earlier notions of a 1 to 16 amplification. However if 102 hours is correct for time from stem cell to metamyelocyte an average T_c of 26 hours is required and this is not compatible with the observed labeling indices [1] and a T_s of 12 hours. Clearly, further experimental and conceptual work needs to be done before one can solve the question of stem cell input and its identity. However, certain ideas about the myelocyte must be reconsidered.

There are two populations of myelocytes based on their sizes.[7] In earlier studies [8] it was considered that the small myelocyte was a more mature version of the larger myelocyte. The fraction of small myelocytes in DNA synthesis is much less than that of the larger myelocytes. The earlier notion that the maturation sequence proceeded from the large myelocyte through the small myelocyte to the metamyelocyte will not provide enough labeled metamyelocytes during the time they are observed to appear. Labeled metamyelocytes can appear only after mitosis of myelocytes which were in S at the time ^{3}HTdR was available. There simply are not enough labeled small myelocytes to provide an influx of 3.3 per cent per hour observed.[1] A whole series of arguments have been presented that suggest the small myelocyte is a G_1 myelocyte.[1] This notion, not yet proved, is compatible with the observed increase in labeling of the small myelocyte and will provide the labeled metamyelocytes that have been observed to appear. If small myelocytes are in G_1, they will be predominately $2n$ cells and this is yet to be ascertained.

The model is drawn as if there were no variance in the system. This would lead to a plateau in metamyelocyte labeling, possibly followed by a decrease in the fraction of labeling at the metamyelocyte as G_1 cells progress to metamyelocyte stage and a secondary increase as the cells labeled in the first myelocyte division proceed into the metamyelocyte compartment about 57 hours after labeling. Such has been observed in canine metamyelocytes.[21] Variance in the system presumably is responsible for smoothing out these step-wise predictions in labeling of the metamyelocyte in man. One could also predict from this model that a continuous perfusion of tritiated thymidine for a period of time T_{G_1} equal to G_1, or roughly 37 hours, should label the entire granulopoietic proliferating compartment.

In the preceding rigid model the generation time (T_C) is about 52 hours, T_S about 12 hours, T_{G_1} about 37 hours. This model is based on observations made on a single patient following in vivo administration of ^{3}HTdR. One can also estimate myelocyte generation time from Donohue's data [20] on absolute marrow cellularity and the GTR.

The myelocyte compartment is in a "steady state" where loss by differentiation into metamyelocytes equals the GTR which equals the MPR. If there were no loss from the myelocyte compartment it would double with each generation [22] so that

$$N = N_o \cdot 2^{t/T_c} = N_o \left(e^{.693}\right)^{t/T_c} \qquad (10)$$

where T_C equals generation time and t equals time of observation and t/T_C equals number of mitoses.

Let's assume, as suggested by others, that under normal conditions the myelocyte compartment is proliferating autonomously, that is, there is no K_{in} from a stem cell compartment. The growth constant or fractional growth rate (λ) is

$$\lambda = \frac{0.693}{T_c} \qquad (11)$$

If one sets aN to be the rate of cell differentiation the net change in N is

$$\frac{dN}{dt} = \lambda N - a N \qquad (12)$$

and in the steady state

$$dN/dt = 0$$

The myelocyte population is 260×10^7 per Kg.[20] The GTR is 6.79×10^7 per Kg.[2] Substituting these values in Equation (6), $\lambda = 0.0264$. Substituting this λ in Equation (10) the T_c equals 26.2 hours. This value is half the T_c of our earlier model. To pursue the notion that the myelocyte compartment might be autonomous similar to the gut epithelium one can estimate the average number of mitoses (n) from t/T_c in Equation (9). The approximate duration of t was estimated earlier to be about 102 hours. Therefore, with T_c of 26.2 hours the average number of mitoses would be 3.8. However it seems unreasonable to ignore myeloblasts and promyelocytes and the general notion of a stem cell input is held to be almost an article of faith. At this time there are several concepts about the structure of granulopoiesis but inadequate data to test the concepts. A more extensive study of the blood neutrophil cycle will provide more information on the stem cell to blood granulocyte time. This with refinement on transit time through the nonproliferating compartment of the marrow will make it possible to provide a better estimate of the stem cell to metamyelocyte time. If this can be established with more certainty one might then be able to construct a better model of the proliferative marrow compartment based on ratios of mitotic figures and labeled cell types.

Earlier in the discussion, arguments were presented to show why it was most logical for nature to increase granulocyte production by adding an additional mitosis and to deliver these cells more quickly by a shortening of the generative cycle and the transit time through the nonproliferating pool in order to present the increased production where it was needed most quickly.

In looking at granulocytopoiesis, it is seen that the proliferating granulopoietic cells mature as they divide. In Table 1 under normal steady state conditions the time from differentiation of a stem cell to its appearance in the peripheral blood is of the order of 276 to 346 hours. It is estimated that stem cell to the metamyelocyte transit time under normal conditions is roughly 146 hours. During this time something happens to turn off further cell cycling. Differentiation is clearly a matter of derepression of those portions of the genetic code that allow the transcription of all the mRNA's required for synthesis of the proteins characteristic of the differentiated cell line. Whether or not all cistrons are simultaneously derepressed by the chemical mediator allowing transcription of all types of mRNA at once is not known. Perhaps the mediator derepresses a single cistron which is linked in sequence to successive cistrons by its protein products derepressing other cistrons in sequence. For example, the primary mediator derepresses *A*. Products of *A* derepress *B*, of *B* derepress *C*, etc. Presumably the protein syntheses are ultimately switched off by concentration of the product—*A* represses *A*, *B* represses *B*, etc. The rate at which differentiation will take place and cytoplasmic hypertrophy will occur is a result of the sum of all the synthetic processes. These protein synthetic rates are linked to the quantity of mRNA, the required ribosomes, tRNA, and substrates. These are controlled either by number of transcription sites or rate of transcription per site.

Two questions are crucial. First, why does any cell enter the generative cycle and divide, and second, what turns off the cycling in this case of the granulopoietic precursors about 146 hours after initial differentiation? Potential answer to the first question comes from the work and notions of Hartmann,[23] Mazia,[24] and Johnson.[25] The first experiment bearing upon this is that of Hartmann [23] who showed that cell divisions in amebae could be postponed indefinitely by repeatedly amputating part of its growing cytoplasm and, thus, never allowing it to reach "critical mass." Mazia [24] felt that in single organisms this was related to size. Johnson [25] has shown the following sequence of events in regenerating liver and compensatory hypertrophy of the kidney. Cells prior to unilaterial nephrectomy or partial hepatectomy have a certain size distribution. The remaining cells are required to do the work of the whole liver or two kidneys. This requires cytoplasmic hypertrophy. Immediately, after the operation there is an increase in RNA and protein synthesis. Cells hypertrophy by production of more mitochondria, more ribosomes, more enzymes, etc. than are required. The larger cells of the usual size distribution attain the "critical mass" sooner. At this time one can see the cells commencing to go into DNA synthesis and after a time equal to DNA synthesis plus G_2 labeled cells are seen in mitosis. DNA synthesis is first initiated in those cells at the upper end of the usual size distribution curve. Most students of cell proliferation have been inclined to think of regulation of cell proliferation in terms of something that specifically regulates DNA synthesis from outside of the cell.

The "critical mass" suggests that there is a cytoplasmic nuclear signal that tells the cell to divide when it has become an inefficient "chemical factory" either because of an imbalance between surface area and volume or diffusion distances are too great or some as yet undisclosed factor. At least in the kidney and the regenerating liver, the evidence is exceptionally strong that cellular hypertrophy is the primary if not the sole determinant for initiation of cell proliferation.

One can question whether or not the critical mass concept is applicable to the proliferating pool of hemopoietc cells in general. Differentiation is the initiation of a sequence of protein syntheses leading to a cell type with multiple characteristics. Transcription sites are derepressed and mRNA synthesis starts. The rate of mRNA synthesis follows the number of transcription sites or rate of transcription per site or both. The synthesis of mRNA, ribosomes, and tRNA in appropriate proportions is essential for efficient protein synthesis. These phemonena result in cytoplasmic and nuclear hypertrophy. The cells double in size and divide and continue to do this—the number of divisions being set by the time that is allowed from differentiation to termination of the cycling. The feedback control of protein synthesis may be related to its concentration. In these circumstances the cell would continue to divide until the concentration of products in the cytoplasm turns off the synthesis of mRNA of all or some types, thus preventing further cellular hypertrophy. It is likely that some transcription continues along with protein synthesis; however, the net degradation and synthesis must become equal so that further hypertrophy does not occur. Rates of mRNA and protein synthesis are not going to be equal in all cells. Therefore, rates of hypertrophy will be unequal. Accordingly, if a substance(s) is being produced in all cells, that is retained in cells, a specific concentration of which represses mRNA transcription, one can visualize that repression will occur when cells have attained different sizes below the "critical mass." This would be a potential explanation for size spectrum of nondividing cells. Furthermore this would be compatible with Stohlman's [26] concept that mean corpuscular hemoglobin concentration switches off cycling in the case of erythropoiesis.

The preceeding concepts imply that cells will hypertrophy at different rates and thus enter S phase after different times. This suggests that G_1 will be quite variable and could contract with greater stimuli for hypertrophy. This is consistent with our notions on how granulocyte production rate may be increased. These notions also suggest that some cells may divide more times than others between differentiation of the stem cell and the time when cycling is suppressed and cells become metamyelocytes. This is of course incompatible with the rigid model we built earlier. This model should serve only as an exercise to appreciate how cells are being produced and conceptually how rates of production can be increased.

Summary

A general scheme of granulopoiesis is presented in Figure 5. Knowledge of the behavior of granulocytes in the blood is based on reasonably solid data. There appears to be a fast feedback loop mediated by LIF that dumps

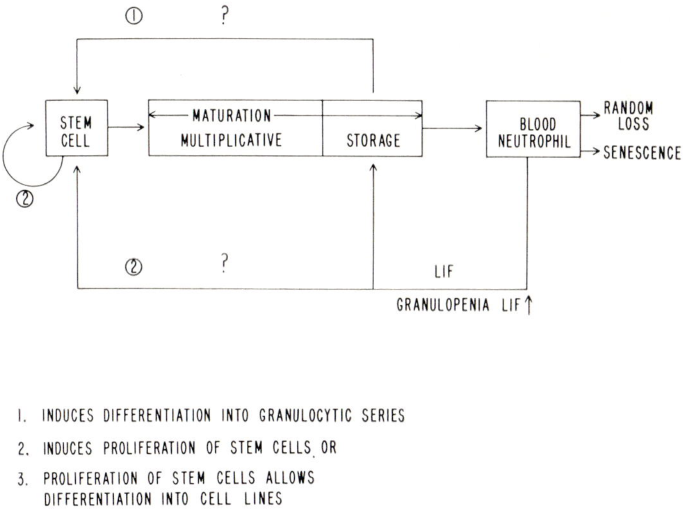

FIG. 5—Schematic presentation of granulopoiesis showing known and potential feedback loops.

granulocyte marrow stores into the blood. The plasma concentration of LIF seems to be inversely related to the concentration of granulocytes in the blood.

The transit time through the storage compartment of the marrow is fairly well established. It can be reduced at least to 48 hours in the presence of severe infection. Upon diminution of the size of the storage compartment by a feedback through either route 1 or 2 in Figure 5 the stem cell input is believed to be increased. However, MPR increases through increased stem cell inputs would take too long. It is hypothesized that the generative cycle of the myelocytic series decreases and that added mitoses will provide more granulocytes more quickly to meet increased peripheral demands. The mechanism of replenishing the stem cell pool is obscure.

Under normal conditions in our present working model the intensity (concentration) of the differentiating stimulus derepresses transcription sites in the stem cells and sets the rate of transcription and/or mRNA quantity in stem cells and probably through all stages of the proliferating pool. This sets the rate of protein synthesis and, thus, cells hypertrophy so that there will be roughly 4 divisions between the stem cell and the metamyelocyte.

Under stimulus of infection the intensity of the chemical stimulus for differentiation is believed to be greater. This in some manner increases the quantity of mRNA and, thus, the rate of cellular hypertrophy. Cells will divide more frequently before the hypothetical substance attains a concentration that inhibits further cycling. This increases the amplification factor of the stem cell and requires a shorter generation time that has not as yet been experimentally verified.

REFERENCES

1. Cronkite, E. P., and Vincent, P. V.: Granulocytopoiesis. Series Haematologica. In press.

2. Cartwright, G. E., Athens, J. W., and Wintrobe, M. M.: The kinetics of granulopoiesis in normal man. Blood 24:780–803, 1964.

3. Fliedner, T. M., Cronkite, E. P., Robertson, J. S.: Granulocytopoiesis. I. Senescence and random loss of neutrophilic granulocytes. Blood 24:402–414, 1964(a).

4. Morley, A. A.: A neutrophil cycle in healthy individuals. Lancet, 1220–1222, 1966.

5. King-Smith, E. A., and Morley, A.: A computer model for the mammalian (human) leukocyte system. Proc. 20th Annual Conference on Engineering in Medicine and Biology, November 1967, Boston, Massachusetts.

6. Stryckmans, P. A., Cronkite, E. P., Fache, J., Fliedner, T. M., and Ramos, J.: DNA synthesis time of erythropoietic and granulopoietic cells in human beings. Nature (London) 211:717–720, 1966.

7. Cronkite, E. P., Fliedner, T. M., Stryckmans, P., Chanana, A. D., Cuttner, J., and Ramos, J.: Flow patterns and rates of human erythropoiesis and granulocytopoiesis. Series Haematogica 5:51–64, 1965.

8. Cronkite, E. P., Bond, V. P., Fliedner, T. M., and Killman, S. A.: The use of tritiated thymidine in the study of haemopoietic cell proliferation. *In* G. E. W. Wolstenholme, and M. O'Connor (Eds.): Ciba Foundation Sympos. on Haemopoiesis. London, J. and A. Churchill, 1960, pp. 70–92.

9. Fliedner, T. M., Cronkite, E. P., Killmann, S. A., and Bond, V. P.: Granulocytopoiesis. II. Emergence and pattern of labeling of neutrophilic granulocytes in humans. Blood 24:683–700, 1964.

10. Bishop, C. R., Athens, J. W., Boggs, D. R., Warner, H. R., Cartwright, G. E., and Wintrobe, M. M.: Leukokinetic studies. XIII. A non-steady state kinetic evaluation of the mechanism of cortisone induced granulocytosis. J. Clin. Invest.. 47:249–260, 1968.

11. Athens, J. W., Haab, O. P., Raab, S. O., Mauer, A. M., Ashenbrucker, H., Cartwright, G. E., and Wintrobe, M. M.: Leukokinetic studies. IV. The total blood, circulating and marginal granulocyte pools and the granulocyte turnover rate in normal subjects. J. Clin. Invest. 40:989–995, 1961.

12. Godwin, H. A., Zimmerman, T. S., Kimball, H. Y., Wolff, S. M., and Perry, S.: The effect of etiocholanolone on the entry of granulocytes into the peripheral blood. Blood 31:461–470, 1968.

13. Athens, J. W., Haab, O. P., Raab, S. O., Boggs, D. R., Ashrenbrucker, H., Cartwright, G. E., and Wintrobe, M. M.: Leukokinetic studies. XI. Blood granulocyte kinetics in polycythemia vera, infection and myelofibrosis. J. Clin. Invest. 44:788, 1965.

14. Galbraith, P. R., Valberg, L. S., and Brown, M.: Patterns of granulocyte kinetics in health, infection and in carcinoma. Blood, 25:683–692, 1965.

15. Maloney, M., and Patt, H. M.: Granulocyte transit from bone marrow to blood. Blood 31:195–201, 1968.

16. Dornfest, B. S., LoBue, J., Handler, E. S., Gordon, A. S., and Quastler, H.: Mechanisms of leukocyte production and release. II. Factors influencing leukocyte release from isolated perfused rat legs. J. Lab. Clin. Med. 60:777–787, 1962.

17. Boggs,. D. R., Cartwright, G. E., and Wintrobe, M. M.: Neutrophilia inducing activity in plasma of dogs recovering from drug-induced myelotoxicity. Amer. J. Physiol. 211:51–60, 1966.

18. Katz, R., Gordon, A. S., and Lapin, D. M.: Mechanisms of leukocyte production and release. VI. Studies on purification of the leukocytosis inducing factor. Res. J. Reticuloendothelial Soc. 3:103–116, 1966.

19. Lisco, H.: Standard Man. Lemont, Ill., Argonne Nat. Lab. Report 4253, 1948.

20. Donohue, D. M., Reiff, R. H., Hanson, M. L., Betson, Y., and Finch, C. A.: Quantitative measurement of the erythrocytic and granulocytic cells of the marrow and blood. J. Clin. Invest. 37:1571–1576, 1958.

21. Maloney, M. A., C. L. Weber, and H. M. Patt: Myelocyte-metamyelocyte transitions. Nature (London), 150–152, 1963.

22. Johnson, H. A.: Some problems associated with the histological study of cell proliferation kinetics. Cytologia (Tokyo) 26:32–41, 1961.

23. Hartman, M.: Arch. Protistenkunde 45:973, 1928.

24. Mazia, D.: The life history of a cell. Amer. Sci. 44:1, 1956.

25. Johnson, H. A.: Liver regeneration and the "critical mass' hypothesis. Amer. J. Path. 57:1, 1969.

26. Stohlman, F., Jr.: Some aspects of erythrokinetics. Seminars Hemat. 4:304–314, 1967.

Studies of Granulocytopoiesis in Abnormal Conditions

By Carter R. Bishop and John W. Athens

Introduction

Knowledge of the kinetics of neutrophilic granulocyte proliferation, maturation and turnover has developed through the use of the DNA labels radiophosphorus [1,2] or ^{3}H-thymidine [3,4] and through the use of the neutrophil cytoplasmic label DF^{32}P.[5,6] From these studies it is evident that the marrow myeloblasts, promyelocytes, and myelocytes are capable of cell division and constitute a proliferative system which produces large numbers of cells. The cells produced then enter a phase of further differentiation, maturation, and storage in the marrow. This maturation phase includes the morphologic cell forms known as metamyelocytes, band or juvenile forms, and segmented or polymorphonuclear neutrophils. In the normal animal, cells from the marrow enter the blood only after completing maturation, and considerable time is required for this process. The blood neutrophils, predominantly segmented forms, leave the blood randomly with a short half disappearance time and enter tissues where they presumably perform their major function as phagocytes. There is general agreement concerning the above scheme of granulocytopoiesis and granulocyte turnover, but there is considerable disagreement on specific details and time parameters of the system.[7,8]

One parameter of the system about which there is disagreement is the time from the myelocyte stage to entrance of the mature polymorphonuclear neutrophil into the blood. This time period is referred to here as the myelocyte to blood transit time. When radiophosphate or tritiated thymidine was administered to patients thought to be hematologically normal, the minimal myelocyte to blood transit time (4 days) was the time that elapsed from injection of the DNA label until labeled cells appeared in the blood. The average values (6 to 9 days) were the times from label injection to the peak of the blood granulocyte radioactivity curves [7] (Fig. 1A). On the other hand, studies in which DF^{32}P was injected intravenously into normal volunteers led to the conclusion that the mean myelocyte to blood transit time is 11 days with a range of 8.5 to 14 days.[8,9] The myelocyte to blood transit time in these studies was taken to be the time elapsed from label injection to the end of Phase II (see Fig. 1B).

Since disease might be expected to alter the myelocyte to blood transit time, the studies reported here were carried out. First, an attempt was made to resolve the apparent difference between estimates of myelocyte to blood transit time by the DNA labeling and cytoplasm labeling technics. To accom-

---

From the Department of Medicine, University of Utah, and the Veterans Administration Hospital, Salt Lake City, Utah.

This investigation was supported by a research grant (AM–004489) and a graduate training grant (AM–5098) from the National Institute of Arthritis and Metabolic Diseases, and grants (FR–00012) and (FR–00064) from the National Institutes of Health, Division of Research Facilities and Resources, Bethesda, Maryland.

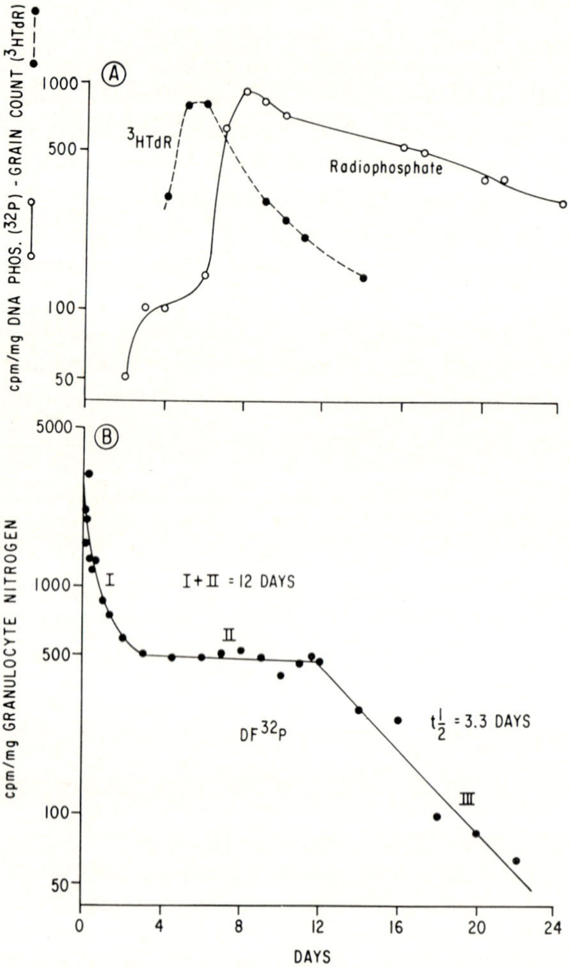

FIG. 1—Blood granulocyte radioactivity curves after intravenous injection of ^{3}H-thymidine or radiophosphorous (A), DF^{32}P (B). The time from the onset of the study to the peak radioactivity in the DNA label curves and the time to the end of Phase II in the DFP curve is the myelocyte to blood transit time. (Replotted from Craddock and Cronkite.)

plish this, DF^{32}P and ^{3}HTdR were given simultaneously to 2 patients and to normal dogs and the myelocyte to blood transit time was measured by the two methods in the same subject at the same time.

In addition, preliminary results of studies attempting to define clinical abnormalities of granulocytopoiesis are described. To date, these have been done primarily by the in vivo DF^{32}P method.

METHODS AND MATERIALS

Subjects

Simultaneous labeling with ^{3}H-Thymidine and DF^{32}P. Two patients were selected for study because each harbored a malignancy that markedly limited life expectancy. A. F. was a 61-year-old man who had been operated on for an oral squamous cell carcinoma. Shortly before this study, he was found to have a metastatic nodule in his lung and a lytic lesion in a dorsal vertebra. He had been given vinblastine 4 months before this study but the drug had been discontinued because of neutropenia. His peripheral blood neutrophil count returned to normal soon thereafter and was 5100 at the time of study.

A. W. was a 47-year-old woman in whom a recurrent adenocarcinoma of the colon was found at laparotomy. There were metastases in her liver but none were found in bone. (There had been a brief response to 5-fluorouracil several months prior to our study, but then the tumor mass in her liver began to enlarge). The last dose of 5-fluorouracil had been given 6 weeks prior to our study. There had been no evidence of bone marrow toxicity. On the fifth day of study, a leukemoid reaction developed and her granulocyte count rose from 4740 on day 1 to 21,200 on day 16; at this time, there were 220 myelocytes per $mm.^3$ in the blood.

Normal adult mongrel dogs weighing at least 25 kilograms were also studied.

Studies to define clinical abnormalities in granulocytopoiesis. Subjects for these studies were patients in the hematology clinic or in the university hospital who were being followed for neutropenia or a chronic infection and a sustained granulocytosis. In addition, normal adult male volunteers with sustained neutrophilia as a result of the ingestion of 40 mg. of prednisone per day were used as subjects.

Isotopic Labeling

The patients selected for simultaneous labeling by both methods were given 400 μc. of $DF^{32}P$ (approximately 1.2 mg.) diluted in 20 ml. of isotonic saline followed immediately by 100 μc. of ^{3}HTdR (thymidine-methyl ^{3}H, 1.9 Ci/mmole), per Kg. of body weight.

Patients and subjects studied by means of the in vivo $DF^{32}P$ method were given an injection of this agent only. Labeling methods used in the dogs was the same except that the dose of $DF^{32}P$ was 1.0 mc. or about 3 mg.[10]

Measurement of Blood Granulocyte Radioactivity

Blood samples were withdrawn at appropriate intervals after label injection and granulocytes were separated by dextran sedimentation, saponin hemolysis, and differential centrifugation as previously described[5] except that instead of isotonic saline an acid citrate solution* was used for the 4 washes. After the fourth wash with acid citrate, the cell button was drained thoroughly and the cells were resuspended in exactly 15 ml. of acid citrate solution. A WBC count and differential count were done on this suspension. Blood granulocyte radioactivity (BGRA) was measured in a liquid scintillation counter and expressed as $CPM/15 \times 10^6$ cells. The BGRA values were plotted on semilogarithmic paper and lines were fitted by the method of least squares.

---

* Citric acid 0.73 Gm. per cent and sodium citrate 2.2 Gm per cent.

Results

Resolution of Previous Differences in Myelocyte to Blood Transit Time

From the results obtained in 2 patients simultaneously labeled with $^{3}HTdR$ and $DF^{32}P$, it is apparent that the time from the start of the study to the peak of the tritiated thymidine curve was about the same as the time to the end of Phase II of the $DF^{32}P$ curve (Fig. 2). In other words, in these 2 patients studied with both methods at the same time the myelocyte to blood transit time was identical. The myelocyte to blood transit times in these patients are shorter (8.3 days and 4.3 days, respectively) than values previously reported for normal subjects (8.5 to 14 days). The 8.3 and 4.3 day transit times are more consistent with results previously reported with $^{3}HTdR$ alone.

In similar double label experiments carried out in 6 normal dogs, the mean myelocyte to blood transit time was about 5 days and was also the same with both methods (Table 1). Although the myelocyte to blood transit time in dogs was less than values reported in normal man, this difference might be expected in the 2 different species and has been reported previously.[10]

The Decremental Slopes of the $^{3}HTdR$ and $DF^{32}P$ Curves

It is noteworthy that the last parts of the $DF^{32}P$ curve (Phase III) and of the $^{3}HTdR$ curve in the 2 doubly-labeled patients (Fig. 2) were identical in

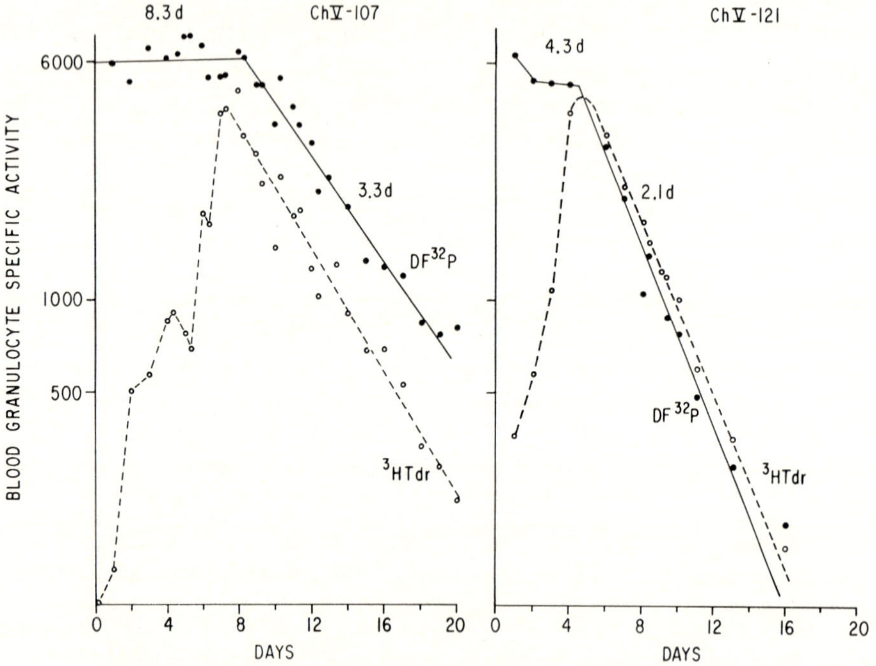

FIG. 2—Blood granulocyte radioactivity curves in 2 patients given ^{3}H-thymidine and $DF^{32}P$ intravenously at the same time.

TABLE 1—*Myelocyte to Blood Transit Time in Normal Dogs as Measured by Simultaneous Labeling with ^3HTdR and $DF^{32}P$*

Subject	Marrow Reserve Transit Time In Days $DF^{32}P$	^3HTdR
179	6.3	5.3
324	4.9	5.4
324	5.0	4.2
328	4.6	4.7
329	4.9	5.3
329	5.2	5.5
Mean	5.2	5.1

slope and both were exponential. Furthermore, the slope identity was maintained over at least 4 half times.

Since myelocytes appear to be labeled uniformly by $DF^{32}P$,[11] the label in subsequent generations of myelocytes would be one-half of that in the former. Thus, the decline in the SA of cells derived from successive generations of myelocytes falls exponentially. It has been postulated that the Phase III of the $DF^{32}P$ curves reflects this process and that the $t\frac{1}{2}$ of Phase III gives a measure of myelocyte generation time. As a result of these several observations, it now seems worthwhile to reexamine the possible usefulness of the later phases of both the ^3HTdR and $DF^{32}P$ curves as measures of events occurring in the marrow myelocyte compartment. Such studies are in progress but are too preliminary to warrant further consideration at this time.

Studies of Granulocytopoiesis in Abnormal States

If one accepts the thesis that after $DF^{32}P$ injection the time elapsed to the end of Phase II measures myelocyte to blood transit time and that the $t\frac{1}{2}$ of Phase III gives a measure of myelocyte generation time, then such measurements carried out in abnormal states may be informative.

To date, a few such measurements have been made in patients with neutropenia, in patients with chronic infections who had stable, elevated granulocyte counts during the study period, and in subjects chronically ingesting prednisone who also had a stable granulocytosis during study. These studies are compared with results obtained previously in normal volunteers (Table 2).

Neutropenic states. Patients with neutropenia were studied by means of the in vitro $DF^{32}P$ method in order to measure total blood granulocyte pool size, blood granulocyte half-disappearance time ($t\frac{1}{2}$) and blood granulocyte turnover rate (GTR).[12] On the basis of these studies, patients with neutropenia due to decreased production (normal $t\frac{1}{2}$ and low GTR) and patients with neutropenia due to increased destruction (short $t\frac{1}{2}$ and increased GTR) were selected for in vivo labeling with $DF^{32}P$.

It can be seen in Table 2 that the myelocyte to blood transit times and the myelocyte generation times were the same in both groups of neutropenic patients and did not differ significantly from values obtained in normal subjects.

TABLE 2—*Comparison of Myelocyte to Blood Transit Time and Myelocyte Generation Time in Normal Subjects, Neutropenic Patients, Patients with Infection, and Subjects Receiving Prednisone*

	No. of Subjects	Myelocyte to Blood * Transit Time (Days)	Myelocyte Generation Time (Days)	Granulocyte Count × 10^3
Normal	21	11.4 †	2.9 †	4.8
		(8.6–14.2)	(1.5–4.3)	(2.3–7.1)
Neutropenia due to	5	10.7	2.9	2.4
decreased production		(9.3–13)	(2.2–3.6)	(5.0–3.5)‡
Neutropenia with	4	11.1	2.3	1.7
increased production		(9–14)	(1.0–4)	(1.4–2.5)‡
Infection	2	11.0	2.5	11.5
		(10.6–11.3)	(2.3–2.6)	(1.9–15.0)
Steroid granulocytosis	5	10.6	2.3	10.8
		(10–12)	(2.1–2.5)	(8.0–14.5)

* Time elapsed between labeling and end of Phase II.
† Figures shown are the mean and absolute range.
‡ One patient in each of these groups was not neutropenic at the time of study though both were neutropenic immediately before study.

Infection. In 2 patients with infection and a stable granulocytosis, measurements of myelocyte to blood transit time and myelocyte generation time were also normal.

Steroid granulocytosis. In 5 subjects with stable granulocytosis as a result of prednisone ingestion, measurements of myelocyte to blood transit time and myelocyte generation time also were normal.

DISCUSSION

The myelocyte to blood transit time as measured with ^{3}HTdR in one patient with a brain tumor and with radiophosphate in 3 convalescing elderly hospital patients was 6 to 9 days (time elapsed to the peak of the curves).[2,4] In contrast, measurement of myelocyte to blood transit time with DF^{32}P in 21 healthy, ambulatory, male volunteers resulted in values ranging from 8.5 to 14 days.[9] Two possible explanations for these different results have been advanced: (1) Since ^{3}HRdR and radiophosphate label cells in the bone marrow that are in DNA synthesis while DF^{32}P labels the cytoplasm of neutrophilic granulocytes at the myelocyte stage and beyond regardless of cell cycle phase, the two methods may not measure the same physiologic time span and would not be expected to give the same myelocyte to blood transit time; (2) on the other hand, Boggs [10] has pointed out that the myelocyte to blood transit time in normal dogs is the same regardless of whether or not the measurement is made with ^{3}HTdR, DF^{32}P, radiophosphate or radiosulfate. For this reason we have argued that the different values reported in man are the result of differences in the subjects studied. Severely ill, bed ridden or convalescent patients might be expected to have granulocyte kinetic parameters different from those of normal healthy subjects. From the 2 patients studied simultaneously with both labels (Fig. 2) it is evident that the two methods give very similar if not identical measurements of myelocyte to blood transit time. It should be noted that in one of the patients (Ch V-107) the myelocyte to blood transit time value falls within the range reported for normal subjects. However, in the second patient (Ch V-121) the myelocyte to blood transit time was only 4.3 days, similar to values found in other patients by Perry et al.[13] and Bond et al.[4] This short transit time appears to reflect a more rapid than normal emptying of marrow cells into the blood since the patient fortuitously developed a leukemoid reaction with a rise in granulocyte count to 21,600/mm.[3] during the study. In spite of this nonsteady state, both labeling methods gave the same transit time value.

The parallel and exponential nature of the decremental slopes of both the $DF^{32}P$ and the $^{3}HTdR$ curves deserves consideration. Although it has been postulated that the slope of Phase III of the in vivo, $DF^{32}P$ curve provides a measure of marrow myelocyte generation time,[9] a similar interpretation of the last portion of $^{3}HTdR$ and radiophosphate curves has not been made. This is perhaps justified in the case of radiophosphate since Ottesen [1] demonstrated prolong availability of radiophosphate after injection and since phosphate reutilization seems likely. However, it probably does not apply to $^{3}HTdR$ which is clearly a "flash" label. Evidence that $^{3}HTdR$ is reutilized in the rat has been presented,[14,15] but from the present curves in man, significant reutilization seems unlikely. If reutilization of $^{3}HTdR$ were significant, one would have to postulate similar reutilization for the very different compound DFP to explain the parallel decremental slopes. In addition, since the exponetial fall off persists for a period of time as long as 3 or 4 half times, significant reutilization is most unlikely. Nevertheless, certain objections to interpreting the decremental slope as a measure of myelocyte turnover and generation time can be raised. The most important is that estimates of myelocyte generation time in man by other technics, namely grain count halving time [16] or per cent labeled mitosis curves,[17] have provided shorter times than those obtained from Phase III of DFP curves. Here again the number of such studies in man are few and they have been carried out in ill patients because of the large doses of radioactivity that must be given. Until measurements of myelocyte generation time by the $DF^{32}P$ method and one or more of the $^{3}HTdR$ methods are carried out simultaneously in several subjects complete resolution of this controversy is unlikely. Such studies are in progress.

If blood granulocyte radioactivity curves obtained after the injection of $^{3}HTdR$ or $DF^{32}P$ do provide reasonable measures of myelocyte to blood transit time and myelocyte generation and turnover time, the results of such studies in diseased patients would be of considerable pathophysiologic interest. On the assumption that this is the case the studies reported in Table 2 and Figure 2 deserve consideration. In subjects studied while in a steady-state, regardless of whether or not they were granulocytopenic or had granulocytosis, the myelocyte to blood transit times and the half disappearance times of Phase III ("myelocyte generation time") were normal, Table 2. This can easily be understood in the case of steroid granulocytosis, a situation in which blood granulocyte turnover rate is known to be normal and the neutrophilia appears to reflect a partial block to egress of granulocytes from the blood.[18] However, myelocyte to blood transit time and Phase III $t\frac{1}{2}$ values were also normal in patients with stable granulocytosis due to infection, a situation in which previous studies have demonstrated increased granulocyte production and turnover, and in patients with granulocytopenia with either increased or decreased granulocyte production. This suggests that in any steady-state situation (granulocytopenia, normal or granulocytosis) the time parameters of marrow granulocyte production are fairly constant. One interpretation of this finding is that under steady-state conditions granulocyte production proceeds with normal cell cycle times, while the quantity of cells produced is a function of the number of "assembly lines" in production. The latter may in turn reflect the number of uncommitted stem cells directed into granulocyte production channels. While this hypothesis is highly speculative, it is consistent with current concepts of the control mechanism for erythrocyte production, and can be subjected to experimental testing.

From patient Ch V-121, Figure 2, it seems clear that in nonsteady states the myelocyte to blood transit time may be short. Sacchetti and co-workers have reported similarly shortened values in patients studied after chemotherapy.[19] Ponassi et al.[20] attempted to bring about a shortening of the myelocyte to blood transit time by giving endotoxin repeatedly during an in vivo $DF^{32}P$ study. The failure to shorten the plateau duration was evidently because of the relatively small demands made on marrow reserves by endotoxin. It will be most interesting to see if studies of more acute nonsteady-state situations will demonstrate alterations in myelocyte generation time in addition to changes in myelocyte to marrow transit time.

Summary

Previously reported differences in myelocyte of blood transit time have been resolved by simultaneously using a DNA label (^{3}HTdR) and a cytoplasmic label ($DF^{32}P$) in the same subjects. In 2 patients and 6 dogs so studied, the estimate of the myelocyte to blood transit time was the same by the two methods.

In addition, the ^{3}HTdR curve fell off exponentially after the peak and was parallel to the Phase III of the $DF^{32}P$ curve. Thus, the estimates of myelocyte generation time obtained from the $t\frac{1}{2}$ of these exponentials, were the same.

Preliminary results of studies of granulopoiesis in abnormal states are reported. No changes in either myelocyte to blood transit time or myelocyte generation time were found in patients studied while they were in a steady state. One patient developed a leukemoid reaction during the study. In this nonsteady state, a shortened myelocyte to blood transit time was demonstrated.

ACKNOWLEDGMENTS

We wish to thank Dr. F. T. Johnson of the Upjohn Co., Kalamazoo, Michigan for the Heparin used in these studies and Mr. C. L. Blair of the E. I. du Pont de Nemours & Co., Wilmington, Delaware for the supply of Mylar.

REFERENCES

1. Ottesen, J.: On the age of human white cells in peripheral blood. Acta. Physiol. Scand. 32:75, 1954.

2. Craddock, C. G., Jr., Perry, S., and Lawrence, J. S.: The dynamics of leukopoiesis and leukocytosis, as studied by leukopheresis and isotopic techniques. J. Clin. Invest. 35:285,1956.

3. Hughes, W. L., Bond, V. P., Cronkite, E. P., Brecher, G., Painter, R. B., Quastler, H., and Sherman, F. G.: Cellular proliferation in the mouse as revealed by autoradiography with tritiated thymidine. Proc. Nat. Acad. Sci. U.S.A. 44:476–483, 1958.

4. Bond, V. P., Fliedner, T. M., Cronkite, E. P., Rubini, J. R., and Robertson, J. W.: Cell turnover in blood and blood forming tissues studied with tritiated thymidine. *In* Stohlman, F., Jr. (Ed.): The Kinetics of Cellular Proliferation. New York, Grune & Stratton, 1959, p. 188.

5. Athens, J. W., Mauer, A. M., Ashenbrucker, H., Cartwright, G. E., and Wintrobe, M. M.: Leukokinetic studies. I. A method for labeling leukocytes with diisopropylfluorophosphate ($DF^{32}P$). Blood 14: 303, 1959.

6. Mauer, A. M., Athens, J. W., Ashenbrucker, H., Cartwright, G. E., and Wintrobe, M. M.: Leukokinetic studies. II. A method for labeling granulocytes *in vitro* with radioactivxe diisopropylfluorophosphate ($DF^{32}P$). J. Clin. Invest. 39:1481, 1960.

7. Perry, S., Godwin, H. A., and Zimmerman, T. S.: Physiology of the granulocyte. Part I & Part II. J.A.M.A. 203:937 & 1025, 1968.

8. Warner, H. R., and Athens, J. W.: An analysis of granulocyte kinetics in blood and bone marrow. Ann. N.Y. Acad. Sci. 113: 523, 1964.

9. Cartwright, G. E., Athens, J. W., and Wintrobe, M. M.: The kinetics of granulopoiesis in normal man. Blood 24:780, 1964.

10. Boggs, D. R., Athens, J. W., Cartwright, G. E., and Wintrobe, M. M.: Leukokinetic studies. IX. Experimental evaluation of a model of granulopoiesis. J. Clin. Invest. 44:643, 1965.

11. Kurth, D., Athens, J. W., Cronkite, E. P., Cartwright, G. E., and Wintrobe, M. M.: Leukokinetic studies. V. Uptake of tritiated diisopropylfluorophosphate by leukocytes. Proc. Soc. Exp. Biol. Med. 107:422, 1961.

12. Bishop, C. R., Athens, J. W.: Leukokinetic studies in neutropenic patients. (Abst.) J. Clin. Invest. 18:8a, 1969.

13. Perry, S., Craddock, C. G., Jr., Lawrence, J. S.: Rates of appearance and disappearance of white blood cells in normal and in various disease states. J. Lab. Clin. Med. 51:501, 1958.

14. Bryant, B. J.: Reutilization of leukocyte DNA by cells of regenerating liver. Exp. Cell Res. 27:431, 1962.

15. Diderholm, H., Fichtelius, K. E., and Linder, O.: Availability time of ^{3}H label after administration of ^{3}H thymidine *in vivo*. Exp. Cell Res. 27:431, 1962.

16. Stryckmans, P., Cronkite, E. P., Fache, J., Fliedner, T. M., Ramos, J.: Deoxyribonucleic acid synthesis time of erythropoietic and granulopoietic cells in human beings. Nature (London) 211:717, 1966.

17. Cronkite, E. P., Bond, V. P., Fliedner, T. M., and Killman, S. A.: The use of tritiated thymidine in the study of haemopoietic cell proliferation. *In* Walstenholme, G. E. W., and O'Connor, M. (Eds.): Ciba Foundation Symposium on Haemopoiesis. Boston, Little, Brown, 1960, p. 70.

18. Athens, J. W., Haab, O. P., Raab, S. O., Mauer, A. M., Ashenbrucker, H., Cartwright, G. E., and Wintrobe, M. M.: Leukokinetic studies. IV. The total blood, circulating and marginal granulocyte pools and the granulocyte turnover rate in normal subjects. J. Clin. Invest. 40:989, 1961.

19. Sacchetti, C., Boccaccio, P., Ponassi, A., Morra, L.: Neutrophilic leukocyte reserve of the bone marrow. 1) "In vivo" detection and quantitation with DFP^{32}. Minerva Nucleare 9:55, 1965.

20. Ponassi, A., Boccaccio, P., Morra, L., Sacchetti, C.: Studio della leucocitosi prodotta dalla somministrazione endovena di endotossine bateriche nell'uomo. (Studies of leukocytosis due to intravenous administration of bacterial endotoxin in man.) Minerva Nucleare 8:333, 1964.

In Vitro and In Vivo Colony Technic in the Study of Granulopoiesis

By K. A. Rickard, R. K. Shadduck, A. Morley, and F. Stohlman, Jr.

So we grew together,
Like to a double cherry, seeming parted,
But yet an union in partition
Two lovely berries moulded on one stem.
Shakespeare, *A Midsummer Night's Dream,* Act 3

As Shakespeare discussed the stem of origin of the two lovely berries we would like to consider how the two systems of spleen colonies and agar colonies relate to stem cells, to each other, and how they assist us in the study of granulopoiesis.

We propose to consider some of the background studies of the in vitro colony system, some of the critical studies on the spleen colony technic, and some of the observations that have been made in our laboratory on the colony stimulating factor and, more importantly, on the cellular origin of in vitro colonies. Our latter observations on the cellular origin of in vitro colonies suggest that the cell responsible for these colonies is a primitive member of the granulocytic series, perhaps a committed myeloid stem cell in a proliferative phase, and as such is distinct from the pluripotential transplantable cell responsible for spleen colonies.

The In Vitro System

Based on the observation of Sanders and Burford [1] and MacPherson and Montagnier [2] on tumor cells transformed by Polyoma virus, Pluznik and Sachs [3] described a method for the cloning of normal mouse mast cells from spleen cell suspensions in a tissue culture in which they used a soft agar medium. In their studies the formation of colonies required the presence of an embryo cell feeder layer, and it was their interpretation that colonies could be initiated by a single "colony-forming unit" and were formed as a result of cell multiplication. Further they noted that there was a linear relationship between the number of cells plated and the number of colonies which developed on the plate. At about the same time Bradley and Metcalf [4] described a similar method for growing cells on agar plates from murine bone marrow. They obtained approximately the same numbers of colonies whether using in their feeder layers 10×10^6 eight-day-old mouse kidney or mouse embryo cells from the seventeenth day of gestation. Their regression data was consistent with the thesis that colonies arise from a single cell. Further, the cellular analysis of colonies so obtained showed that early during the growth period the dominant cell type was a mononuclear cell, but eventually cells akin to polymorphonuclear leucocytes developed.

From St. Elizabeth's Hospital, Tufts University School of Medicine, Boston, Massachusetts.
Supported by grant HE 5600 and HE 7542 from National Heart Institute.
Dr. Rickard is a Fellow of the Leukemia Society of America, Inc.

The In Vivo and the In Vitro System

In these early papers both Pluznik and Sachs [3] and Bradley and Metcalf [4] were quick to draw an analogy between the system for cloning cells in vitro and the classic studies described earlier by Till and McCulloch [5] who showed that transplanted murine hemopoietic tissue may grow as distinct colonies in the spleens of heavily irradiated syngeneic hosts. It is important, however, to note that as yet there has been no evidence for the development of erythroid cells with the in vitro cloning technic, whereas the in vivo technic of Till and McCulloch [5] yields erythroid and megakaryocytic colonies as well as myeloid colonies.

Becker et al.,[6] using a chromosome marker, concluded that the in vivo colonies developed from a single cell. More recently Wu et al.[7] have also applied the chromosome marker technic to the in vitro colonies and have observed that a similar marker is obtained in both spleen and in vitro colonies. This not only suggests that the colonies arise from a single cell, but that the cells giving rise to colonies in agar are closely related to the hemopoietic cells which give rise to spleen colonies after transplantation. Lewis and Trobaugh [8] noted that the hemopoietic repopulating cells responsible for spleen colonies undergo renewal and differentiation and can be subsequently retransplanted, thus indicating that they are indeed stem cells. It is of interest that subcultures of in vitro colonies have been reported [9] although the secondary colonies do not appear to be as well defined as the primary colonies.

The Colony Stimulating Factor

Based on the superiority of kidney feeder layers in inducing colonies, Bradley and Metcalf [4] raised the possibility that the kidney may be active by virtue of secretion or release of a specific granulopoietic stimulatory substance analogous to erythropoietin for the red cells. Senn et al.,[10] using mouse kidney tubules in a feeder layer, were able to grow colonies from human bone marrow. More recently Robinson et al.[11] have demonstrated the development of agar colonies from normal and leukemic human bone marrow grown on feeder layers of human peripheral granulocytes. Pluznik and Sachs [12] have described an inducing factor for clones of mast cells in conditioned medium from nonirradiated embryo cell monolayers. Bradley and Sumner [13] have also used medium conditioned by mouse embryo monolayers and find it slightly superior to medium conditioned by kidney monolayers. Like the latter workers, we have shown that the colony stimulating factor in conditioned medium is potentiated by dialysis (Fig. 1) and is filtrable. Ichikawa et al.[14] reported that the inducer in conditioned medium was thermostable but was not destroyed by incubation with trypsin, chymotrypsin, pronase or collagenase but was precipitated by 33–50 per cent ammonium sulfate. Stanley and Metcalf,[15] on the other hand, reported that urine which promoted colony growth when heated to 90° for 30 minutes or exposed to high concentrations of pronase lost its colony stimulating ability. They sug-

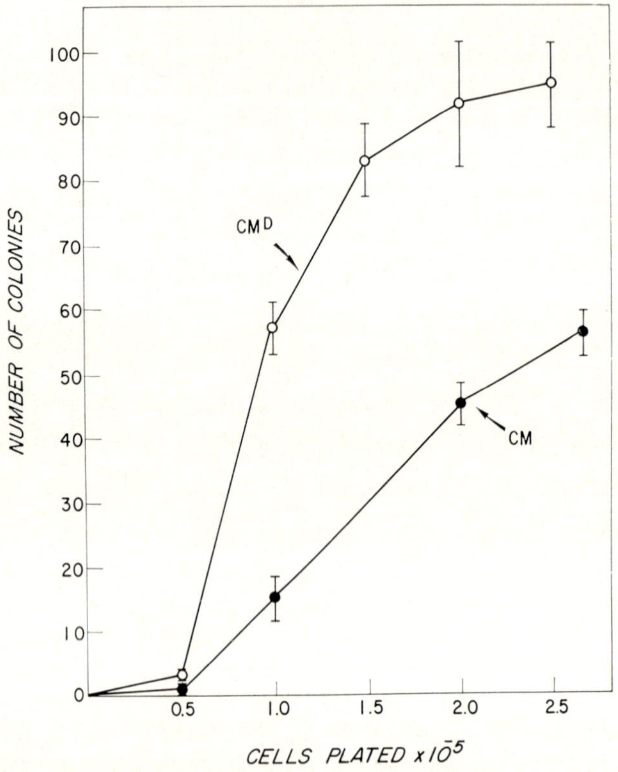

FIG. 1.—The effect of dialysis of conditioned medium. CM-conditioned medium prepared from embryo cell monolayers and used at a dose of 0.1 ml. per plate. CM^D-conditioned medium prepared from embryo cell monolayers and dialyzed at 4 C. against four changes of distilled water in 48 hours, then used at a dose of 0.1 ml. per plate.

gested that the colony stimulating factor was a glycoprotein with a molecular weight of 190,000. From these observations it might be suggested that the stimulating factor provided by conditioned media differs from that of urine or alternatively that a stabilizing factor has been lost during excretion.

Of particular interest in the recent observations of Paran et al.[16] who have shown that mature normal rat granulocytes inhibit the activity of the inducer in conditioned medium. We have been able to confirm these observations.[17] This might be interpreted as suggesting that control mechanisms regulating the growth and development of in vitro colonies include a feedback loop or induction inhibitor which is released by mature granulocytes or macrophages.

Colony stimulating activity has also been reported from the serum of a variety of animals and man. Robinson et al.[18] observed that 90 per cent of sera from AKR mice with spontaneous or transplanted lymphoid leukemia and some 40–50 per cent of sera from normal or pre-leukemic AKR mice stimulated colony formation by C 57BL bone marrow cells. This active factor like that described in the urine was filtrable, nondialyzable, and heat and ether labile. Metcalf and Foster [19] have also demonstrated that serum from Swiss mice with viral induced lymphoid leukemia stimulated in vitro granulocyte or mononuclear colonies. In two surveys conducted in Buffalo and Melbourne [20] it was noted that colony stimulating activity was rare in sera from normal persons but was detected in 15 to 70 per cent of sera of patients with leukemia and allied disorders, proliferative diseases of leukopoiesis, and in patients with nonbacterial infections. These same workers [21] observed that unconcentrated urine from some 50 per cent of normal people, as well as patients with a variety of disease states, exhibit colony stimulating activity. Finally, Marsh [22] described the growth of colonies from dog marrow in the in vitro agar system in the absence of a feeder layer or conditioned media but in the presence of fetal calf serum which may have supplied the stimulating material.

The Cellular Origin of In Vitro Colonies

From the above it is clear that there is a great deal of information available regarding the factors necessary for stimulation of in vitro colonies. The identity of the marrow cell or cells responsible for colony production, however, has received less attention. As previously noted, the possibility that the in vitro colonizing cell is similar to that responsible for spleen colony formation in the transplanted lethally irradiated mouse has been suggested. Further, Robinson et al.[23] demonstrated an exponential decrease in the percentage of cells in the bone marrow which were capable of in vitro colony formation after increasing doses of whole-body irradiation. This correlated with a fall in small lymphocytes which raised the possibility that the lymphocyte is the cell responsible for in vitro colonization. As the authors indicate, this is by no means conclusive evidence that the small lymphocyte or a subclass thereof is in fact the cell which grows in tissue culture. Bennett et al.[24] have shown that the marrow cells from genetically anemic W^v/W^v mice and the nonanemic $W^v/+$ carrier not only form colonies of granulocytic and mononuclear cells in vitro but that the proliferative potential appears to be normal as is their ability to differentiate. In contrast, cells from the marrows of these mice have an impaired ability to form splenic colonies when transplanted into heavily irradiated recipients. Those colonies which are formed are significantly smaller than those formed by normal mouse marrow. Post-transfusion polycythemia is associated with an increase in the frequency of the in vitro colony-forming cells but not that of the in vivo progenitor cells of the marrow. These findings suggest that the in vitro colony-forming cells (CFC) are distinct from the in vivo CFU. Moreover, it would appear that the CFU is a more primitive pluripotential cell than is true for the in vitro colony-forming cell. Metcalf [25] noted that after venesection there was a decrease in the concentration of CFC/10^5 bone marrow cells. From this he concluded that the cell responsible for in vitro colony formation was not erythroid in origin.

The Effect of Hydroxyurea on CFU and CFC

In our laboratory we have studied the effect of hydroxyurea on the development of colonies using both in vivo and in vitro technic. Marrow samples were obtained from 12–14-week-old randomized female CF1S mice immediately before or 2 hours after the administration of hydroxyurea (900 mg./Kg.). Cell suspensions were prepared as previously described.[26] Aliquots of the cell suspensions were injected into heavily irradiated (800 r) syngeneic mice and the animals sacrificed on the eighth post-transplantation day. Colonies were counted by two independent observers. Aliquots of the same suspensions were grown in vitro on E1010-agar plates,[19] using conditioned medium [13] prepared from mouse embryo monolayers as a stimulant.

The bone marrows from donors treated with hydroxyurea (OHU) 2 hours prior to sacrifice had a significant reduction in the numbers of colonies grown on agar plates (Table 1). The reduction amounted to approximately 55 per cent over those from control animals. In contrast there was a much smaller reduction in the total number of colonies formed after transplantation of the marrow into the heavily irradiated recipients (Fig. 2). In three separate experiments shown here there was a reduction varying from 15 to 25 per cent. In a fourth experiment, which we have recently concluded, there was no reduction. Given the technical problems with the CFU technic, such a modest reduction in CFU's must be interpreted with some measure of caution. It is clear, however, that the difference in the degree of reduction of colony formation by OHU differed significantly between the two systems.

Hydroxyurea is a cytotoxic agent which appears to be specifically lethal for cells during S. In addition there may be a hold up of cells in G_1 in tissue culture but this has not been borne out in the in vivo experiments, presumably because of the rapid metabolism of the agent. Carry over of hydroxyurea into the tissue culture in the present experiments, while a possibility, does not appear to be likely due to the rapid inactivation of the agent. From these sets of experiments it would appear that a significant number of the cells responsible for formation of granulopoietic colonies in vitro are in active cycle, indeed ~50 per cent were in S at the time of administration of hydroxyurea. In contrast, the majority of the pluripotential cells responsible for transplantation colonization were in either G_0 or a prolonged G_1 which is in concert with the observations of Becker et al.[27] and more recently Morse et al.[28]

Table 1—*Effect of OHU on In-Vitro Colony-Forming Cells (CFC's)*

Control	OHU (0.9 Gm./Kg.)	
Mean CFC's Per Tibia	Mean CFC's Per Tibia	P Values
7745 ± 300	3424 ± 158	<.001
4831 ± 393	2101 ± 368	<.001
4592 ± 153	2351 ± 314	<.001
1512 ± 152	670 ± 84	<.001

Hydroxyurea (OHU) was given to donor mice 2 hours prior to sacrifice. Values are expressed as total numbers of colony forming cells per tibia. ±S.E.

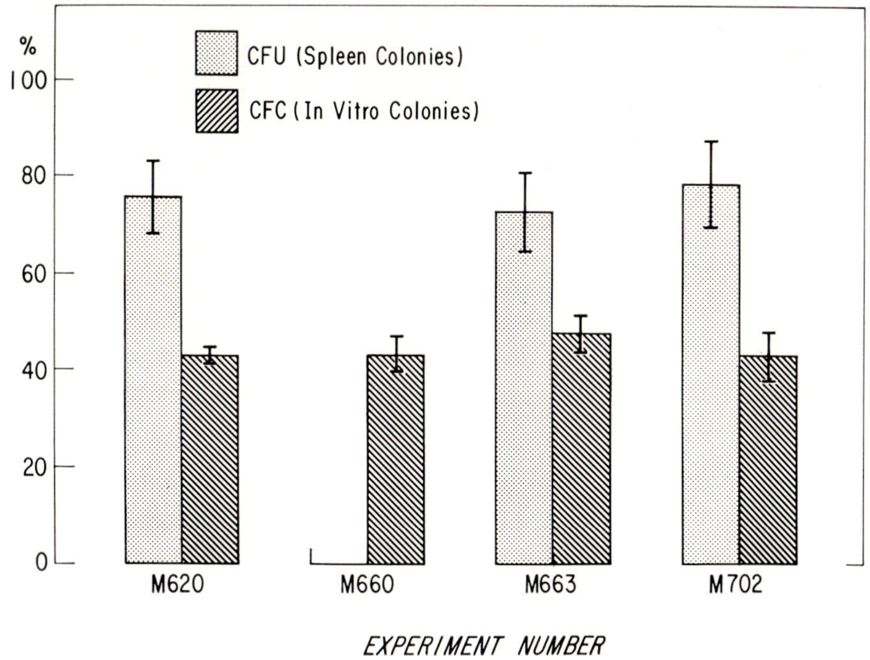

FIG. 2—The effect of treating bone marrow donors with hydroxyurea (900 mg./Kg.) on in vitro (agar) and in vivo (spleen) colonies. The total number of in vitro colony-forming cells and in vivo colony-forming units per tibia of donor mice after OHU are expressed as percentages of the values obtained for control mice. ± S.E.

The CFU and CFC in the Shielded Limb of Irradiated Mice

To further investigate the notion that the in vitro colony-forming cell is a primitive member of the granulocytic cell line—a committed precursor cell, rather than the transplantable CFU—we studied the marrow from irradiated mice in which one hind limb was shielded at the time of irradiation (ILS). In ILS animals the shielded limb exhibits increased granulopoiesis due to the physiologic demand of irradiation induced neutropenia. Five days after receiving 800 r there is a marked increase in the size of the myeloblast—promyelocyte compartment in the shielded limb (Fig. 3).

This granulocytic hyperplasia presumably is mediated through an increase in the rate of differentiation of stem cells into this compartment. Coincident with the granulocytic hyperplasia, the number of in vitro colony-forming cells/tibia had strikingly increased, being more than doubled in 3 experiments in the shielded limb of the ILS animal (Fig. 4). The increased numbers of in vitro colonies appeared to be directly correlated with the aforementioned granulocytic hyperplasia and presumably reflected an increase in the size of the committed myeloid stem cell compartment. It is of interest that the size of the pluripotential compartment in the ILS-treated animal as judged by transplantation was insignificantly increased in the shielded marrow (Table 2). The latter presumably reflects differentiation of cells into the committed myeloid compartment or migration from the shielded limb. The doubling

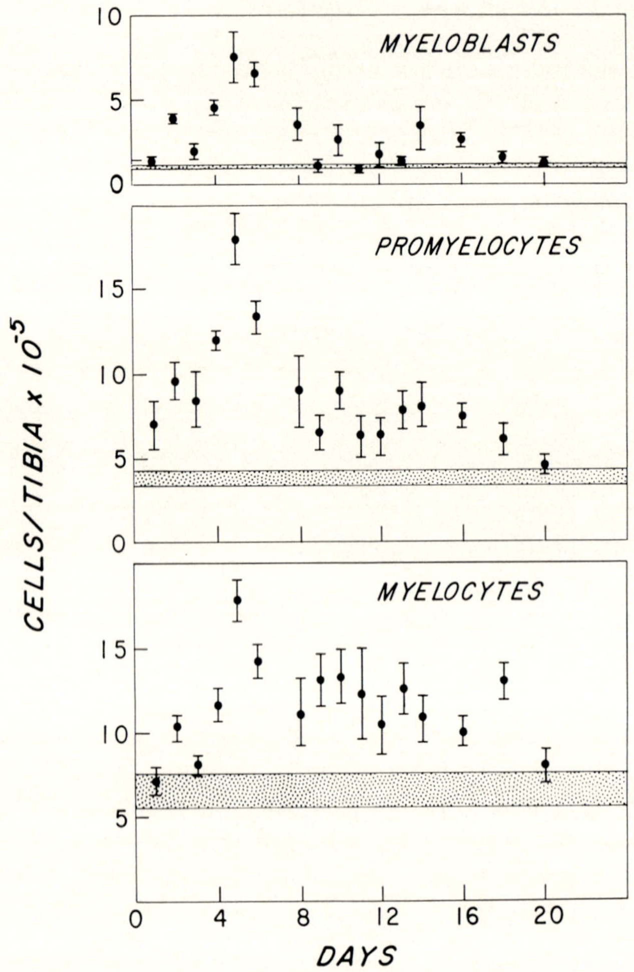

Fig. 3—The number of myeloblasts, promyelocytes, and myelocytes in the shielded limb of ILS mice in relation to days post irradiation. ± S.E. (Reprinted from Morley and Stohlman, Reference 30.)

Table 2—*Colony-Forming Units (CFU's) in Shielded Limb of ILS Mice*

Control	ILS
Mean CFU's Per Tibia	Mean CFU's Per Tibia
2320 ± 179	3260 ± 299
2620 ± 260	2700 ± 280
3110 ± 480	3460 ± 400

Values are total numbers of splenic CFU per tibia in control mice and, the shielded limb, of mice 5 days after receiving 800 r. ± S.E.

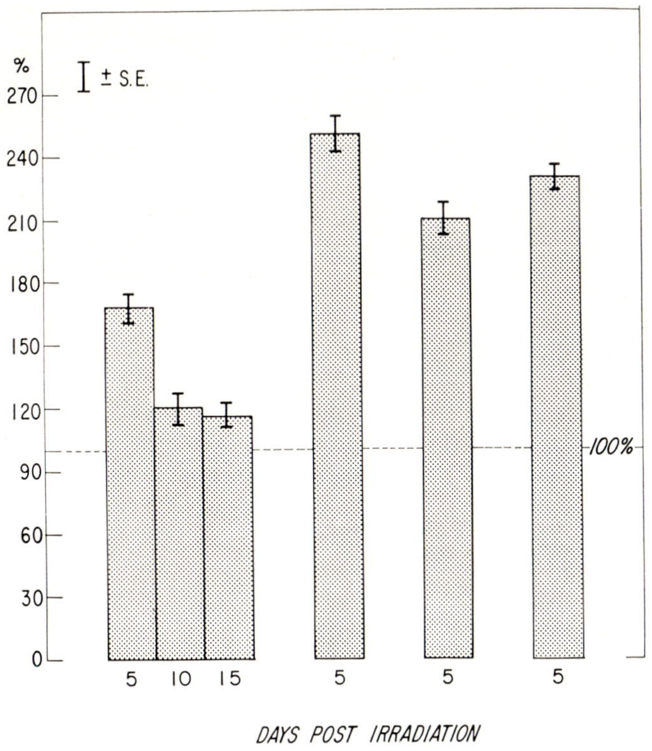

FIG. 4—The total number of in vitro colony-forming cells per tibia in the shielded limb of ILS mice are expressed as percentages of the values obtained for control animals. The results from four experiments are shown in the figure.

of CFC with a constant number of CFU provides further cogent evidence against the thesis that the in vitro colony-forming cell and the transplantable CFU are identical. Moreover, it appears to be entirely compatible with the notion that the CFC is in fact a precursor cell already committed to myeloid differentiation.

The In Vitro Growth Potential of Fetal Liver

We have previously demonstrated[29] that in CF_{1S} strain of mice 7 to 8 times the numbers of fetal liver cells must be transplanted to yield the same number of spleen colonies that are observed when adult bone marrow is used. The plating efficiency between the two differ, 9.6 per cent for fetal liver versus 14 per cent for adult marrow. From this one might project that the frequency of the pluripotential transplantable cells in fetal liver is no more than ~⅙ of that observed in adult bone marrow. However, when fetal liver cells from a similar suspension were grown with the in vitro agar technic the frequency of CFC's in the fetal liver approached that observed with adult marrow (Fig. 5).

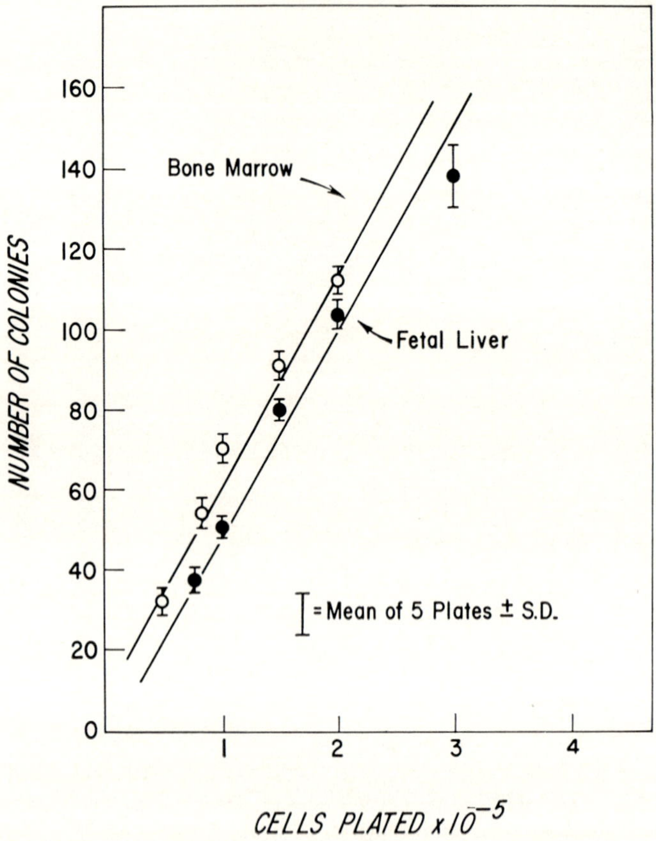

FIG. 5—The results of one experiment in which the numbers of colonies obtained from adult bone marrow and 16½-day-old fetal liver suspensions are plotted against cell numbers plated. Dialyzed conditioned medium was used as the colony stimulating factor.

In 4 experiments in which the in vitro growth potential of fetal liver cells was compared with bone marrow, the ratio of CFC/10^5 cells in fetal liver suspensions to bone marrow varied from 0.70–0.82. These observations lend further support to a committed rather than a pluripotential cell as the cell of origin of the in vitro colonies. It is of interest, that although the 16½ day old fetal liver is predominantly an erythroid organ, the frequency of the myeloid precursor cell as assessed by the agar colony technic approaches that of adult marrow.

SUMMARY

We have considered in this paper the relationship between the in vitro and the in vivo colony technics and how they assist us in the study of granulopoiesis. We have also shown firstly, that on the basis of a reduction in bone marrow in vitro colony-forming cells with hydroxyurea and secondly, because of the increase in bone marrow in vitro colony-forming cells in the shielded limb of irradiated leg shielded mice and thirdly, because of the comparable relationship between the numbers of in vitro colony-forming cells in fetal liver and bone marrow that the cell compartment giving rise to the in vitro colony is a primitive member of the granulocyte series, not the transplantable colony-forming unit, but more likely a committed myeloid stem cell which is in cycle.

ACKNOWLEDGMENTS

We are grateful to Mr. D. Howard for his skilled assistance and to Mrs. E. Aguilo for the preparation of glassware during the course of these studies.

REFERENCES

1. Sanders, F. K., and Burford, B. O.: Ascites tumors from BHK 21 cells transformed *in vitro* by polyoma virus. Nature (London) 201:786–789, 1964.

2. Mac Pherson, I., and Montagnier, L.: Agar suspension culture for the selective assay of cells transformed by polyoma virus. Virology 23:291–294, 1964.

3. Pluznik, D. H., and Sachs, L.: The cloning of normal mast cells in tissue culture. J. Cell. Physiol. 66:319–324, 1965.

4. Bradley, T. R., and Metcalf, D.: The growth of mouse bone marrow cells *in vitro*. Aust. J. Exp. Biol. Med. Sci. 44:287–300, 1966.

5. Till, J. E., and McCulloch, E. A.: Direct measurement of the radiation sensitivity of normal mouse bone marrow. Radiat. Res. 14:213–222, 1961.

6. Becker, A. J., McCulloch, E. A., Till, J. E.: Cytological demonstration of the clonal nature of spleen colonies derived from transplanted mouse marrow cells. Nature (London) 197:452, 1963.

7. Wu, A. M., Siminovitch, L., Till, J. E., McCulloch, E. A.: Evidence for a relationship between mouse hemopoietic stem cells and cells forming colonies in culture. Proc. Nat. Acad. Sci. U.S.A. 59:1209–1215, 1968.

8. Lewis, J. P., Trobaugh, F. E.: Haematopietic stem cells. Nature (London) 204: 589–590, 1964.

9. Metcalf, D., Foster, R.: Behavior on transfer of serum stimulated bone marrow colonies. Proc. Soc. Exp. Biol. Med. 126: 758–762, 1967.

10. Senn, J. S., McCulloch, E. A., Till, J. E.: Comparison of colony forming ability of normal and leukaemic human marrow in cell culture. Lancet ii:597–598, 1967.

11. Robinson, W. A., Pike, B., Lampel, J.: Human bone marrow colony growth in agar-gel. Fourth Leucocyte Culture Conference. Hanover, New Hampshire. Abstract 75, 1969.

12. Pluznik, D. H., Sachs, L.: The induction of clones of normal mast cells by a substance from conditioned medium. Exp. Cell. Res. 43:553–563, 1966.

13. Bradley, T. R., Sumner, M.: Stimulation of mouse bone marrow colony growth *in vitro* by conditioned medium. Aust. J. Exp. Biol. Med. Sci. 46:607–618, 1968.

14. Ichikawa, Y., Pluznik, D. H., Sachs, L.: In vitro control of the development of macrophage and granulocyte colonies. Proc. Nat. Acad. Sci. U.S.A. 56:488–495, 1966.

15. Stanley, E. R., Metcalf, D.: Partial purification and some properties of the factor in normal and leukaemic human urine stimulating mouse bone marrow colony growth *in vitro*. Aust. J. Exp. Biol. Med. Sci. 47:467–483, 1969.

16. Paran, M., Ichikawa, Y., Sachs, L.: Feedback inhibition of the development of macrophage and granulocyte colonies. II. Inhibition by granulocytes. Proc. Nat. Acad. Sci. U.S.A. 62:81–87, 1969.

17. Rickard, K. A., Morley, A.: Unpublished observations.

18. Robinson, W., Metcalf, D., Bradley, T. R.: Stimulation by normal and leukemic mouse sera of colony formation *in vitro* by mouse bone marrow cells. J. Cell. Physiol. 69:83–92, 1967.

19. Metcalf, D., Foster, R.: Bone marrow colony stimulating activity of serum from mice with viral induced leukemia. J. Nat. Cancer Inst. 39:1235–1245, 1967.

20. Foster, R., Metcalf, D., Robinson, W. A., Bradley, T. R.: Bone marrow colony stimulating activity in human sera. Brit. J. Haemat. 15:147–159, 1968.

21. Robinson, W. A., Stanley, E. R., Metcalf, D.: Stimulation of bone marrow colony growth *in vitro* by human urine. Blood 33: 396–399, 1969.

22. Marsh, J. C.: Colony growth of dog bone marrow *in vitro*. Fourth Leucocyte Culture Conference. Hanover, New Hampshire. Abstract 69, 1969.

23. Robinson, W. A., Bradley, T. R., Metcalf, D.: Effect of whole body irradiation on colony production by bone marrow cells *in vitro*. Proc. Soc. Exp. Biol. Med. 125:388–391, 1967.

24. Bennett, M., Cudkowicz, G., Foster, R. S. Metcalf, D.: Hemopoietic progenitor cells of W anemic mice studied *in vivo* and *in vitro*. J. Cell. Physiol. 71:211–226, 1968.

25. Metcalf, D.: The effect of bleeding on the number of *in vitro* colony forming cells in the bone marrow. Brit. J. Haemat. 16: 397–407, 1969.

26. Morley, A., Shadduck, R. K., Howard, D., Stohlman, F., Jr.: The effect of heat damaged red cells on stem cells. Radiat. Res. 41:70, 1970.

27. Becker, A. J., McCulloch, E. A., Simonovitch, L., Till, J. E.: The effect of differing demands for blood cell production on DNA synthesis by hemopoietic colony forming cells of mice. Blood 26:296–308, 1965.

28. Morse, B. S., Rencricca, N. J., Stohlman, F., Jr.: The relationship of erythropoietin effectiveness to the generative cycle of the erythroid precursor cell. Blood. In press.

29. Kubanek, B., Rencricca, N. J., Porcellini, A., Howard D., Stohlman, F., Jr.: The pattern of stem cell repopulation in heavily irradiated mice receiving transplants of fetal liver. Blood. 35:64, 1970.

30. Morley, A., and Stohlman, F., Jr.: Studies on the regulation of granulopoiesis I. The response to neutropenia. Blood. 35:312, 1970.

Colony Growth of Human Bone Marrow Cells in Vitro

By William A. Robinson and Beverley L. Pike

Introdution

A method of bone marrow colony growth in agar-gel medium has recently been described.[1,2] In this system single cells can be stimulated to divide, mature, and give rise to granulocytic and mononuclear cell colonies in the presence of an appropriate stimulus.[3] A number of materials have been used to initiate and sustain colony growth, the most effective being kidney and whole mouse embryo "feeder layers," [1,4] "conditioned medium" from these cells,[2,5] and human and animal serum and urine.[6,7,8] While these stimuli have been quite effective, using murine bone marrow as target cells, they have not been effective in initiating colony growth from human bone marrow in our hands. Human bone marrow colony growth has been reported [9] in this system using feeder layers of mouse kidney tubule cells, but the colonies obtained were quite small and the numbers low when compared with murine bone marrow in the same system.

We have recently found that excellent human bone marrow colony growth can be obtained when these cells are grown on feeder layers of normal human peripheral white blood cells. The colonies obtained are similar in size, number, and morphology to those described using murine bone marrow. The technic, which is simple and reliable, is detailed herein.

Material and Methods

Preparation of Feeder Layers

Human peripheral blood is collected from normal volunteers in tissue culture Vacutainer tubes * containing 2 drops of 1:1000 heparin.† After collection, the blood is allowed to stand for 2 hours at room temperature. The plasma containing the white blood cells and platelets is then removed with a Pasteur pipet. The number of nucleated WBC in the plasma is counted in a hemocytometer. The usual number is about 20×10^6/ml. with a ratio of one nucleated WBC to one RBC. For routine studies feeder layer are prepared from cells which have not been washed from plasma. As will be seen, however, the cells appear to be the essential component. The medium used to prepare both the underlays and overlays is outlined in Table 1. It consists of a modified preparation of McCoy's 5A medium ‡ to which has been added 15 per cent fetal calf serum and a variety of other nutrient elements. Human serum is particularly avoided since

From the Division of Hematology, Department of Medicine, University of Colorado Medical Center, Denver, Colorado.

Supported by grants from the American Cancer Society (E–498) and the National Cancer Institute, National Institutes of Health (I ROI CA 11305–DI).

* No. 3208 NT, Becton, Dickinson and Company, Rutherford, N. J.

† Liquaemin, Sodium "10," Organon Inc., West Orange, N. J.

‡ Biogen Laboratories, Denver, Colorado.

TABLE 1—*Composition of Medium Used for Human Bone Marrow Colony Growth*

McCoy's 5A medium single strength	800 ml.
Fetal calf serum	150 ml.
Sodium bicarbonate (7.5% solution)	6 ml.
Sodium pyruvate (100 mM solution)	10 ml.
Eagles MEM vitamins (100 ×)	4 ml.
Eagles MEM amino acids (50 ×)	8 ml.
Eagles MEM nonessential amino acids (100 ×)	4 ml.
Eagles MEM glutamine (200 mM)	4 ml.
L-serine (21 mg./ml.)	0.4 ml.
L-asparagine (10 mg./ml.)	1.6 ml.

Underlays: 9 parts of medium are mixed with 1 part of 5 per cent agar solution—the final agar concentration being 0.5 per cent.

Overlays: 9 parts of medium are mixed with 1 part of 3 per cent agar solution—the final agar concentration being 0.3 per cent.

* Filtered through 0.45 μ Millipore filter prior to use.

it may contain factors which alter colony growth in this system.[7] A constant source of fetal calf serum, a single batch obtained from Biogen Laboratories, has been utilized throughout these experiments. McCoy's 5A medium, prepared as above, is mixed in a 9:1 concentration with boiled 5 per cent agar ¶ to give a final agar concentration of 0.5 per cent. The medium must be warmed to 37 C. prior to adding the agar to prevent gelling. The agar is added to the medium at a temperature just below the boiling point. The mixture is then allowed to cool to approximately 40 C. To this mixture is then added human peripheral blood cells in plasma to a final concentration of 1×10^6 cells per ml. One ml. aliquots of this are pipetted into 35 mm. plastic Falcon * petri dishes. The agar medium is then allowed to gel at room temperature. Overlays may be pipetted directly onto the underlays once they have gelled or the plates may be stored at 37 C. in a humidified incubator for up to 14 days without loss of colony stimulating activity.

In studies of the effect of plasma versus plasma plus cells or cells alone, the amount of plasma incorporated into the 1 ml. underlays was the amount which contained 1×10^6 cells.

Preparation of Bone Marrow Overlays

All human bone marrow for culture is obtained from the posterior iliac crest using a Westerman-Jensen needle. The specimen for culture is collected by aspirating 5–10 ml. of the bone marrow into a 10 ml. glass syringe rinsed with 1:1000 heparin just prior to use. The aspirate is transferred to a test tube and allowed to stand at room temperature for 2 hours as for peripheral blood. The plasma containing white blood cells, platelets, and small fat particles is removed with a Pasteur pipete. This is sedimented in a bench centrifuge and washed three times in McCoy's 5A medium. No attempt is made to break apart the fat particles as these are removed during the washing procedure. McCoy's

¶ Difco-Bacto-Agar.

* Falcon Plastics, Los Angeles, California.

5A medium is then mixed in a 9:1 concentration with a 3 per cent boiled agar as above. Washed bone marrow cells are added to this mixture in a concentration of 2×10^5 cells per ml, and plated on top of the previously prepared underlays. After gelling at room temperature, the plates are incubated in a humidified incubator containing a constant flow of 7.5 per cent CO_2 in air.

Control Plates

Control plates are prepared in all experiments using underlays which consist only of 1 ml. agar-medium mixture without added peripheral blood cells. Overlays containing human bone marrow cells are from the same mixture as above.

Microscopic Studies

Colony counts are done routinely at days 12 to 15 of incubation with a dissecting microscope at times 16 magnification. For microscopic studies colonies are removed from the agar using a finely drawn Pasteur, flattened with cover slips on glass slides, and then stained by floating 0.45 per cent Aceto-ocein stain under the cover slips as previously described.[3]

Results

Colonies are first recognizable on plates at day 4–5 of incubation. The rate of colony growth (Fig. 1) is considerably slower than with murine bone marrow. There is a slow increase in colony size between days 4 and 10 after which there is a rapid increase until about day 15–20 when colonies reach maximum size of 500–1500 cells. This is in contrast to murine marrow colonies where maximum size is attained at day 10–12 of culture.[3]

Colonies are almost exclusively compact in character (Fig. 2). The loose cluster arrangement noted with murine bone marrow [3] is only occasionally seen and usually in the very late stages of culture. The cellular development within colonies is similar to that previously described for murine granulocytic colonies.[3] The sequence of appearance of various forms is, however, spread over a longer period of time. At day 7 of incubation colonies are composed almost exclusively of large mononuclear cells, mainly myelocytes. By day 12 the predominant cells are metamyelocytes and bands (Fig. 3) with a few large mononuclear cells and occasional mitotic figures. By day 15–20 colonies are almost exclusively mature granulocyte forms (Fig. 4). Culture beyond this point usually results in cellular death and degeneration. The large mononuclear cells appearing in murine bone marrow colonies at this point [3] are only occasionally seen. Likewise the mononuclear cell colonies seen in murine systems from initiation of culture can only rarely be found. Granulocytic colonies comprise 95 per cent or more of those grown from normal human bone marrow.

All marrows in the present study were also plated on concentrated normal human urine and on control plates without a stimulus in the underlay. Occasional colony growth was noted on these plates, but the colonies were very small (25–100 cells) when compared with those grown on peripheral WBC.

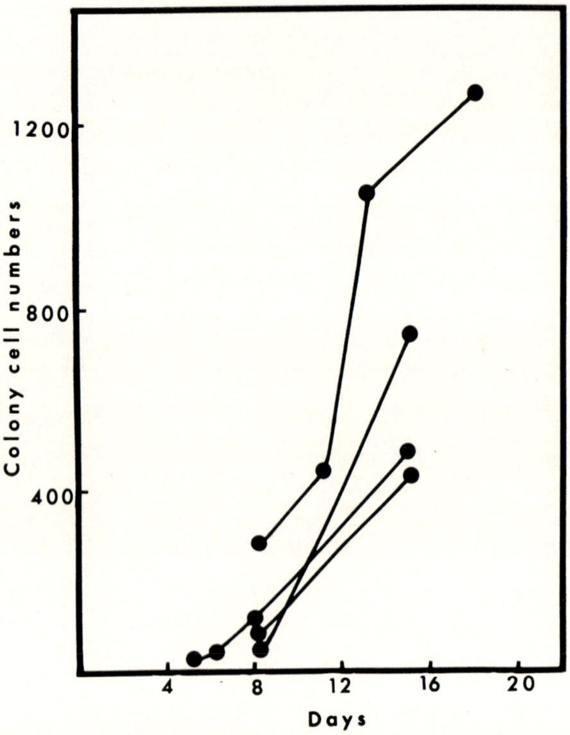

FIG. 1—Growth rate of colonies from normal human marrow grown on WBC underlays. Each line represents an individual marrow at varying times. Each point represents the mean of eight plates. Studies of colony size beyond the time shown were not done.

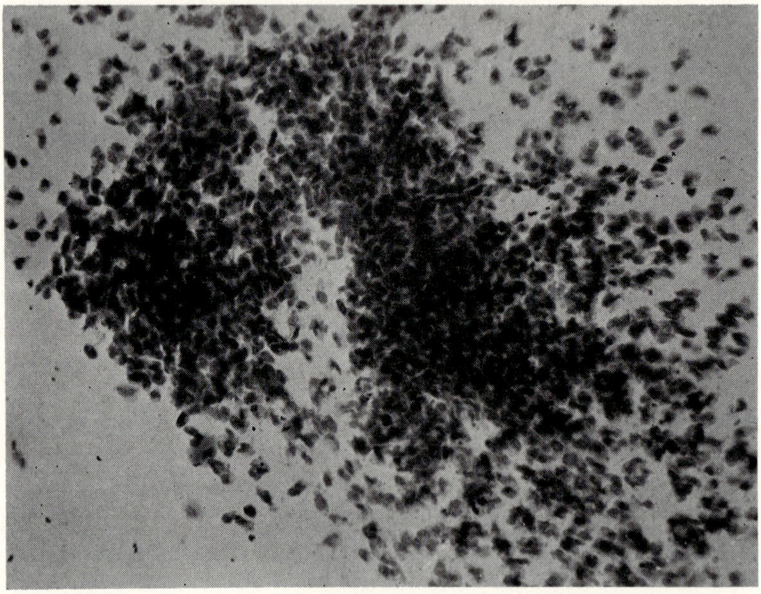

FIG. 2—Typical compact colony from normal human bone marrow (NB.) at day 10 of incubation. The apparent cleft is artefact resulting from removal of the colony from the culture plate for photography. (Aceto-orcein × 150.)

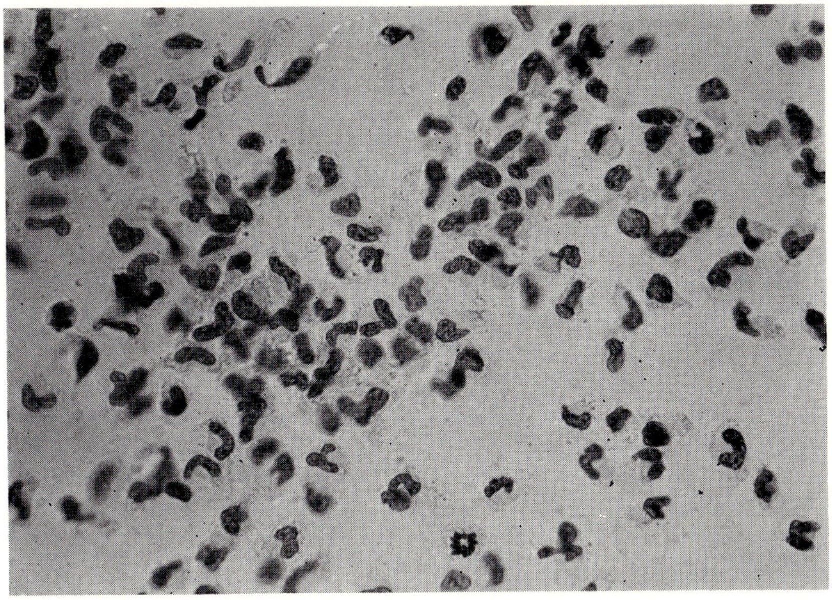

FIG. 3—Portion of a colony from normal human bone marrow (NB) at day 12 of incubation. Note predominance of metamyelocyte and band forms with an occasional mononuclear cell and a single mitotic figure. (Aceto-orcein × 400.)

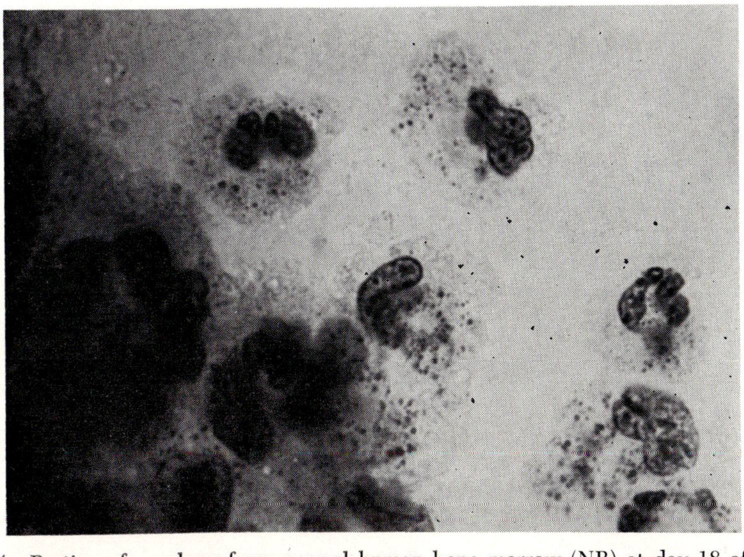

FIG. 4—Portion of a colony from normal human bone marrow (NB) at day 18 of culture showing typical lobulated granulocyte forms comprising colony. (Aceto-orcein × 1200.)

In general, those bone marrows which showed the greatest growth on WBC feeder layers showed the best growth on control plates. We have speculated that this might be due to contamination of the bone marrow specimen with peripheral blood cells leading to auto stimulation. Growth on plates with human urine was only slightly better than with no stimulus.

The number of colonies grown from human bone marrow, both normal and from various disease states, is shown in Table 2. Normal human marrow yielded 32–105 colonies with a mean of 74. This is about 25 per cent of the yield of colony-forming cells obtained from murine bone marrow under optimum conditions.[3] It must be remembered, however, that the bone marrow collection process is much different and results in dilution of the specimen with nucleated peripheral blood cells which are included when counting prior to plating. For this same reason no comparison can be made between the colony counts of individual bone marrows in the present study. In order to make such comparisons valid, differential counts must be performed on the plated specimens and end-stage cells eliminated from the counts. This has been done and will be the subject of a latter communication. In this study the colony counts from various human disease states varied widely. For the reasons cited above, no definite statements can be made concerning this finding. Of greatest interest in this group, however, is the finding that marrow from most patients with acute granulocytic leukemia grew extremely well, often covering the plate with a solid sheet of cells. We have also found that peripheral blood cells from these patients grow well in this system although the colonies are smaller than with normal human marrow cells. Both bone marrow and peripheral blood cells from patients with acute granulocytic leukemia show evidence of maturation during the latter stages of culture. This finding will also be the subject of a later communication.

It has not been determined whether or not the peripheral white blood cells used in feeder layers are unique in this regard. It would be surprising if they were in light of the fact that murine bone marrow can be stimulated to grow on a wide variety of cell sources.[4] It has been shown, however, that it is the WBC themselves and not the plasma that are the essential component of the present system (Table 3). Attempts to make conditioned medium from WBC have, however, been thus far unsuccessful. It has also been found that feeder layers of immature peripheral blood cells from patients with acute granulocytic leukemia are not capable of stimulating colony development (Table 4). Further studies of the feeder layer potential of other human cell sources is underway.

There is no evidence that the stimulation of colony growth is analagous with mixed leukocyte cultures. Although leukocyte typing of WBC and bone marrow donors has not been done, it has been shown that bone marrow grows equally well on WBC feeder layers from autologous or unrelated homologous donors. Heterologous WBC from dogs, mice, and sheep will also stimulate human bone marrow colony growth, but the colonies are smaller and fewer in number. It would also appear that intact WBC are not necessary in feeder layers to obtain colony growth. Feeder layer plates may be stored at 37 C. for

TABLE 2—*Colony Formation by Human Bone Marrow*

Patient	Diagnosis: Clinical	Diagnosis: Pathologic	Number of Colonies
W.R.	Normal	Normal	51 ± 11
B.P.	Normal	Normal	32 ± 9
J.K.	Normal	Normal	88 ± 10
H.S.	Normal	Normal	102 ± 17
N.B.	Normal	Normal	52 ± 7
A.J.	Psoriasis	Normal	105 ± 18
A.D.	Reticulum cell sarcoma	Normal	83 ± 10
W.M.	Lymphosarcoma	Normal	49 ± 8
A.J.	Cirrhosis	Normal	105 ± 18
L.B.	Hemolytic anemia	Lymphoid metaplasia	14 ± 9
W.D.	Blood loss anemia	Erythroid hyperplasia	75 ± 5
E.H.	Anemia, cause undetermined	Erythroid hyperplasia	75 ± 13
L.F.	Refractory anemia	Megaoloblastic changes, erythroid and granulocytic	0
M.F.	Pancytopenia cause undetermined	Panhyperplasia	20 ± 4
R.E.	Pancytopenia cause undetermined	Panhypoplasia	15 ± 4
H.L.	Secondary polycythemia	Erythroid hyperplasia	132 ± 23
M.M.	Secondary polycythemia	Erythroid hyperplasia	51 ± 8
M.S.	Anemia, cause undetermined	Erythroid hypoplasia	43 ± 5
E.B.	Carcinoma, prostate	Granulocytic hyperplasia	34 ± 4
B.F.	Hodgkin's disease	Granulocytic hyperplasia	45 ± 3
A.L.	Hodgkin's disease	Granulocytic hyperplasia	20 ± 4
E.B.	Chronic granulocytic leukemia	Granulocytic hyperplasia	45 ± 4
J.S.	Chronic lymphatic leukemia	Lymphoid hyperplasia	18 ± 4
R.L.	Subacute granulocytic leukemia	Maturation arrest granulocytic	4 ± 1
M.N.	Acute granulocytic leukemia	Acute granulocytic leukemia	0
A.C.	Acute granulocytic leukemia	Acute granulocytic leukemia	64 ± 20

TABLE 2 (*Continued*)

Patient	Diagnosis: Clinical	Diagnosis: Pathologic	Number of Colonies
A.H.	Acute granulocytic leukemia	Acute granulocytic leukemia	Too numerous to count
D.K.	Acute granulocytic leukemia	Acute granulocytic leukemia	Too numerous to count
P.D.	Acute granulocytic leukemia	Acute granulocytic leukemia	Too numerous to count
R.B.	Acute granulocytic leukemia	Acute granulocytic leukemia	Too numerous to count
B.B.	Acute lymphatic leukemia	Acute lymphatic leukemia	0
C.S.	Acute stem cell leukemia	Acute stem cell leukemia	0
J.St.	Pancytopenia	? Early granulocytic leukemia	115 ± 20

TABLE 3—*Colony Stimulating Activity of Peripheral WBC Feeder Layers with and without Plasma*

Underlay (Feeder Layer)	Number of Colonies/2 × 10⁵ Normal Human Bone Marrow
* WBC alone	38 ± 6
* WBC in plasma	45 ± 3
† Plasma alone	4 ± 2

* 1×10^6 nucleated cells/plate.
† Amount of plasma containing 1×10^6 WBC/plate.

TABLE 4—*Colony Formation by Human Bone Marrow on Normal and Leukemic WBC Underlays*

Bone Marrow Overlay	Number of Colonies: Normal WBC	Number of Colonies: Leukemic WBC *
M.H.	22 ± 4	0
F.E.	75 ± 11	1 ± 1
B.F.	45 ± 9	3 ± 1
R.P.	46 ± 9	10 ± 3

Underlays—1×10^6 normal or leukemic cells.
* Leukemic cells obtained from 2 patients with acute granulocytic leukemia, washed free of plasma prior to use.

at least 14 days without loss of colony stimulating activity. Examination of the cells remaining at 14 days shows only ghosts and debris, suggesting that the factors necessary for colony growth have been liberated into the medium. This finding has not been explained in the light of our inability to prepare conditioned medium from WBC grown in liquid culture.

Discussion

The major purpose of the present work has been to detail a method of obtaining human bone marrow granulocyte colony growth in vitro. The system we have described accomplishes this aim if murine systems are used as the present standard. The colony growth obtained in the present studies is similar in many respects to that obtained using mouse bone marrow. There are however distinct differences between the two. The growth rate of human bone marrow colonies is considerably slower than that from the mouse and plates must be incubated for longer periods. The colonies obtained are almost exclusively granulocytic with only rare mononuclear colonies. The nutrient and stimulatory requirements of human bone marrow are apparently quite different from those of murine bone marrow.

The use of McCoy's 5A medium in the present studies is an outgrowth of our inability to grow human marrow in Eagle's medium, so effective in supporting murine colony growth. Numerous previous attempts to use the latter to grow human marrow were totally unsuccessful. The components added to McCoy's 5A medium noted in Table 1 are largely empirical. We have not as yet determined the absolute necessity of each although we have determined that basic McCoy's 5A medium is much less effective in supporting human bone marrow colony growth than the formula noted. Systematic study of the nutrient requirements in the present system is under way. The use of a single strength medium also bears comment. This is merely a result of convenience since it is available locally in this form. Technically it is much easier to use a double strength medium so that the concentration of agar can be markedly reduced for ease of handling. We, therefore, do not advocate the use of a single strength medium but use it ourselves as a matter of convenience.

Likewise it has not been demonstrated that the peripheral WBC used in feeder layers are unique in this regard. It may be that other human cell sources may serve this function equally well. Initial attempts in this laboratory to use adult human kidney, liver, and skeletal muscle as feeder layers have, however, been unsuccessful. It is interesting to speculate that there might be a hitherto unrecognized positive feedback between mature dying granulocytes and renewed production. Whatever the nature of the stimulatory action of WBC it does not appear to depend on intact living cells since 14-day-old plates containing only cellular debris and ghosts are as effective in stimulating colony growth as those freshly prepared. Stimulation of colony growth must, therefore, be dependent upon factors released into the medium. Attempts to define these factors have not been successful. We have tried on numerous occasions to obtain colony growth from liquid medium in which peripheral WBC have been grown for 1 to 14 days without success. The inability to obtain growth in this manner may result from technical problems, namely damage of the target bone marrow cells by degradative enzymes also present in the conditioned medium. This explanation fails, however, to explain why a similar phenomenon does not take place in agar. At present we can only say that the uniqueness of peripheral WBC in this system lies in their ready availability and the ease with which they can be prepared for routine use.

The stimulatory action of peripheral WBC would appear to be dependent upon the number of granulocytes present. As noted, peripheral blood cells from patients with acute granulocytic leukemia have little colony stimulating activity. Peripheral blood from patients with lymphocytosis and granulocytopenia also lack colony stimulating activity.

We should like to emphasize again that in the present studies no comparison can be made between the colony forming ability of marrows from various human disease states. Dilution of specimens with peripheral blood renders such comparisons invalid unless careful differential counts are done. We have attempted here to study a broad spectrum of human disease states to test the system. Studies of the nature of the colony-forming cell and numbers in various disease states are now under way and will be the subject of later communications. One point, however, bears comment and that is the finding of large numbers of colonies growing from marrows of most patients with acute granulocytic leukemia. This is in contrast to one previous report in which no growth was obtained from similar patients when their cells were grown on mouse kidney tubule feeder layers.[9] The reasons for this are not clear but probably reflect differences in the systems used. It should also be noted that all patients in the acute leukemia group in the present studies were untreated. This system should prove particularly valuable for those studying the proliferative capacity and chemotherapeutic response of leukemic cells, both from the bone marrow and peripheral blood.

The technic described here is simple and reliable. It should provide an important method for studying human bone marrow growth potential in normal persons and various disease states, and for the study of factors influencing and regulating white blood cell production.

Summary

A system for growing granulocytic colonies from human bone marrow aspirates in agar-gel medium is described. 2×10^5 nucleated bone marrow cells are plated on feeder layers of 1×10^6 normal human peripheral WBC. Colonies are visible on plates at day 4–5, reaching maximum size (500–1500 cells) at day 15–20. Normal human bone marrow yields colonies which are almost exclusively granulocytic.

REFERENCES

1. Bradley, T. R., and Metcalf, D.: The growth of mouse bone marrow cells in vitro. Aust. J. Exp. Biol. Med. Sci. 44:287, 1966.

2. Pluznik, D. H., and Sachs, L.: The induction of clones of normal mast cells by a substance from conditioned medium. Exp. Cell. Res. 43:553, 1966.

3. Metcalf, D., Bradley, T. R., and Robinson, W. A.: Analysis of colonies developing in vitro from mouse bone marrow cells

stimulated by kidney "feeder layers" and leukemic serum. J. Cell. Physiol. 69:93, 1967.

4. Bradley, T. R.: Aspects of bone marrow colony growth in vitro. Aust. J. Exp. Biol. Med. Sci. 46:335, 1968.

5. Bradley, T. R., and Sumner, M. A.: Stimulation of bone marrow colony growth in vitro by conditioned medium. Aust. J. Exp. Biol. Med. Sci. 46:607, 1968.

6. Robinson, W. A., Metcalf, D., and Bradley, T. R.: Stimulation by normal and leukemic mouse sera of colony formation in vitro by mouse bone marrow cells. J. Cell. Physiol. 69:83, 1967.

7. Foster, R., Metcalf, D., Robinson, W. A., and Bradley, T. R.: Bone marrow colony stimulating activity in human sera. Brit. J. Haemat. 15:147, 1968.

8. Robinson, W. A., Stanley, E. R., and Metcalf, D.: Stimulation of bone marrow colony growth in vitro by human urine. Blood 33:396, 1969.

9. Senn, J. S., McCulloch, E. A., and Till, J. E.: Comparison of colony forming ability of normal and leukemic human marrow in cell culture. Lancet 2:597, 1967.

Prospects for New Directions in Therapy of Acute Lymphoblastic Leukemia

ALVIN M. MAUER, BEATRICE C. LAMPKIN, AND TAKESHI NAGAO

DURING THE PAST SEVERAL YEARS information concerning the nature of the leukemic cell population in acute leukemia has been growing steadily. Of particular interest as regards treatment of acute leukemia have been studies of the proliferative characteristics of these cells. Most chemotherapeutic agents available today have their maximal effect during some portion of the mitotic cycle, most often in the DNA synthesis period. Appropriate use of these drugs does depend to some degree, therefore, on a knowledge of the proliferative characteristics of the cell population under treatment.

Furthermore, some features of the leukemic cell population in acute lymphoblastic leukemia have been uncovered which explain the response characteristics of this disease to treatment. These features also provide some basis for planning new approaches to therapy with currently available drugs with the hope of providing a better therapeutic index. The goal to be achieved, of course, is complete eradication of the abnormal cell population with sparing of normal cells.

Also implicit in our present knowledge of the kinetics of cell population in acute lymphoblastic leukemia is the need to learn more about control mechanisms for population growth of these cells. Indeed some control of leukemic cell growth does appear to be operative in this disease and further knowledge of these mechanisms might well lead to new avenues of therapy.

The purpose of this paper will be to briefly outline current concepts concerning leukemic cell kinetics in acute lymphoblastic leukemia. Then some speculations concerning possible therapeutic approaches based on these concepts will be presented. Some of these approaches are immediately applicable for testing or are being tested currently. Others will require more information before their feasibility is known. However, all are being presented with the aim of stimulating the development of new and potentially testable hypotheses about the therapy of acute lymphoblastic leukemia.

CURRENT CONCEPTS OF CELL KINETICS IN ACUTE LYMPHOBLASTIC LEUKEMIA

Recent reviews of the available studies of cell kinetics in acute leukemia, the concepts derived therefrom, and some over-all consideration of therapeutic approaches are available.[1,2,3] From early concepts of a uniformly rapidly dividing population of cells with exponential growth characteristics, a concept of a more complex kinetic situation has arisen. The population is comprised

---

From the Department of Pediatrics, University of Cincinnati, The Children's Hospital, and The Children's Hospital Research Foundation, Cincinnati, Ohio.

This investigation was supported by Public Health Service research grant CA–04826.

Dr. Lampkin is a recipient of a Special Fellowship Award from the National Institutes of Health.

of some cells in an active proliferative phase and other cells in a resting phase. After one or more mitotic divisions, cells in the active proliferative phase, characterized by large cell size and fine nuclear chromatin pattern on marrow smears, enter a resting phase and are now smaller with a denser nuclear chromatin pattern.[4,5,6] The proportion of cells in proliferating and resting states is not a constant one, but varies from patient to patient and in the same patient during different phases of his disease.[7] It is not known whether or not cells in the resting state are truly in a G_0 phase or merely an extended G_1 phase. In fact the exact biochemical definition of these two resting phases is an inexact one at present.

Of critical importance for considerations of therapy and the pathogenesis of this disease is the proliferative potential of the resting cell. Gavosto and co-workers [8] have demonstrated that the large proliferating cell population is not self-maintaining and must be continuously replaced by new cells entering the compartment to prevent depletion. These replacement cells have been demonstrated to come in large part, if not exclusively, from the cells in a resting phase recycling through a phase of proliferative activity.[9] These studies,[9] done under conditions of a steady state, have been recently confirmed by Gabutti and co-workers [10] for a population in which the effect of perturbation by chemotherapy was studied. Under these conditions, small resting cells were also seen to reenter a proliferative phase.

The significance of these findings for therapeutic considerations would seem to be twofold. Firstly, because of their proliferative potential, resting cells must be eliminated along with actively dividing cells to achieve cure of the disease. Secondly, because there is no need to evoke continued leukemic transformation of normal cells to explain the self-sustaining growth characteristics of leukemic cells, a more hopeful outlook for cure exists upon eradication of the leukemic cell population from the patient.

Studies of generation time for the dividing leukemic cells have not proved to be helpful as yet for considerations of therapy. The generation time is longer for most leukemic cell populations [1,7] than average normal marrow cell generation times or may in a few patients approach normal values.[6] The usual DNA synthesis time is about 20 hours, post-mitotic rest phase an hour or two, and the time for mitosis 2 to 4 hours. The shortest total generation time is usually 60 to 70 hours. Variability in the G_1 phase is evident, however, and longer times may be found in the same population of cells. The distinction between a longer G_1 phase and a true G_0 state, as mentioned before, has not been resolved. It is quite evident, however, because of the mixed nature of the leukemic cell population that generation time and doubling time for population growth are not equivalent. Virtually nothing is known of cell death rates under steady state conditions.

Application of Cell Kinetic Patterns to Therapy

Of immediate importance is the recognition that a population of cells is present, relatively resistant to current chemotherapy but capable of returning to a proliferative phase to reinstitute disease activity. These resting cells are potentially capable of surviving induction therapy, persisting into remission, developing drug resistance, and reentering proliferative activity to institute disease relapse. In one patient [9] about two-thirds of the total cell population went through a proliferative, drug sensitive phase during a 9 day period. This recycling of cells through a drug sensitive phase explains the success of therapy directed solely at dividing cells. However, the resting cells did not appear to be homogeneous, and the smallest cells appeared to have the longest resting times; it may be these cells which escape drug effect.

Three general approaches to increasing therapeutic effectiveness seem possible. Firstly, the leukemic cell population may be capable of manipulation in order to increase the number of cells exposed to drugs during a drug sensitive phase. Secondly, if more could be learned of proliferating and resting cells, a satisfactory therapeutic result might be achieved by shifting the balance in favor of the resting state. Finally, as more is learned concerning the relationship of host and tumor with regard to immunologic tolerance of the host for the tumor cell population, combinations of chemotherapy and immunotherapy might be possible.

Cell Population Manipulation

The aims of such programs would be to give the available chemotherapeutic agents at a time of maximal sensitivity of the cell population. Three potential programs are possible now.

Diurnal variation. A diurnal variation of proliferative activity is present in normal human marrow cell populations [11] and it has recently been proposed that chemotherapy be planned in an effort to take advantage of what variation in proliferative activity might occur in tumor cell populations.[12] Unfortunately, the variation that occurs in proliferative activity of leukemic cell populations is insufficient to expect any therapeutic advantage. In fact, no significant or predictable changes in the percentage of cells in DNA synthesis could be found.[7]

Synchronization of mitotic activity. A therapeutic advantage certainly could be obtained by synchronizing a portion of the leukemic cells in DNA synthesis, a sensitive phase of the cell cycle. Synchronization of proliferative activity can be achieved by a number of methods for cell populations under in vitro culture conditions. Recently, successful attempts to achieve partial synchronization of leukemic cells in vivo have been reported.[13,14]

The technic used in those two studies has been to give a single intravenous injection of cytosine arabinoside in a dose of 5 mg. per Kg. DNA synthesis promptly stops or is markedly slowed. However, the effect is temporary and recovery of most of the cells occurs from 24 to 96 hours later, usually at 48 hours. During the time of drug effect, cells continue to enter the DNA synthesis phase from G_1, and are, likewise, caught by the cytosine arabinoside block. Therefore, when the drug effect is lifted, not only do those cells initially blocked by the drug resume DNA synthesis, but also all those cells entering into the phase during the subsequent period of time. Thus, all of the affected cells

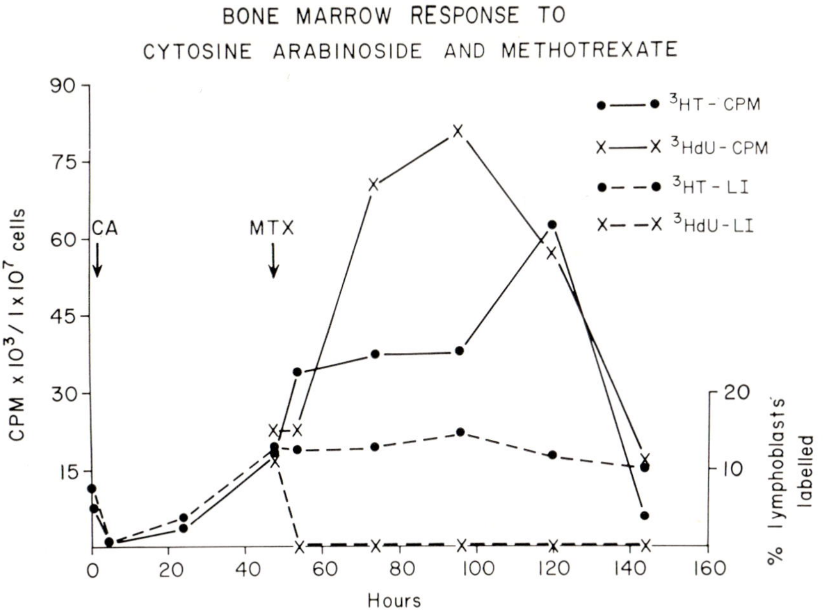

FIG. 1—Results of administration of cytosine arabinoside (CA) as a synchronizing agent followed by methotrexate (MTX) as a secondary agent. Cytosine arabinoside was given intravenously in a dose of 5 mg./Kg. body weight. Methotrexate was given intravenously in a dose of 1 mg./Kg. body weight. Serial bone marrow samples were taken as indicated. The percentage of cells labeled with tritiated thymidine (^{3}HT-LI) and tritiated deoxyuridine (^{3}HdU-LI) was determined by a one hour incubation with aspirated marrow samples followed by preparation of autoradiographs. The whole cell incorporation of tritiated thymidine (^{3}HT-CPM) and tritiated deoxyuridine (^{3}HdU-CPM) was measured by a one hour in vitro incubation with aspirated marrow samples followed by determination of cell radioactivity as counts per minute (CPM) per 10^7 cells.

now resume DNA synthesis synchronously, providing a greater number of sensitive target cells for administration of a chemotherapeutic agent.

Initial studies of synchronization were done with vincristine as the secondary chemotherapeutic agent.[14] Figure 1 shows the result of a study when methotrexate in a dose of 1 mg. per Kg. body weight was given as the secondary agent. After injection of cytosine arabinoside, no incorporation of either tritium labeled thymidine or deoxyuridine could be demonstrated by autoradiography at 4 hours. Some return of DNA synthesis was seen at 24 hours and a degree of synchronization occurred at 48 hours. Methotrexate was then given. Incorporation of deoxyuridine as measured by autoradiography was promptly blocked and this effect persisted during the subsequent period of observation. However, the percentage of cells still capable of incorporating thymidine did not significantly change, indicating that the drug had affected DNA synthesis but not altered cell viability to a significant degree. Of particular interest is the observation that incorporation rates of deoxyuridine and thymidine as measured by liquid scintillation counts of washed whole cell preparations (indicated by the solid lines) actually increased during the period of methotrexate effect.

Certainly synchronization of the cells in a sensitive phase of the cell cycle can be achieved. Further studies are needed to determine the proper use of this technic to achieve maximal effectiveness. There is indication from preliminary studies that normal marrow cells may recover more rapidly from the cytosine arabinoside effect. Thus, some added therapeutic advantage may be possible in treating a synchronized malignant cell population at a time when randomization of proliferative activity has already been achieved by the normal cell population.

Recruitment of cells to proliferative activity. Factors as yet poorly understood determine the balance between the proliferating and resting portions of the leukemic cell population. Recent studies by Chan and co-workers [15,16] have indicated that, during extracorporeal irradiation of blood in patients with acute leukemia, a surprising and prompt increase in proliferative activity of marrow leukemic cells occurred. Gabutti and her co-workers [10] found return of resting cells to proliferative activity after administration of methotrexate. Thus, perturbation of the leukemic cells to effect recruitment of resting cells to proliferative activity is possible. We also have observed an unaccountably large proportion of cells entering DNA synthesis after cytosine arabinoside synchronization, as if recruitment of resting cells may occur also in this circumstance in some patients.

Efforts at increasing cell kill from conventional chemotherapeutic agents by increasing the proportion of susceptible cells through recruitment seems indicated. Although Chan and his co-workers [16] reported failure in 2 patients in whom combined extracorporeal irradiation followed by chemotherapy at a time of heightened marrow proliferative activity failed to induce remission, both patients were refractory to chemotherapy already. Better results might be achieved in acute lymphoblastic leukemia where the cell population is sensitive to available agents. Attention could also be given to such responses as achieved by Gabutti and co-workers;[10] and by more appropriate timing of drug administration, greater numbers of susceptible cells might be affected. Clearly, further studies of the phenomenon are in order.

Suppression of Proliferative Activity

Too little is known concerning the control mechanisms operative in maintaining the balance between proliferating and resting cells to speculate meaningful about possible therapeutic approaches. The indications are that at least one determining factor might be the size or concentration of the leukemic cell population [7] as it appears to be in animal tumor models.[17–20] One deterant to the study of related mechanisms in acute lymphoblastic leukemia is the lack of a technic by which these cells can easily be grown outside of the patient. However, study of control mechanisms operative in leukemic cell growth might provide information useful in suppressing growth by shifting the balance to the resting state. Clearly, further studies are needed here too.

Host Resistance

It is beyond the scope of this presentation to summarize recent attempts to approach therapy of acute leukemia by means of potentiating host immune responses. The end result of this approach is to achieve "graft rejection" of tumor cells by the host. It is clear, however, that success of this approach depends on reduction of tumor mass as much as possible with preservation of an intact immune system. To achieve this end, reduction of tumor mass by means other than the massive chemotherapeutic assualt currently in use may allow better results through a more intact and responsive immune system.

Summary

From this presentation it is hoped that the potential for application of our current concepts of the pattern of cell proliferation in acute leukemia can be realized. Some of the proposed avenues for further therapeutic approachs mentioned can be tested now. Other potential avenues will require first filling in critical remaining gaps in our knowledge of these cells. Further studies will no doubt reveal other avenues as yet unanticipated from these studies.

REFERENCES

1. Killman, S.-A.: Acute leukemia: The kinetics of leukemic blast cells in man. An analytical review. Ser. Haemat. I:38, 1968.

2. Mauer, A. M., Saunders, E. F., and Lampkin, B. C.: Possible significance of nonproliferating leukemic cells. *In* Human Tumor Cell Kinetics, National Cancer Institute Monograph No. 30, pp. 63–78, 1969.

3. Henderson, E. S.: Treatment of acute leukemia. Seminars Hemat. 6:271, 1969.

4. Gavosto, F., Pileri, A., Bachi, C., and Pegoraro, L.: Proliferation and maturation defect in acute leukaemia cells. Nature (London) 203:92, 1964.

5. Killman, S. A.: Proliferative activity of blast cells in leukemia and myelofibrosis. Morphological differences between proliferating and nonproliferating blast cells. Acta Med. Scand. 178:263, 1965.

6. Mauer, A. M., and Fisher, V.: Characteristics of cell proliferation in four patients with untreated acute leukemia. Blood 28:428, 1966.

7. Saunders, E. F., Lampkin, B. C., and Mauer, A. M.: Variation of proliferative activity in leukemic cell populations of patients with acute leukemia. J. Clin. Invest. 46:1356, 1967.

8. Gavosto, F., Pileri, A., Gabutti, V., and Mascra, P.: Non-self-maintaining kinetics of proliferating blasts in human acute leukaemia. Nature (London) 216:188, 1967.

9. Saunders, E. F., and Mauer, A. M.: Reentry of nondividing leukemic cells into a proliferative phase in acute childhood leukemia. J. Clin. Invest. 48:1299, 1969.

10. Gabutti, V., Pileri, A., Tarocco, R. P., Gavosto, F., and Cooper, E. H.: Proliferative potential of out-of-cycle leukaemic cells. Nature (London) 224:375, 1969.

11. Mauer, A. M.: Diurnal variation of proliferative activity in the human bone marrow. Blood 26:1, 1965.

12. Cardoso, S. S., and Carter, J. R.: Circadian mitotic rhythm as a guide for the administration of antimetabolites. Proc. Soc. Exp. Biol. Med. 131:1403, 1969.

13. Lampkin, B. C., Nagao, T., and Mauer, A. M.: Drug effect in acute leukemia. J. Clin. Invest. 48:1124, 1969.

14. Lampkin, B. C., Nagao, T., and Mauer, A. M.: Synchronization of the mitotic cycle in acute leukaemia. Nature (London) 222:1274, 1969.

15. Chan, B. W. B., Hayhoe, F. G. J., and Bullimore, J. A.: Effect of extracorporeal irradiation of the blood on bone marrow activity in acute leukaemia. Nature (London) 221:972, 1969.

16. Chan, B. W. B., Bullimore, J. A., and Hayhoe, F. G. J.: Extracorporeal irradiation of the blood in acute myeloid leukaemia: A clinical and *in-vitro* culture study. Brit. J. Haemat. 17:309, 1969.

17. Mendelsohn, M. L.: Chronic infusion of tritiated thymidine into mice with tumors. Science 135:213, 1962.

18. Mendelsohn, M. L.: Autoradiographic analysis of cell proliferation in spontaneous breast cancer of the C3H mouse. III. The growth fraction. J. Nat. Cancer Inst. 28:1015, 1962.

19. Baserga, R.: Mitotic cycle of ascites tumor cells. Arch. Path. (Chicago) 75:156, 1963.

20. Frindel, E., Malaise, E. P., Alpen, E., and Turiana, M.: Kinetics of cell proliferation of an experimental tumor. Cancer Res. 27:1122, 1967.

DISCUSSION

CORNELIUS ROSSE: The lymphoid cell population of bone marrow is a more complex cell system, at least in the guinea pig, than it would appear from Dr. Mauer's concept. The nonproliferating pool of marrow lymphocytes corresponds to the pachychromatic small lymphocytes and the proliferating pool to the larger leptochromatic cells known as transitional cells. Transitional cells themselves consist of at least two populations—one with a short generation time and another with a much more prolonged generation time. These two populations are morphologically distinguishable and correspond to the basophilic and pale transitional cells respectively as described earlier by Rosse and Yoffey. It was shown that after flash label with ^{3}H-thymidine more than 60 per cent of the basophilic cells incorporated label in contrast to about 15 per cent of the pale cells. Thus, when considering input into the proliferating pool, one must take into account not only the small lymphocytes but also the slowly proliferating portion of transitional cells. We, in fact, presented such a model (Moffatt, Rosse, and Yoffey).

There is as yet no direct evidence that during the process of normal hemopoiesis the small lymphocyte of the bone marrow can enlarge. Osmond showed that this does occur in vitro in response to PHA and pokeweed mitogen. In vivo enlargement of recirculating lymphocytes has been demonstrated by Gowans and associates under immunologic challenge.

I would like to add that, being a student of Yoffey, it gives me great gratification to see at this conference such a wide acceptance of his views and hypotheses formulated many years ago. I have to point out, however, that all the evidence for the stem cell function of marrow lymphocytes presented at this conference is indirect and this conclusion is reached by inference only. Direct evidence on the stem cell function of lymphocytes is still to be awaited.

REFERENCES

C. Rosse, and J. M. Yoffey: The morphology of the transitional lymphocyte in guinea-pig bone marrow. J. Anat. 102:113, 1967.

D. J. Moffatt, C. Rosse, and J. M. Yoffey: Identity of the haemopoietic stem cell. Lancet 547, 1967.

D. G. Osmond: Culture of lymphocyte-rich fractions of bone marrow with phytohemagglutinin and pokeweed mitogen. Anat. Rec. 160:403, 1968.

G. L. Gowans: The fate of parental strain small lymphocytes in F_1 hybrid rats. Ann. N. Y. Acad. Sci. 99:432, 1962.

ALVIN M. MAUER: It would appear that the cell population in normal marrow known by the collective name of marrow lymphocytes is indeed a complex one. The population of marrow leukemic lymphoblasts in acute lymphoblastic leukemia, on the other hand, seems to be some what more homogeneous, at least as to its derivation and functional aspects. The model developed from the available data for the marrow leukemic lymphoblast is the simplest one which fits and is advanced for the puposes of developing working hypotheses. Further studies are in order to test the validity of this model and add to the concepts implicit therein.

A Hypothesis Concerning the Relationship Between Normal and Leukemic Hemopoiesis in Acute Myeloid Leukemia

By Sven-Aage Killmann

Clinically, acute leukemia is characterized by an excess of cells of the blastic variety and a lack of functionally useful blood cells. In this paper some thoughts on the relationship between these two aspects of acute leukemia are presented.

Acute myeloid leukemia (AML) appears to be a disease of the common hemopoietic stem cell [1] as will also be discussed later. There are neither similar clues as to the cell of origin of acute lymphatic leukemia (ALL), nor clarity as to whether or not this disease starts in bone marrow lymphoid cells or in some extramedullary lymphatic tissue. AML and ALL differ in several respects: cytology,[2] age distribution,[3] response to therapy at the clinical [4] and cellular [5,6] level, ploidy of abnormal stem lines,[7,8] and absence of preleukemic conditions in ALL. In view of these differences, it is emphasized that the present considerations apply only to AML.

In theory, the deficit of normal hemopoiesis in AML could be explained by any one of the following mechanisms:

1. Only leukemic stem cells remain, whereas all normal stem cells have been destroyed.
2. Normal and leukemic stem cells coexist.
    - A. Normal stem cells are present in adequate numbers but,
        - (a) they are blocked from feeding into differentiating hemopoiesis, or
        - (b) they feed into differentiating hemopoiesis but the resulting cells die.
    - B. Normal stem cells are present in inadequate numbers.

The first possibility would appear to be a very sombre one, since the very obvious implication is that the disease could only be remedied by transplantation or biochemical repair. Superficially, it seems that this possibility can easily be dismissed for the very simple reason that, even in advanced AML, a few normally appearing neutrophil and erythroid precursors can be seen in the marrow, and some functionally useful cells are still produced. However, at this junction, the question must be asked how normal these cells really are. On closer scrutiny, many minor aberrations can be found in the apparently normal cells of untreated AML: varying degrees of megaloblastoid changes in the erythroid precursors, nucleocytoplasmic dissociation in neutrophil precursors, and binucleate or hypersegmented neutrophils. Atypical eosinophils are frequently encountered which have their granules accumulated in groups, whereas other cytoplasmic regions are completely free of granules.[1] Similarly, the dimi-

From the Division of Hematology, Department of Medicine A, Rigshospitalet, University Hospital of Copenhagen, Copenhagen, Denmark.

nution of red cell H-substance and blood group changes in AML and in preleukemic disorders [1,9] indicate disturbed erythroid differentiation. Such changes could be interpreted in two ways: (1) they represent damage to normal cell lines from the leukemic environment; (2) the cells with such changes are part of the leukemic cell line but have, nevertheless, differentiated almost normally. This view of course involves that leukemicness—in terms of differentiation block—is not an all-or-none phenomenon but a matter of gradation. Several observations favor the second interpretation. Brecher has noted Auer-bodies in segmented granulocytes in AML (9a). In chromosome studies of AML with abnormal stem lines we found in our laboratory that the percentage of erythroid mitoses often clearly exceeds the percentage of normal karyotypes in the bone marrow.[10,11] As in chronic myeloid leukemia (CML),[12–15] this is good circumstantial evidence that chromosome abnormalities are also present in erythoid precursors, i.e., that a fraction of the leukemic cells differentiate into erythropoiesis. The results thus strongly suggest that AML is not a strict white cell disorder but rather a process that concomitantly involves neutropoiesis, erythropoiesis, and—by extrapolation—thrombocytopoiesis, probably at the level of a common stem cell. This would put AML into the same category as CML, i.e., a neoplasia of a multipotent hemopoietic stem cell. That the defect of AML and CML reside in the same cell is consistent with the very high incidence of AML ("blastic crisis") in CML. The apparent capability of some AML cells to differentiate complicates interpretations: it is not possible to be sure that a normal-looking cell really is normal; it could as well be a basically leukemic cell which after all has succeeded in differentiating. Possibly, this does not occur at random but rather is a feature of some subclones of the leukemic cells, which one could think of as "leukemia minor" strains. Support for this concept comes from some observations by Müller [16] who studied 2 patients with blastic crisis of CML by ultramicrospectrophotometry. In one case, bone marrow myeloblasts were hypodiploid; in the other, they were hyperdiploid. However, in both cases the granulocytes in the blood were strictly diploid. This suggests that only the remaining diploid strains were able to differentiate whereas the aneuploid strains were so severely disturbed that they could not differentiate beyond the blast level. It appears that we are facing the ugly fact that we have no assurance that there are any normal hemopoietic stem cells left in AML. Some relief is, of course, obtained from the fact that remissions are seen in AML. Still, this does not represent a formal proof since hemopoiesis in remission might just reflect a temporary take-over of leukemia minor strains instead of the "leukemia major" strains which, as far as differentiation is concerned, are in large measure arrested at the blast stage.

After these very dire aspects, some comfort may be obtained if one considers the karyotype pattern during relapse and remission.[16–25] In relapse, the karyotypes are characterized by aneuploidy and abnormal stem lines in about half of the cases and by blurred chromosome morphology whereas during remission ploidy and chromosome structure are perfectly normal. During subsequent relapse, the original pattern reappears. The normal karyotype pattern during remission strongly suggests but does not prove that, at this time, normal stem cells are operative. The reason that the normal karyotypes do not prove normality is that approximately 50 per cent of AL in relapse have normal karyotypes.[7,8] However, an argument for normality during remission is that at this time, chromosome morphology is clearly defined rather than blurred. Still, it may be disputed whether or not a subjective criterion as distinctiveness versus blurredness is sufficient to differentiate between normality and leukemia. Nevertheless, taking all into consideration, the most likely possibility at the present time is that during remission hemopoiesis is really maintained by normal stem cells. At the same time it appears rather likely, however, that the total number of effective normal stem cells in AML is reduced or inhibited in some way; as a clinician, at least, one is impressed with the fact that for what appears to be the same degree of leukemic cell killing (marrow hypoplasia), remissions occurs much less readily in AML than in ALL.

In summary of the above, it will be assumed that in AML normal stem cells still exist, albeit perhaps in reduced numbers. The problem then is why and how normal hemopoiesis is suppressed in AML. Competition between cell lines is not a likely reason, considering that marrow compensation for hemolysis does not result in granulocytopenia; moreover, it seems that neither relative nor absolute cell production in AML exceeds normal granulocyte production.[26] A high granulocytic chalone concentration as has been reported in rat chloroleukemia [27] might explain neutropenia but not anemia and thrombocytopenia, and presumably it would quickly be reduced following reduction of the tumor cell mass. It appears that one has to look for some more specific leukemic-cell to normal-cell interaction. In principle, this might take place at the level of differentiating cells, with leukemic cells interfering with normal cell development and proliferation leading to the death of the normal cells, or at the level of stem cells where leukemic cells would inhibit normal stem cells from becoming differentiated cells. Interaction between leukemic and differentiating hemopoietic cells cannot be dismissed but will not be considered here. Instead, the possibility will be examined that the suppression of normal hemopoiesis in AML could be due to a competition between leukemic and normal stem cells.

A salient point in this context is whether or not leukemia is unicellular in origin. In CML the evidence for this is strong, based on the very high incidence of the Ph^1 chromosome [13,28,29] and the enzyme-genetic studies of Fialkow et al.[30,31] Pertinent data bearing directly on AML are not available. However, AML and the blastic phase of CML are very similar not only clinically and cytologically but also with respect to their non-Ph^1 chromosome abnormalities. One may actually consider CML as being the most clearly characterized preleukemic (with respect to AML) condition and the one which carries the highest risk of developing into AML. This point of view is also supported by an interesting case observed by Zuelzer's group.[32] Because of the many links between AML and CML, the most likely prediction is that AML also arises in a single cell. Perhaps the most general formulation is that myeloid leukemia—"myeloid" referring to marrow and not to the neutrophil series—arises in a single common hemopoietic stem cell. If the progeny of this cell retains the capacity of differentiation for some time, conditions like the noncharacteristic preleukemic cytopenias, Ph^1-positive or Ph^1-negative CML, subacute myeloid leukemia, or erythroleukemia develop; gradually, differentiation potential is lost, and the condition sooner or later terminates in AML. If the progeny of the altered stem cell has no or very little differentiating capacity, the well known sudden onset type of AML ensues.

The problem now is: how is a single cell capable of gaining such complete dominance and suppressing normal hemopoiesis? Usually, it is postulated that the leukemic cells have a growth advantage. Although this is a possibility it may still be worthwhile to consider other mechanisms. One such alternative hypothesis will be briefly discussed here.

Figure 1 shows two possible stem cell schemes. In 1A, all stem cells divide and have an equal chance to remain stem cells or to become differentiating cells. In scheme 1B, one stem cell has become dormant in generation 2; it is assumed that it can resume proliferation if appropriately stimulated. The sister cell continues to proliferate and maintains the cell line. It will be seen that, for an equal production of differentiating cells, a scheme like 1B has the advantage over scheme 1A of retaining a number of stem cells at a short

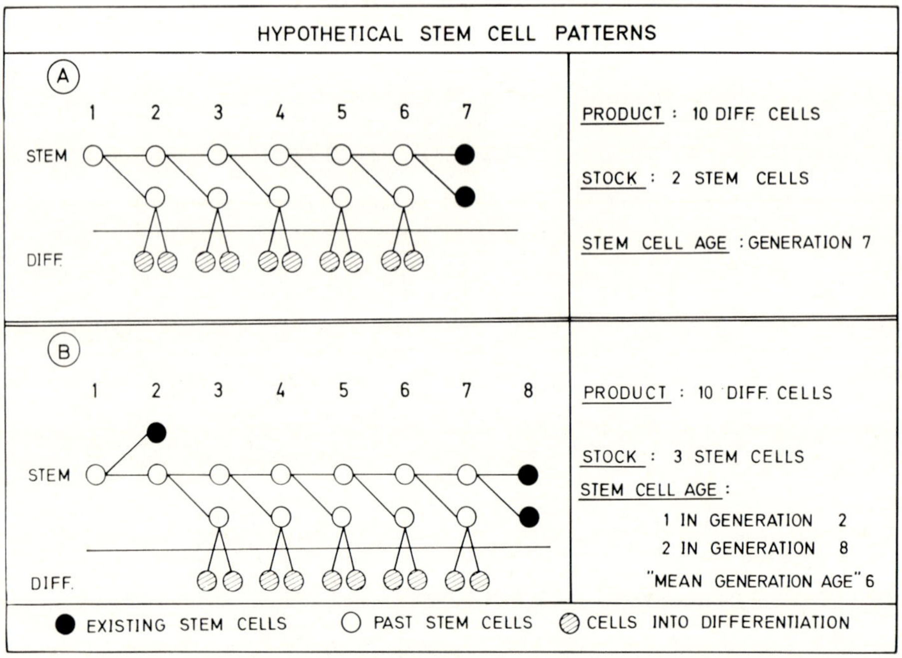

FIG. 1—Two possible stem cell schemes. In A, all stem cells continue to divide. In B, a fraction of the stem cells enters a dormant state at an early time. For an equivalent production of cells destined for differentiation, B has the advantage over A that some stem cells remain at a short lineage from the fertilized ovum and hence have a good preservation of the genetic blueprint. Open circles symbolize past development of cells, filled in circles are existing stem cells, and hatched circles represent cells which have entered the pathway toward differentiating cells. Figures indicate generation number.

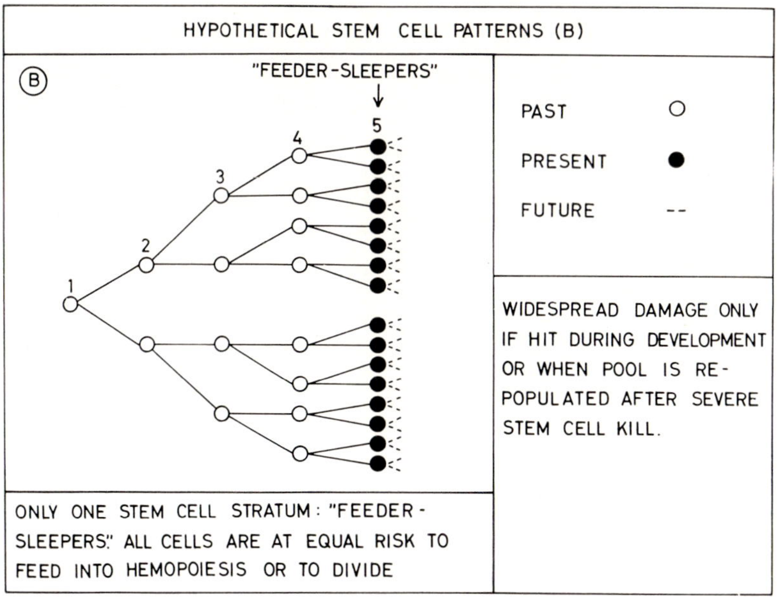

FIG. 2—Hypothetic stem cell pool. All cells are equal—they are at equal distance from the very first stem cell and have equal chances of further destiny: entry into differentiating hemopoiesis or replication in stem cell pool. Past stages in the development of the population are indicated by open circles, cells in present existence by filled in circles. Dotted lines symbolize capacity of division. For details, see text.

lineage from the fertilized ovum which favors the conservation of the genetic information.

Figure 2 is another representation of the stem cell proliferation pattern depicted in Figure 1A. The pool of existing stem cells are at equal distance from their origin and have equal chance of further destiny, i.e., replication in stem cell pool or going down the pathway that ultimately leads to differentiating hemopoiesis. These cells are referred to as "feeder-sleepers" since they are assumed to feed hemopoiesis but to turn over slowly.[1] It is seen that alteration in a single cell of this pool will be of infinitesimal importance provided the event takes place when the pool is fully developed, barring the possibility of growth advantage. However, single cell damage can spread in the pool and become manifest in a significant fraction of the pool and the cell lines derived from it if the single cell change takes place during development of the pool or is present during repopulation of the pool after severe stem cell loss.

Figure 3 depicts a stem cell pool such as may emerge from scheme 1B. A pattern like this has also been proposed by Kay.[33] Some cells have entered a dormant state at a certain distance from the fertilized ovum; an important feature is that these "sleepers" are assumed to be blocked from going directly into the pathway which leads to differentiating hemopoiesis. Genetically, it would be quite wasteful to spend such a cell to produce, say, a few dozen red cells. The maintenance of hemopoiesis takes place from more advanced

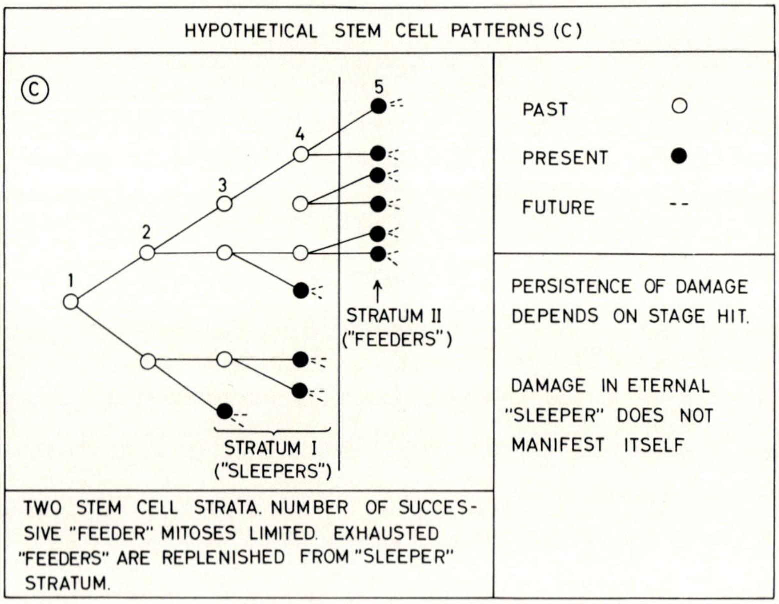

FIG. 3—Hypothetic stem cell pool. All cells are not equal. Some are feeders which are actively proliferating; their function is to feed differentiating hemopoiesis and maintain the feeder pool. The vertical line symbolizes a block which prevents sleepers from feeding into differentiated hemopoiesis. When the feeder stratum is exhausted it is replenished from the sleeper stratum which sets up a new feeder clone. Otherwise, sleepers are nondividing (dormant). Sleepers may be at different distance from the very first stem cell. Symbols are explained in text of Figure 2. For details, see text.

clones, the "feeders," which also are self-perpetuating. If the feeder pool is exhausted or damaged, it is replenished from the sleeper pool. The system thus is one of clonal succession.[33] The scheme postulates a sharp distinction between feeders and sleepers but, of course, a more gradual transition between the two functional states is possible; the feeders might just be a fraction of stem cells which respond more readily to mitotic stimuli and have a longer pedigree. One may ask how such a wide distribution, with regard to length of cell pedigree, could develop. A possible mechanism would be that stem cells divide at random; the resulting skewed distribution of pedigree lengths and clone sizes could become even more accentuated if passage through one cell cycle disposed for initiation of another cell cycle.

The main virtue of this system is that it would permit a reserve of stem cells with a short pedigree. However, it also has its disadvantages when one considers the consequences of a change in a single cell. There are several possibilities. A single cell alteration in the feeder clone would not be observable, again barring growth advantage. Likewise, a change in an eternal sleeper would not express itself. However, a single cell change in a sleeper would become manifest if this cell was called upon to set up a new feeder clone. The prevalence of abnormal feeders and, hence, of hemopoietic cells would in large measure depend on the number of sleeper clones participating in repopulating the feeder pool. The abnormality could gain complete dominance if the feeder pool was substituted from a single abnormal sleeper or was replenished exclusively from the progeny of a single abnormal sleeper. At first glance such an exclusive recruitment from a single sleeper clone may appear unlikely. However, a possible mechanism already alluded to, would be that sleepers are triggered into division at random. Division of a sleeper would increase the relative and absolute size of the particular clone, and hence increase the chance of another division taking place in the same clone, etc. A hypothetical mechanism that might further enhance this avalanche principle is the following: any stem cell division facilitates a subsequent division. This can also be rephrased so that the sensitivity to a particular division stimulus increases with cell lineage. This would mean that the more divisions a cell has undergone the more likely is it to divide again; it would make advanced sleepers the most suitable candidates for taking over the proliferative feeder function of supplying the hemopoietic cell lines. That sleepers are advanced means that they have a rather long pedigree. In view of the exponential growth pattern of sleepers this means that an advanced sleeper clone is large, assuming, of course, that no significant cell death took place while the clone was developing. Whether or not resulting from one or both of the outlined mechanisms, the avalanche principle might result in quite substantial dominance of a single sleeper clone within the sleeper pool. Consequently, repopulation of the feeder pool could be completely dominated by a single clone. The persistence of such a ruling clone would depend on its capacity for continued division (degree of stemness) and the loss of cells it incurs due to feeder function and possible intraclonal cell death at the feeder and/or sleeper level. When a clone has been spent, another clone will expand; in this way one can visualize that the pattern of feeder pool replacement will approach a clonal succession. It is implicit in the idea of clonal succession that stem cells have a finite stemness. Yet, the considerable stem capacity of stem cells is attested to not only by experience with hemopoietic cell transplantation but also by the fact that, in CML, hemopoiesis may be run by the Ph^1 positive clones for years.[28]

Turning now again to leukemia, an attempt will be made to explain some clinical aspects of AML within the framework of the clonal succession theory. From the present scheme, the etiologic and clinical linkage between AML and aplastic anemia[34–39] would be explained on the basis that it is the loss of normal stem cells (aplasia) which increases the chance for a leukemic clone to grow out; the same would apply to the development of paroxysmal nocturnal hemoglobinuria (PNH) strains in aplastic anemia.[40] It is of interest to note in this context that PNH may terminate in AML[41–43] and that PNH may occur in myelofibrosis[44–46] which is also linked with AML.

Radiation induced leukemia could be due to normal stem cell kill with consequently increased chances of an abnormal sleeper to become manifest, radiation mutation, or both. The rather variable incubation time of radiation leukemia[47] would be expected from the scheme. From the scheme one would also predict that, occasionally, the leukemic event would take place in a clone of rather short life expectancy; this would be expected to result in transitory leukemia. Cases of this type, also exhibiting chromosome abnormalities, are on record in mongoloid children.[48–50] For the same reason, a transient leukemia might superimpose itself on marrow aplasia; examples of this may be the cases of untreated acute leukemia which terminate in aplasia without evidence of leukemia at autopsy.[51]

A remarkable feature of cytotoxic therapy of AML is that a phase of severe aplasia appears to be a prerequisite for remission to take place but far from guarantees that remission will occur. Figure 4 is a diagrammatic representation of AML. The single cell change that initiated leukemia took place in generation 2. The operating feeder clone is leukemic, i.e., the disease is in relapse. Normal sleepers are present but, since the feeder pool is full, they are blocked from advancing to the feeder pool and thus cannot express themselves; therefore, normal hemopoietic cells are not produced. By applying cytostatic or cytocidal chemotherapy to the system the actively proliferating leukemic feeders are eradicated and sleepers are called upon to repopulate the feeder pool, thus setting the stage for a possible clone change. Whether or not a clone change will take place will depend on the relative position (cell number) of the existing sleeper clones. In the example of Figure 4, the most advanced sleeper clone (generation 5) is a clone of normal cells; if the feeder pool was eradicated, the normal generation 5 sleepers would replenish the feeder pool,

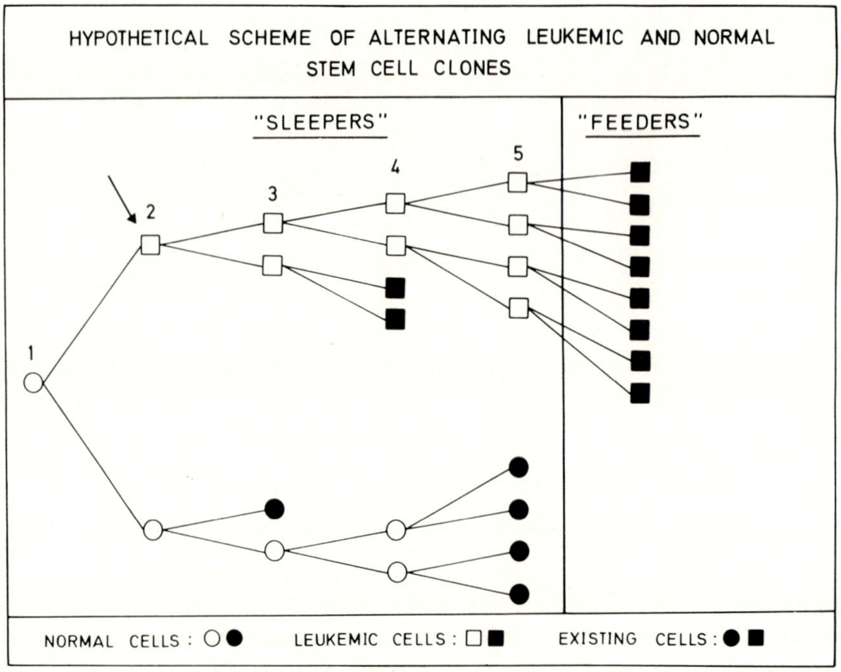

FIG. 4—Diagrammatic representation of a hypothetic clonal succession of leukemic and normal stem cells in acute myeloid leukemia. The leukemic event took place in one cell in generation 2 (arrow). For details, see text. Circles represent normal stem cells, squares are leukemic cells. Open symbols indicate past development, filled in symbols represent existing cells.

leukemic sleepers would be prevented from becoming feeders, and remission would take place. Had the most advanced sleeper clone been leukemic, relapse would have continued. In the example of Figure 4, remission will go on until the normal clone has been spent. Then a new clonal succession will take place, in this case from the leukemic generation 4. Consistent with this scheme, when AML releapses the original abnormal stemline(s) reappear(s).[8,22,23] The scheme would also account for the suddenness with which relapse may take place and useful hemopoiesis collapse; it might explain the variable length of remissions as a function of clone size and stemness. This theory [1] has gained some acceptance.[52] There is, however, one serious objection to the present attempt of explaining the remission/relapse pattern of AML on the basis of clonal succession: from the scheme, one would predict that cytotoxic maintenance therapy would shorten remissions because it would accelerate clonal succession, and inherent to clone change is the risk of relapse. In contrast to ALL, there is no evidence that maintenance treatment increases the duration of remissions in AML but, on the other hand, remission of AML apparently is not shortened by such therapy.[53] However, the situation of maintenance treatment may be a very complex one, e.g., due to possible differences in sensitivity between leukemic and normal cells, or by influencing immunologic processes that perhaps could be involved in marrow damage [54] which in turn might favor the evolution of leukemic cell strains. Another argument against a pure clonal succession is the coexistence of a double red cell population in PNH.[55]

For the time being, it is difficult to envisage clinical situations which could contribute further to support the present theory or refute the feeders and sleepers as mere "dreamers." It appears, however, that the problem could be approached experimentally: in principle, irradiated animals should be injected with equal parts of marrow cells with different karyotypes, e.g., $T_6+T_6T_6$ into a T_6 negative recipient. Variable dominance of one karotype in the bone marrow of different animals or change of karyotype dominance with time in the animals would support the theory. It might also be of interest to study the T_6/T_6T_6 distribution during serial transplantations, using a $T_6+T_6T_6$ injected chimera as the initial donor. A somewhat similar experiment on serial transplantation of bone marrow cells from a heavily irradiated mouse has been reported; [56] when the originally irradiated mouse had recovered from radiation, three cell clones could be distinguished on the basis of radiation-induced marker chromosomes. During subsequent passages, the relative clone sizes varied remarkably. This would be consistent with the sleeper-to-feeder theory. It would be of interest to repeat this experiment with a source of nonirradiated cells, e.g., with the T_6 system as outlined above.

Summary

1. AML is proposed to be a neoplasia which resides in the common hemopoietic stem cell.

2. In part, the progeny of the diseased stem cell is able to differentiate.

3. To explain some clinical aspects of AML, a clonal succession of normal and leukemic stem cells is suggested.

REFERENCES

1. Killmann, S.-Aa.: Acute leukemia: Development, remission/relapse pattern, relationship between normal and leukemic hemopoiesis and the "sleeper-to-feeder" stem cell hypothesis. Ser. Haemat. I,1:103, 1968.

2. Hayhoe, F. G. J., Quaglino, D., and Doll, R.: The Cytology and Cytochemistry of Acute Leukemias. London, H. M. Stationary Office, 1964.

3. Wintrobe, M. M.: Clinical Hematology (ed. 6). Philadelphia, Lea & Febiger, 1967.

4. Holland, J. F.: Progress in the treatment of acute leukemia. *In* Dameshek, W., and Dutcher, R. M. (Eds.): Perspectives in Leukemia. New York, Grune & Stratton, 1968, p. 217.

5. Lampkin, B. C., Nagao, T., and Mauer, A. M.: Drug effect in acute leukemia. J. Clin. Invest. 48:1124, 1969.

6. Ernst, P., and Killmann, S.-Aa.: Effect of antileukemic drugs on cell cycle of human leukemic blast cells in vivo. Acta Med. Scandinav. 186:239, 1969.

7. Sandberg, A. A., Takagi, N., Sofuni, T., and Crosswhite, L. H.: Chromosomes and causation of human cancer and leukemia. Cancer 22:1268, 1968.

8. Jensen, M. Krogh: Chromosome Studies in Acute Leukemia (Thesis). Copenhagen, Munksgaard, 1969.

9. Salmon, Ch.: A tentative approach to variations in ABH and associated erythrocyte antigens. Ser. Haemat. II, 1:3, 1969.

9a. Brecher, G.: Personal communication.

10. Jensen, M. Krogh, and Killmann, S.-Aa.: Chromosome studies in acute leukemia. Evidence for chromosomal abnormalities common to erythroblasts and leukaemic white cells. Acta Med. Scandinav. 181:47, 1967.

11. Jensen, M. Krogh, and Killmann, S.-Aa.: Evidence for chromosome abnormalities in the erythroid precursors in acute myeloid leukemia. Submitted to Acta Med. Scandinav.

12. Trujillo, J. M., and Ohno, S.: Chromosomal alteration of erythropoietic cells in chronic myeloid leukemia. Acta Haemat. (Basel) 29:311, 1963.

13. Whang, J., Frei, E., III, Tjio, J. H., Carbone, P. P., and Brecher, G.: The distribution of the Philadelphia chromosome in patients with chronic myelogenous leukemia. Blood 22:664, 1963.

14. Clein, G. P., and Flemanns, R. J.: Involvement of the erythroid series in blastic crisis of chronic myeloid leukemia. Further evidence for the presence of Philadelphia chromosome in erythroblasts. Brit. J. Haemat. 12:754, 1966.

15. Rastrick, J. M.: A method for the positive identification of erythropoietic cells in chromosome preparations of bone marrow. Brit. J. Haemat. 16:185, 1969.

16. Müller, D.: Histochemische Untersuchungen an akuten und chronisch-myeloischen Leukämien II. Der DNA-Gehalt leukämischer Zellen. Blut 13:152, 1966.

17. Baikie, A. G., Jacobs, P. A., McBride, J. A., and Tough, I. M.: Cytogenetic studies in acute leukemia. Brit. Med. J. I, 1564, 1961.

18. Fitzgerald, P. H., Adams, A., and Gunz, F. W.: Chromosome studies in adult acute leukemia. J. Nat. Cancer Inst. 32:395, 1964.

19. Reisman, L. E., Mitani, M., and Zuelzer, W. W.: Chromosome studies in leukemia. I. Evidence for the origin of leukemic stem lines from aneuploid mutants. New Eng. J. Med. 270:591, 1964.

20. Sandberg, A. A., Ishihara, T., Kikuchi, Y., and Crosswhite, L. H.: Chromosomal differences among the acute leukemias. Ann. N.Y. Acad. Sci. 113:663, 1964.

21. Kiossoglou, K. A., Mitus, W. J., and Dameshek, W.: Chromosomal aberrations in acute leukemia. Blood 26:610, 1965.

22. Jensen, M. Krogh: Chromosome studies in acute leukemia. III. Chromosome constitution of bone marrow cells in 30 cases. Acta Med. Scandinav. 182:629, 1967.

23. Reisman, L. E., Zuelzer, W. W., and Thomsen, R. I.: Further observations on the role of aneuploidy in acute leukemia. Cancer Res. 24:1448, 1964.

24. Baikie, A. G., Court Brown, W. M., Jacobs, P. A., and Milne, J. S.: Chromosome studies in human leukaemia. Lancet II, 425, 1959.

25. Sandberg, A. A., Ishihara, T., Crosswhite, L. H., and Hauschka, T. S.: Chromosomal dichotomy in blood and marrow of acute leukemia. Cancer Res. 22:748, 1962.

26. Killmann, S.-Aa.: Acute leukemia: The kinetics of leukemic blast cells in man. Ser. Haemat. I, 3:38, 1968.

27. Rytömaa, T., and Kiviniemi, K.: Control of DNA duplication in rat chloroleukaemia by means of the granulocytic chalone. Europ. J. Cancer, 4:595, 1968.

28. Pedersen, B.: Cytogenetic Evolution in Chronic Myelogenous Leukaemia (Thesis). Copenhagen, Munksgaard, 1969.

29. Whang-Peng, J., Canello, G. P., Carbone, P. P., and Tjio, J. H.: Clinical implications of cytogenetic variants in chronic myelocytic leukemia. Blood 32:755, 1968.

30. Fialkow, P. J., Gartler, S. M., and Yoshida, A.: Clonal origin of chronic myelocytic leukemia in man. Proc. Nat. Acad. Sci. U.S.A. 58:1468, 1967.

31. Fialkow, P. J., Lisher, R., Detter, J., Giblett, E. R., and Zavala, C.: 6-phosphogluconate dehydrogenase: Hemizygous manifestation in a patient with leukemia. Science 163:194, 1969.

32. Mastrangelo, R., Zuelzer, W. W., and Thompson, R. I.: The significance of the Ph^1 chromosome in acute myeloblastic leukemia: Serial cytogenetic studies in a critical case. Pediatrics 40:834, 1967.

33. Kay, H. E. M.: How many cell-generations? Lancet II:418, 1965.

34. Dameshek, W., and Gunz, F.: Leukemia. New York, Grune & Stratton, 1958.

35. Dreyfus, B., and Bessis, M.: Leucémies aiguës précédées d'une aplasie myéloïde. Sang 25:787, 1954.

36. Court-Brown, W. M., and Doll, R.: Leukaemia and aplastic anaemia in patients irradiated for ankylosing spondylitis. Med. Res. Council Spec. Rep. Ser. No. 295, London, H. M. Stationary Office, 1957.

37. Garriga, S., and Crosby, W. H.: The incidence of leukemia in families of patients with hypoplasia of the marrow. Blood 14: 1008, 1959.

38. Hayhoe, F. G. J.: Leukemia. Research and Clinical Practice. London, Churchill, 1960.

39. Cronkite, E. P., Moloney, W., and Bond, V. P.: Radiation leukemogenesis. An analysis of the problem. Amer. J. Med. 28: 673, 1960.

40. Lewis, S. M., and Dacie, J. V.: The aplastic anemia-paroxysmal nocturnal hemoglobinuria syndrome. Brit. J. Haemat. 13: 236, 1967.

41. Jenkins, D. E., Jr., and Hartmann, R. C.: Paroxysmal nocturnal hemoglobinuria terminating in acute myeloblastic leukemia. Blood 33:274, 1969.

42. Holden, D., and Lichtman, H.: Paroxysmal nocturnal hemogloblinuria with acute leukemia. Blood 33:283, 1969.

43. Kaufmann, R. W., Schechter, G. P., and McFarland, W.: Paroxysmal nocturnal hemoglobinuria terminating in acute granulocytic leukemia. Blood 33:287, 1969.

44. Dameshek, W., and Fudenberg, H.: Paroxysmal nocturnal hemoglobinuria. Atypical manifestations suggesting an immunologic disease. Arch. Intern. Med. (Chicago) 99:202, 1957.

45. Hansen, N. E., and Killmann, S.-Aa.: Paroxysmal nocturnal hemoglobinuria. Acta Med. Scandinav. 184:525, 1968.

46. Hansen, N. E., and Killmann, S.-Aa.: PNH and PNH-traits in patients with myelofibrosis. Submitted to Blood.

47. Cobb, S., Miller, M., and Wald, N.: On the estimation of the incubation period in malignant disease. J. Chronic Dis. 9:385, 1959.

48. Schunk, G. J., and Lehmann, W. L.: Mongolism and congenital leukemia. J.A.M.A. 155:250, 1954.

49. Ross, J. D., Moloney, W. C., and Desforges, J. D.: Ineffective regulation of granulopoiesis masquerading as congenital leukemia in a mongoloid child. J. Pediat. 63: 1, 1963.

50. Engel, R. R., Hammond, D., Eitzman, D. V., Pearson, H., and Krivit, W.: Transient congenital leukemia in seven infants with mongolism. J. Pediat. 65:303, 1964.

51. Brittin, G. M., Brecher, G., Steinfeld, J. L., and Shimkin, M. D.: Unusual end stages of untreated acute leukemia. Amer. J. Path. 43:93, 1951.

52. Henderson, E. S.: Treatment of acute leukemia. Seminars Hemat. 6:271, 1969.

53. Acute Leukemia Group B, unpublished; communicated by Dr. N. Nissen, Copenhagen.

54. Krantz, S. B., and Koo, V.: Studies on red cell aplasia II. Report of a second patient with an antibody to erythroblast nuclei and a remission after immunosuppressive therapy. Blood 34:1, 1969.

55. Dacie, J. V.: The Hemolytic Anemias. Congenital and Acquired (ed. 2, part 4). London, Churchill, 1967.

56. Micklem, H. S., and Loutit, J. F.: Tissue Grafting and Radiation. New York, Academic Press, 1966, p. 86.

Megakaryocytopoiesis

By T. T. Odell, Jr. and C. W. Jackson

Knowledge concerning the nature of the maturation or differentiation process in megakaryocytes has expanded considerably during the past few years. As a result of recent investigations, some older concepts of megakaryocytopoiesis have had to be revised. Let us review the findings that led to our present concepts, and also indicate some of the uncharted areas.

Cronkite and his co-workers [1] first noted that only a small proportion of the recognizable megakaryocyte population in a patient's hemopoietic tissue was labeled soon after a single injection of tritiated thymidine (^{3}HTdr), a specific precursor of DNA. This observation showed that DNA synthesis is restricted to a small part of the population or, put in another way, that the DNA synthetic period makes up only a short part of the maturation time of recognizable megakaryocytes. Feinendegen et al.,[2] using rats, then showed that the initially labeled cells were representative of the youngest recognizable megakaryocytes, and Ebbe and Stohlman [3] confirmed this finding. Moreover, mitotic figures were seen only in these youngest megakaryocytes. Cell division (cytokinesis) was never observed. With time, the more mature elements of the megakaryocyte line were also labeled, thereby indicating a progression of cells from the least mature to the most mature compartments. In fact, the sequential appearance of label in cells that had been classed as a maturing series supported the morphologic classification system. This classification system, developed by Feinendengen et al.,[2] divided the recognizable megakaryocytes into three groups: the first group was composed of the immature cells with a nongranulated, basophilic cytoplasm and evenly stippled chromatin in the nucleus, corresponding generally to what are often called megakaryoblasts; in the third group the cells were larger, with many granules in the copious cytoplasm, a higher cytoplasm-to-nucleus ratio and more dense, unevenly distributed chromatin in the nucleus (megakaryocytes); the second group of cells were intermediate in appearance (promegakaryocytes). Averaging the results of Feinendegen and those of Ebbe, the frequency of megakaryocytes in these morphologic stages was about 17 per cent in Group I, 22 per cent in Group II, and 62 per cent in Group III. In this paper the recognized population of megakaryocytes will be divided and referred to as immature or young, corresponding to group I in the foregoing system, and mature, corresponding to Groups II and III. In addition, the naked nuclei that remain after platelets have been released will also be mentioned. About 20–25 per cent of the recognized megakaryocytes are immature, about 70–75 per cent mature, and about 5 per cent naked nuclei.

Since both initial labeling with ^{3}HTdr and mitotic figures were seen only in immature megakaryocytes, it became evident that the development of

---

From the Biology Division, Oak Ridge National Laboratory, Oak Ridge, Tennessee.

Research sponsored by the U. S. Atomic Energy Commission under contract with the Union Carbide Corporation.

polyploidy does not continue throughout the whole maturation period, a concept that had generally been held before these discoveries were made. Rather, development of polyploidy is restricted to cells that are classified morphologically as young or immature.

With the goal of learning more about the process of megakaryocyte maturation, we undertook to measure the amount of DNA in individual megakaryocytes of rats, using the two-wavelength microspectrophotometirc method of Ornstein [4] and Patau.[5] At this time Garcia [6] in rabbits and de Leval [7] in guinea pigs reported ploidy values of 8, 16, 32, and 64N in megakaryocytes using the microspectrophotometic method. Our results in rats were similar, indicating that replication of DNA in megakaryocytes occurs synchronously rather than by diploid set of chromosomes, since interclasses of DNA amounts were not apparent.[8] In addition, we determined the frequency of megakaryocytes in each ploidy class, observing the following percentages: 1.6 per cent in 4N cells, 10.5 per cent in 8N, 71.2 per cent in 16N, 17.1 per cent in 32N, and 0.1 per cent in 64N. Much earlier Japa,[9] using the technic of counting the number of lobes of megakaryocyte nuclei in squash preparations of human marrow, had found a similar distribution of ploidy values (4N, 2.5 per cent; 8N, 25.5 per cent; 16N, 53 per cent; 32N, 18 per cent; 64N, 1 per cent).

The labeling index results, the percentage distribution of megakaryocytes in morphologic stages, and the determinations of cell frequency in ploidy classes suggested a new hypothesis of megakaryocyte maturation: that the development of polyploidy by DNA replication takes place in immature megakaryocytes, after which DNA synthesis permanently stops and the cells continue their differentiation toward mature platelet-producing megakaryocytes.[8] Moreover, it was postulated that DNA synthesis may cease at either the 4, 8, 16, or 32N ploidy level in any one cell followed by cytoplasmic differentiation, with the further implication that platelet production and megakaryocyte death may occur at any of these ploidy levels (Table 1).

We investigated this hypothesis by determining the ploidy value of individual cells whose morphologic stage was known. Smears of rat bone marrow were

TABLE 1—*Megakaryocyte Maturation*

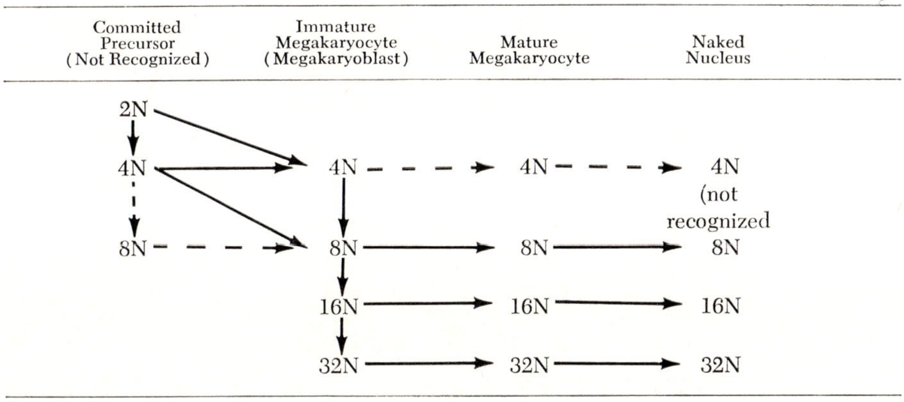

first stained with Giemsa to classify the megakaryocytes morphologically, and then they were restained with the Feulgen procedure to measure the DNA of individual cells by microspectrophotometry. By recording the position of the Giemsa-stained cells on the slides, it was possible to find the same cells and measure their DNA content. In addition, the donor animals were injected with ^{3}HTdr 30 minutes before they were killed, so that we could distinguish the cells that were in DNA synthesis at the time of collection. Cell labeling was recorded in autoradiograms of the Giemsa-stained smears.

The results (Table 2) showed that both the immature and the mature megakaryocytes in the recognized population include cells with ploidy values of 8, 16, and 32N.[10] Since the more mature megakaryocytes no longer replicate DNA and therefore cannot change ploidy class, these findings support the postulation that megakaryocytes mature at several ploidy levels. This raises the question of the nature of the mechanism that stops DNA synthesis and that, moreover, accomplishes this at different ploidy levels in different cells. It has been suggested that accumulation of hemoglobin, a protein unique to red blood cells, inhibits further DNA replication in that cell line,[11] and a similar function has been suggested for thrombosthenin, the unique contractile protein of blood platelets.[12] However, it seems possible that thrombosthenin may appear too late in the differentiation process of megakaryocytes to have the function of inhibiting DNA synthesis.[13] Both light and electron microscope studies indicate that the complex structural organization of megakaryocyte cytoplasm that leads to platelet production is seen in more mature megakaryocytes.[14]

The occurrence of nonendoreplicating, differentiating megakaryocytes with different ploidy values implies that platelets are derived from megakaryocytes of differing ploidy, and raises the question of whether or not various functional parameters of platelets may vary depending on the ploidy of their progenitors.

The percentage distribution of cells of the several ploidy classes within maturation stages was also calculated (Table 2). It was found that the proportion of 8N cells is markedly smaller among mature than among immature cells, whereas the proportions of 16 and 32N cells are somewhat greater in the older cells. These results are consistent with the model, since the expectation was that many of the young 8N cells would replicate DNA to supply the young 16N and 32N cells while others would move on into the mature differentiating compartment as 8N cells; most 16 and 32N megakaryocytes would continue to differentiate into mature cells.

The finding that 8N cells make up only about one-third of the population

TABLE 2—*Ploidy Distribution of Immature and Mature Megakaryocytes*

	Megakaryocytes (%)		
N *	Immature	Mature	Naked Nuclei
8	32	15	7
16	59	68	78
32	9	18	15

* N signifies one haploid chromosome complement.

TABLE 3—*Initial Labeling Index (30 Minutes After ^{3}HTdr) and Megakaryocyte Size in Ploidy Classes*

	Ploidy Class				
	All	4N	8N	16N	32N
Per cent labeled (young megakaryocytes)	32	73	51	22	14
Size of young megakaryocytes (relative)			1.06	1.78	2.74
Size of mature megakaryocytes (relative)			1.57	2.74	4.70

of immature megakaryocytes—even though they presumably supply all of the young 16N cells, which make up about 60 per cent of this stage, as well as the 32N cells and a few mature 8N cells—can also be reconciled by taking into consideration the relative times spent by the various cells in the young compartment. Labeling index results have shown that about 25–30 per cent of the immature megakaryocytes are initially labeled and, therefore, in DNA synthesis. Since the DNA synthesis period (S) comprises about 80 per cent of the generation cycle of megakaryocytes,[15] approximately 35 per cent of the young recognizable megakaryocytes are in the replication cycle. Thus, the remaining 65 per cent presumably are in a postreplicating stage. It appears, therefore, that what we designate morphologically as the immature stage may consist of two subcompartments, one consisting of endoreduplicating cells and a second comprising cells in a post-S phase. Since many of the 8N cells apparently replicate DNA and become 16N, only a relatively small proportion of the 8N cells would be expected to enter the second subcompartment, and thus the average turnover time of all young 8N megakaryocytes would be shorter than that for 16N and 32N megakarycoytes, most of the latter going on to differentiate rather than replicating DNA again. Therefore, although the proportion of 8N cells seen at any one instant is about one-third the population of young megakaryocytes, the number passing through this compartment per unit of time is relatively greater because of the shorter average time spent there as compared with 16 and 32N cells.

In addition, the labeling index of the several ploidy classes of the immature megakaryocytes 30 minutes after injection of ^{3}HTdr (Table 3) shows that the proportion of cells in the replication phase is high among cells of low ploidy and decreases in the higher ploidy classes. These results are consistent with the interpretation that many cells of lower ploidy classes are engaged in replication to produce cells of higher ploidy and relatively few enter the second, post-S subcompartment of the immature population. In fact, the findings suggest that almost all of the 4N cells belong to the replicating population. On the other hand, a large proportion of the higher ploidy cells do proceed to the second subcompartment.

Looking only at the initially labeled cells (cells in DNA synthesis when the ^{3}HTdr was given), we found that 54 per cent belong to the 8N class, 42 per cent to the 16N class, and 4 per cent to the 32N class. It is therefore apparent that the 16N and 32N replicating cells can be supplied from replicating 8N cells.

The sizes of megakaryocytes in the several ploidy classes (8, 16, and 32N) within the maturation stages have also been determined by making photographic enlargements on uniform-weight photographic paper of 35 mm. pictures of cells, cutting them out, and weighing them. The result (Table 3) showed that the higher the ploidy value was among cells of a maturation stage, the larger was the cell. Morever, cells with the same ploidy had a markedly greater average size among mature megakaryocytes than among immature ones. The larger size of cells with higher ploidy in the immature compartment is doubtless due in large part to increase in nuclear size; cytoplasm is relatively sparse in all immature cells. On the other hand, the increase in size of mature cells is due to an increase in the cytoplasmic mass, and it indicates a very active synthesis of cytoplasmic components. It is also apparent from electron micrographs that various cytoplasmic structures and organelles are being laid down during this time, such as the demarcation membranes that mark the future boundaries of the platelets.[14] The findings that the size distribution curve of blood platelets appears to be unimodal while megakaryocytes attain different sizes at maturity in relation to their ploidy level suggest that the number of platelets produced by a megakaryocyte may be related to its size, and therefore in turn to its degree of polyploidy.

There are many unanswered questions concerning the relations between DNA, RNA, and protein synthesis and the differentiation and maturation processes of megakaryocytes. One stumbling block in the way of studying subcellular or molecular aspects of megakaryocytes is that no technic is presently available for collecting large numbers of megakaryocytes free of other kinds of cells, much less for collecting megakaryocytes of particular maturation stages or ploidy classes. As a result it is not possible to study these cells by usual biochemical technic. However, some information at this level can be obtained by using autoradiographic and cytochemical technic.

As we have seen, the maturation of megakaryocytes includes two processes, the development of a polyploid condition and cytoplasmic differentiation. The development of polyploidy begins with a committed diploid cell; this can be thought of as the beginning of megakaryocytopoiesis, and thus the point at which to begin the timing of megakaryocyte maturation. Since individual megakaryocytes reach full maturity at different ploidy levels, the degree of polyploidy achieved will be one determinant of the maturation time of an individual cell. It would take two generation cycles to go from a committed diploid cell to a recognized 8N cell, three for 16N cells, and four for 32N cells. Since each generation cycle in rat megakaryocytes is about 9.3 hours,[15] it would take 18.6, 27.9, and 37.2 hours, respectively. In addition, the differentiation time after DNA synthesis has permanently stopped appears to take about 42 hours, on the basis of labeling index studies,[2,3,16,18] so that this time can be added to those already given. The range of maturation time is therefore 60.6–79.2 hours, from diploid cell to mature platelet-producing megakaryocyte. Estimates have usually been made, however, for only the recognized megakaryocyte population; and, since recognized megakaryocytes are mostly at least 8N in ploidy, the duration of two generation cycles (or 18.6 hours) should be subtracted from the foregoing values to make them comparable to other estimates. In these terms, the range would be 42 to 60.6 hours.

Another finding, from labeling index studies, consistent with these estimates was that 50 per cent of the naked nuclei showed the presence of label about 54 hours after injection of ^{3}HTdr, another indication of the average maturation time of recognized megakaryocytes.

The rather meager information available for man suggests that maturation of megakaryocytes in humans takes a somewhat longer time than in rats and mice. In a human patient, maximum labeling of megakaryocytes was attained 5 days after injection of tritiated thymidine; [1] the effects of radiation on the peripheral platelet count can also be interpreted to indicate a maturation time of 5 or more days for megakaryocytes in man.[17]

Since the diploid and probably most of the tetraploid cells that supply the megakaryocyte population are generally not recognized, they have to be studied indirectly. Results of various experiments in which the system has been perturbed suggest additional interpretations of megakaryocyte maturation. For example, it was found that megakaryocyte numbers and the mitotic index of megakaryocytes had just begun to increase 24 hours after induction of thrombocytopenia by platelet-specific antiserum, and the rise thereafter was rapid.[18] Presumably, during this lag period between induction of thrombocytopenia and the first observable changes in the megakaryocyte population, the low platelet count first caused release of a humoral regulatory agent, which then stimulated precursor cells to produce more megakaryocytes. Since the cell generation cycle of megakaryocytes lasts about 9.3 hours, it would take nearly 20 hours to go from an unrecognized 2N cell to a recognized 8N

TABLE 4—*Concept of Megakaryocytopoiesis*

	Compartment	Characteristics
Unrecognized	Uncommitted stem cell (diploid)	Renews next compartment, especially upon increased demand.
	↓	
Unrecognized	Committed precursor cell (diploid)	Regulated by "thrombopoietin," the level of which is determined by the numbers of circulating platelets. a. Renews itself by cell division (includes pool of "resting" cells, G_1 or G_0). b. Matures to recognized megakaryocyte (megakaryoblast) by 1 or 2 endomitotic divisions.
	↓	
Recognized	Immature megakaryocyte (megakaryoblast)	Ploidies of 4, 8, 16, 32, 64N. 1. Endoreduplicating subcompartment. Endomitosis restricted to this compartment of recognized population. 2. Post-DNA-synthesis subcompartment.
	↓	
Recognized	Mature megakaryocyte	Ploidies of (4), 8, 16, 32, 64N. Differentiates to mature platelet-producing megakaryocyte. Growth and differentiation of cytoplasm (no endomitosis).

megakaryocyte. Therefore, the hypothetical humoral agent may become active soon after induction of thrombocytopenia. The 20-hour time from induction of thrombocytopenia to appearance of new megakaryocytes also suggests that these cells may be derived from a "resting pool" of committed diploid precursor cells.

In conclusion, a concept of megakaryocytopoiesis based on the foregoing considerations is presented in Table 4. In this concept, an uncommitted stem cell pool renews a committed precursor cell pool, perhaps at a very slow pace under normal conditions and more rapidly under conditions of unusual demand. The committed diploid cell both renews its own compartment and also sends cells along the course of polyploidy development. After reaching one of several possible degrees of polyploidy, these cells then cease to replicate DNA and undergo a final differentiation period (in which changes in the cytoplasm are especially apparent) until the stage of a mature platelet-producing cell is reached. The whole process is controlled by one or more thrombopoietic agents that act primarily on a committed diploid precursor cell.

REFERENCES

1. Cronkite, E. P., Bond, V. P., Fliedner, T. M., Paglia, D. A., and Adamik, E. R.: Studies on the origin, production and destruction of platelets. *In* Johnson, S. A., Monto, R. W., and Horn, R. C., Jr. (Eds.): Blood Platelets. Boston, Little, Brown, 1961.
2. Feinendegen, L. E., Odartchenko, N., Cottier, H., and Bond, V. P.: Kinetics of megakaryocyte proliferation. Proc. Soc. Exp. Biol. Med. 111:177, 1962.
3. Ebbe, S., and Stohlman, F., Jr.: Megakaryocytopoiesis in the rat. Blood 26:20, 1965.
4. Ornstein, L.: The distributional error in microspectrophotometry. Lab. Invest. 1: 250, 1952.
5. Patau, K.: Absorption microphotometry of irregular-shaped objects. Chromosoma 5: 341, 1952.
6. Garcia, A. M.: Feulgen-DNA values in megakaryocytes. J. Cell Biol. 20:342, 1964.
7. De Leval, M.: Dosages cytophotométriques d'ADN dans des mégacaryocytes normaux de Cobaye. C. R. Soc. Biol. (Paris) 158:2198, 1964.
8. Odell, T. T., Jr., Jackson, C. W., and Gosslee, D. G.: Maturation of rat megakaryocytes studied by microspectrophotometric measurement of DNA. Proc. Soc. Exp. Biol. Med. 119:1194, 1965.
9. Japa. J.: Study of morphology and development of megakaryocytes. Brit. J. Exp. Path. 24:73, 1943.
10. Odell, T. T., Jr., and Jackson, C. W.: Polyploidy and maturation of megakaryocytes. Blood 32:102, 1968.
11. Stohlman, F., Jr., Lucarelli, G., Howard, D., Morse, B., and Leventhal, B.: Regulation of erythropoiesis. Medicine 43: 651, 1964.
12. Paulus, J.-M.: Multiple differentiation in megakaryocytes and platelets. Blood 29: 407, 1967.
13. Ebbe, S.: Megakaryocytopoiesis and platelet turnover. Ser. Haemat. 1:65, 1968.
14. Paulus, J.-M.: Etude ultrastructurale et microphotométrique de la maturation mégacaryocytaire. Nouv. Rev. Franc. Hemat. 8:394, 1968.
15. Odell, T. T., Jr., Jackson, C. W., and Reiter, R. S.: Generation cycle of rat megakaryocytes. Exp. Cell Res. 53:321, 1968.
16. Odell, T. T., Jr., Burch, E. A., Jr., Jackson, C. W., and Friday, T. J.: Megakaryocytopoiesis in mice. Cell Tissue Kinet. 2:363, 1969.
17. Odell, T. T., Jr.: Radiation effects on the megakaryocyte and platelet system. *In* Manual on Radiation Hematology. International Atomic Energy Agency. In press.
18. Odell, T. T., Jr., Jackson, C. W., Friday, T. J., and Charsha, D. E.: Effects of thrombocytopenia on megakaryocytopoiesis. Brit. J. Haemat. 17:91, 1969.

Responses of the Thrombocytopoietic System to Platelet Depletion and Irradiation

By Shirley Ebbe

A small proportion of cells in the compartment of recognizable megakaryocytes synthesize DNA, but cell division does not occur.[1,2] Cellular proliferation must occur at the level of the unrecognizable precursors, and it would be anticipated that responses to homeostatic mechanisms and to cytotoxic agents would also occur at the precursor cell level. Some of the properties of these unrecognizable precursors may, therefore, be revealed by the responses of the recognizable components to perturbations of the thrombocytopoietic system. This report will summarize some of the changes which were observed when the system was stimulated by depletion of circulating platelets or depressed by sublethal doses of radiation.

Materials and Methods

Experiments were done in Sprague-Dawley rats or CF_{1s} mice (Carworth Farms). Platelet counts were done by phase microscopy.[3] The methods used have been detailed in previous publications as follows.

1. Morphologic classification of megakaryocytes and labeling with tritiated thymidine (^{3}HTdR).[2]
2. Platelet depletion in rats by exchange transfusion with platelet-poor homologous blood or rabbit anti-rat platelet serum.[4]
3. Estimation of relative sizes of rat megakaryocytes.[4]
4. Preparation and administration of rabbit anti-mouse platelet serum and sublethal irradiation of mice.[5]

Results and Discussion

Platelet Counts

Figure 1 shows the platelet counts of rats for 9 days after circulating platelets were depleted by exchange transfusion to ~9 per cent of normal. The observed values were compared to the theoretical recovery from this degree of thrombocytopenia if production and destruction remained normal during this period. Normal production rates were based on the average normal platelet count of 1.1×10^6/mm.^{3} in these rats and either a platelet life span of 4 days [6] or the random destruction of 40 per cent of circulating platelets daily.[7] It can be seen that platelet counts increased almost exactly as predicted for the first 2 days after platelet depletion. This period of normal production tended

From St. Elizabeth's Hospital and Tufts Medical School Boston, Massachuetts.

Supported in part by grant AM–08263 and Research Career Development Award 1–K3 AM–8634 from the National Institute of Arthritis and Metabolic Diseases, by a Research Grant from the American Cancer Society (Massachusetts Division), Inc., and Grant No. T–505 from the American Cancer Society, Inc.

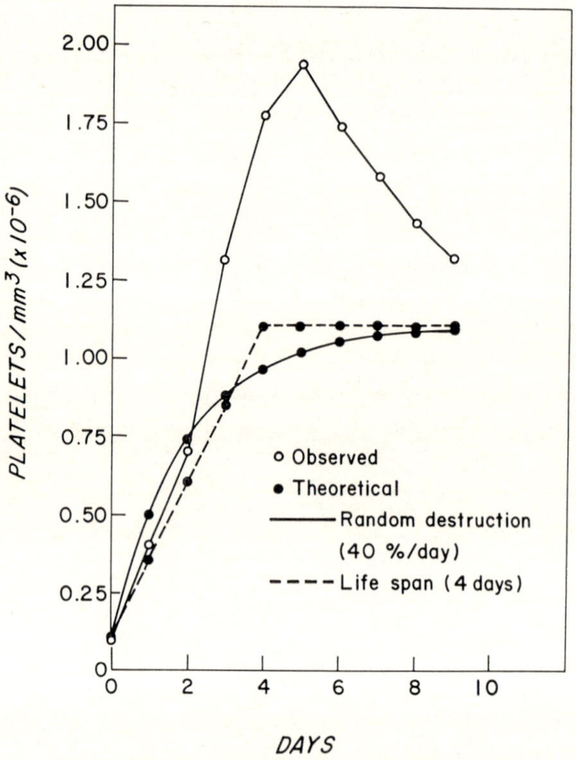

FIG. 1—Rat platelet counts after depletion of circulating platelets to ~9 per cent of normal by exchange transfusion. Each "observed" point is the average of 6–16 rats, except for the first which is the average of 155 rats; S.E.M. <20,000/mm.^{3} for each point. (From Reference 29).

to rule out the presence of any substantial reservoir of platelets which were available for immediate delivery to the circulation and suggested, rather, that platelet production was dependent on an orderly maturation and disintegration of megakaryocytes. In this respect, control of platelet production could not be likened to that of granulocytes for which there appears to be a large reserve of cells which are released into the circulation within a few hours after induction of neutropenia.[8]

During the third day after depletion, however, the magnitude of the increment in concentration of circulating platelets indicated that production had increased. Comparable findings in rats have been published by other investigators,[9,10] and a lag period of 3 to 4 days between thrombocytopheresis and development of thrombocytosis has been described in dogs.[11] The increase in platelet production and subsequent development of compensatory thrombocytosis provided cogent evidence for a homeostatic mechanism by which thrombocytopenia stimulated platelet production. Some of the effects of this mechanism on megakaryocytopoiesis were seen in the bone marrow during the 2 days preceding the increase in platelet production.

Megakaryocyte Turnover

In smears of bone marrow, stained with Wright's and Giemsa stains, morphologic criteria were used to subdivide the compartment of recognizable megakaryocytes into three stages of maturation: megakaryoblast (Stage I), basophilic megakaryocyte (Stage II), and granular megakaryocyte (Stage III).[2,12] In this classification, the naked nuclei which are left after formation of platelets were not included. Generally, there were about 19 per cent Stage I, 25 per cent Stage II, and 56 per cent Stage III.[2] When a single injection of ^{3}HTdR was given to normal rats, the only recognizable megakaryocytes which became labeled immediately were about 20–30 per cent of Stage I cells. Cell division did not occur within the compartment, but the immediate precursors, which were morphologically not recognizable, labeled heavily with ^{3}HTdR.[1,2] Therefore, the rates at which the percentage of labeled Stage I cells increased and labeled cells appeared in the subsequent maturation stages gave measures of the rates of differentiation of precursor cells into the recognizable compartment and maturation of megakaryocytes through the three morphologic stages, respectively.

When ^{3}HTdR was administered immediately before or after onset of acute platelet depletion, there was no difference from normal in the pattern of labeling of megakaryocytes during the first post-thrombocytopenic day.[13] An acceleration of megakaryocyte turnover was apparent, however, when labeling with ^{3}HTdR was accomplished a day after production of thrombocytopenia. During the second post-thrombocytopenic day, the rate of increase of labeled cells in Stage I was increased, indicating accelerated differentiation of precursor cells into the compartment of recognizable megakaryocytes. At the same time, maturation through the compartment was accelerated, as judged from the rate at which the initially labeled cells matured into Stages II and III. Coincident with these changes in turnover rate there was no change in the proportion of cells in each of the stages of maturation. Therefore, it was concluded that accelerated influx was balanced by accelerated maturation during the second day of the megakaryocytic response to a single thrombocytopenic stimulus. Other workers [14] have also reported an increased rate of labeling of megakaryocytes when ^{3}HTdR was given a day after induction of thrombocytopenia by antiplatelet serum. They, however, found an increase in the relative numbers of immature megakaryocytes which appeared to begin about one day after induction of thrombocytopenia and was followed by an increase in more mature cells. Hence, they attributed the apparent shortening of megakaryocyte transit time, in part, to an added input of labeled cells, and, in part, to an actual decrease in maturation time of megakaryocytes. A strict comparison between the two studies can not, however, be made since thrombocytopenia was maintained by repeated injections of antiserum in one,[14] and, in the other,[13] the animals were permitted to recover from a single thrombocytopheresis or antiserum injection.

For the population size to change when the maturation rate was accelerated, influx from the precursor pool would have to increase out of proportion to the maturation rate. During the period of adjustment, a shift in the differential count of megakaryocytes would have occurred. This was not observed at any time during 8 days after acute thrombocytopenia in rats.[13] The differential count is an insensitive measure of megakaryocytopoiesis, but the absence of any change indicating that no large alteration in the size of the megakaryocyte compartment occurred during the 2 day period preceding an increase in platelet production. Matter et al.[9] reported an increase in platelet production to about 4 times normal on the third day following a single platelet depletion; associated with this they found only a 25 per cent increase in numbers of megakaryocytes. Thus, the population size remained relatively constant even though there was an increased influx of megakaryocytes. This seeming paradox was explained by the accelerated maturation which offset the increased influx. The ability of the megakaryocyte compartment to expand in response to more chronic stimulation is not completely clear. Harker[15] reported the number of megakaryocytes to be about 1.4 times normal after daily thrombocytopheresis for 4 days, but they had increased to about 3.3 times normal after 10 days of continued platelet depletion. Odell et al.[14] found about 2.5 times the normal number of megakaryocytes after 2.5 days of thrombocytopenia maintained by repeated injections of antiplatelet serum and little additional change in the succeeding 1.5 days.

Megakaryocyte Size

A second change in megakaryocytes which was observed during the 2 post-thrombocytopenic days when platelet production remained normal was an increase in size of individual megakaryocytes.[4] During the first 9 hours, sizes of megakaryocytes, as estimated from bone marrow smears, remained normal. With moderate thrombocytopenia (platelet counts of ~40 per cent or ~8 per cent of normal) macrocytosis was maximal in Stage I at 24 hours, but was present also in Stage II; 48 hours after induction of thrombocytopenia, macrocytosis had diminished somewhat in Stage I, but had become maximal in the Stage III, platelet-producing megakaryocytes. When more severe thrombocytopenia (platelets <0.1 per cent of normal) was produced by heterologous antiplatelet serum macromegakaryocytosis was apparent in all three maturation stages 24 hours later. However, at that time it had reached maximum proportions in Stage I while Stages II and III proceeded to become larger during the ensuing 24 hours. This earlier appearance of macrocytosis in mature megakaryocytes may be related to the more rapid rates of differentiation and maturation in these severely thrombocytopenic rates.[13] The degree of macrocytosis achieved was comparable in the three maturation stages. With platelet counts reduced to ~40 per cent or ~8 per cent of normal, megakaryocyte sizes increased to 140–150 per cent of normal. After platelets were reduced to <0.1 per cent of normal, megakaryocytes became 170–190 per cent the size of normal. Macrocytosis disappeared from the compartment of recognizable megakaryocytes in the same sequence it had appeared Stage I returned to normal size first, Stage III last.

The sequential appearance and disappearance of macromegakaryocytosis indicated that the stimulus for increase of cell size was probably effective on the precursor cells rather than directly on the recognizable megakaryocytes. This conclusion was substantiated by the development of comparable degrees of macrocytosis in all three stages of maturation.

Events related to the thrombocytopoietic system which followed a single episode of acute platelet depletion in rats can be summarized as follows.

1. For 2 days, platelet production remained normal.
2. Megakaryocytopoiesis in the recognizable compartment remained normal for at least 9 hours.
3. Twenty-four hours after onset of thrombocytopenia, macrocytosis of cells of Stage I and II was present.
4. Between 24 and 48 hours, megakaryocyte turnover was accelerated.
5. By 48 hours, macrocytosis was maximal in the Stage III cells.
6. After 48 hours, platelet production increased.

The mechanism underlying macromegakaryocytosis has not been clarified, but it may relate to stimulation of nuclear replication and endomitosis. Harker [15] has described an increased number of nuclei in large megakaryocytes, but quantitative determination of nuclear DNA content has not been reported. It seems reasonable that cells with large amounts of DNA would form more nuclear lobes than those with lesser amounts. However, the multiple nuclei of megakaryocytes appear to arise by a process of segmentation rather than being formed at the time of endomitotic replication of DNA, so there may not be an exact correlation between number of segments and DNA content. Megakaryoblasts (Stage I) are the only cells of the series which synthesize DNA,[1,2,16] hence, full polyploidy is achieved before cytoplasmic maturation ensues. Bessis [12] describes the megakaryoblast as a cell with a single nucleus which undergoes segmentation during maturation to the basophilic megakaryocyte (Stage II). In concert with this observation is the finding by deLeval and Paulus [17] that ^{3}HTdR is incorporated only into nonsegmented megakaryocytic nuclei in guinea pigs. The data of Odell and Jackson [16] indicate that the cells which label initially with ^{3}HTdR in the rate are already polyploid. Clarification of the relationships between DNA content, nuclear lobe count, and megakaryocyte size obviously awaits further investigations.

The relative importance of macromegakaryocytosis as compared to increased turnover rate in determining the final increase in numbers of platelets produced is another point which is not clear. There is a considerable amount of evidence that newly formed platelets are larger than those that have aged in the circulation [18–20] and, also, that macrothrombocytes may be produced in association with accelerated platelet production.[9,12,18,21,22] It is conceivable, therefore, that macromegakaryocytes may produce larger platelets rather than large numbers of platelets.

The delay between onset of thrombocytopenia and onset of demonstrable changes in megakaryocytes is probably determined by at least two factors. First, there may be a delay in elaboration and accumulation of a critical concentration of a stimulant, or thrombopoietin. Second, a certain amount of time may be required for the stimulated precursor cells to reach the compartment of recognizable megakaryocytes.

Irradiation and Acute Platelet Depletion

When thrombocytopoiesis was depressed in normal mice by exposure to sublethal doses of X-radiation (200–400 r) the platelet counts showed a characteristic response.[5] For 4 days, normal levels of circulating platelets were maintained. During the subsequent 4 to 5 days, the platelet levels fell to reach minimum values on the eighth or ninth postirradiation day. Thereafter, they gradually returned to normal values. The initial plateau of 4 days corresponded to the time that normal platelet production was maintained by radioresistant cells. The recognizable megakaryocytes which do not proliferate would be expected to be resistant to radiation. However, the maturation time for megakaryocytes in this strain of mice appeared to be only about 3 days or less when determined with ^{3}HTdR.[23] Prolongation of the initial plateau beyond the maturation time of recognizable megakaryocytes suggested that an immediate precursor, with a maturation time of about a day, may also be radioresistant.

An alternative explanation would be that the splenic platelet pool was adequate to maintain normal platelet levels for about a day after production by the marrow ceased. To test this hypothesis, splenectomized mice were exposed to 300 r, and, as shown in Figure 2, their platelet counts responded

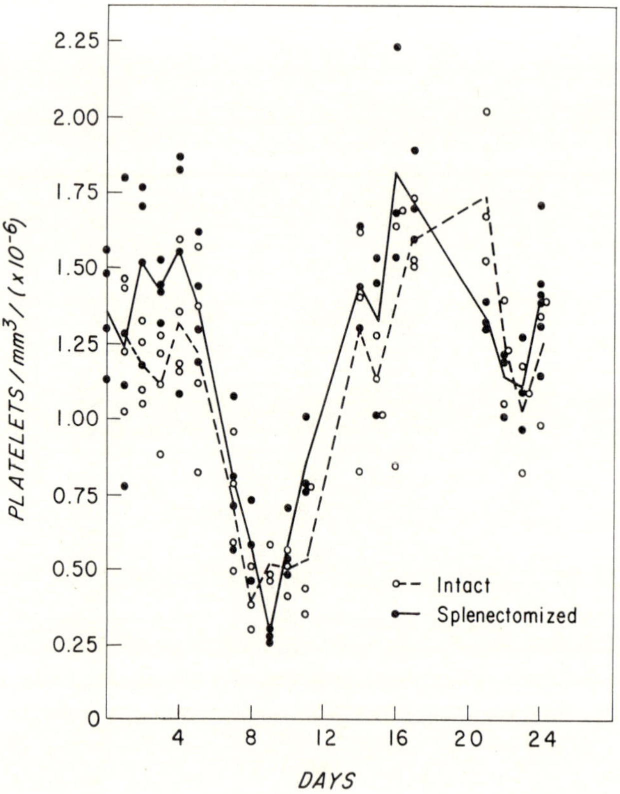

FIG. 2—Platelet counts of intact and splenectomized mice after exposure to 300 r. Individual values are shown by the open and closed circles; the lines connect average values.

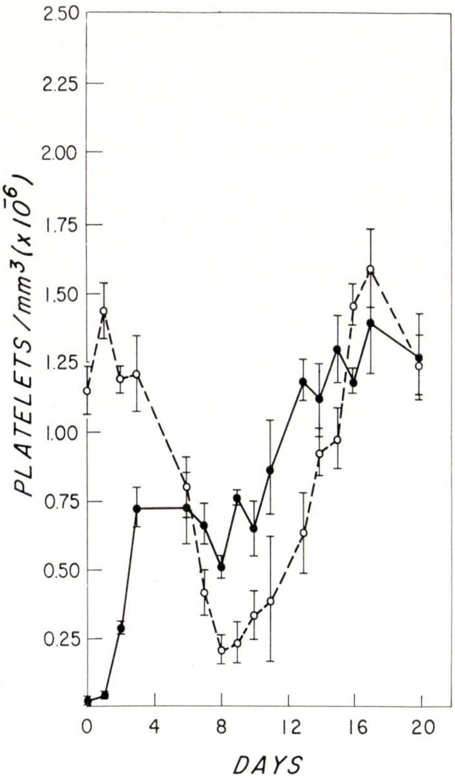

FIG. 3—Platelet counts of mice after exposure to 300 r. Closed circles and solid lines represent mice which were made thrombocytopenic 1–2 hours before irradiation by a single injection of antiplatelet serum. Open circles and dashed lines are control irradiated mice. Each point is the average of 3–6 mice, and vertical bars indicate ± 1 S.E.M. (From Reference 5).

to radiation in the same way as did those of the intact control mice. Therefore, it appeared probable that the bone marrow was able to maintain normal thrombocytopoiesis for a longer time after exposure to radiation than would be predicted from the time that cells spend only in the compartment of recognizable megakaryocytes.

It was demonstrated that erythropoiesis is stimulated in sublethally irradiated animals which were acutely anemic or hypoxic during the immediate pre- or postirradiation period.[24] This modification of the radiation-induced erythroid suppression was later shown to be due to the humoral factor, erythropoietin.[25] The effects of acute and chronic platelet depletion on the response of the thrombocytopoietic system to irradiation have more recently been studied.[5] Figure 3 shows the effect of acute platelet depletion by a single injection of antiplatelet serum in the immediate preirradiation period on the course of radiation-induced thrombocytopenia. In those mice which were acutely thrombocytopenic at the time of irradiation, the radiation-induced thrombocytopenia was neither as severe nor as prolonged as in controls. A similar effect was seen in the mice whose platelet counts are shown in

Figure 4. They were made thrombocytopenic by a single injection of antiplatelet serum immediately after exposure to 300 r. In the animals so treated, the radiation-induced thrombocytopenia was also less severe and of shorter duration than in irradiated controls.

These results showed modification of the thrombocytopoietic response to sublethal radiation by acute platelet depletion during the immediate pre- or postirradiation period. Therefore, they provide further evidence to indicate that there is a homeostatic mechanism by which the level of circulating platelets influences production of platelets.

Since acute depletion after irradiation appeared to stimulate megakaryocytopoiesis, a significant number of cells which are responsive to thrombopoietin apparently are in a state in which they are not heavily damaged by radiation. This conclusion was substantiated by demonstration of more rapid recovery from radiation-induced thrombocytopenia when the stimulus of immunothrombocytopenia was introduced as late as a day after exposure to 300 r.[5] The level of precursor cell which was affected by the thrombocytopenic stimulus can not be determined. It may be a cell already committed to megakaryocytopoiesis or, possibly, the pluripotential stem cell.

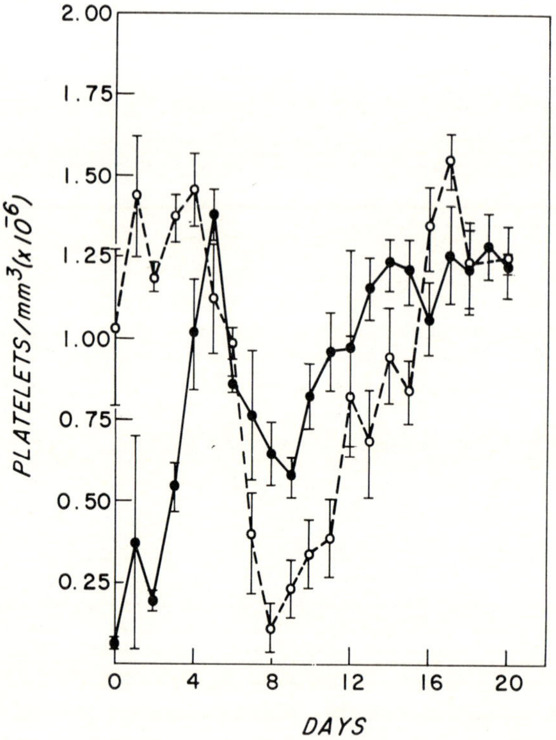

Fig. 4—Platelet counts of mice after exposure to 300 r. Closed circles and solid line represent mice which were made thrombocytopenic by a single injection of antiplatelet serum within an hour after exposure to x-ray. Open circles and dashed line are irradiated controls. Each point is the average of 3–6 mice, and vertical bars indicate ± 1 S.E.M. (From Reference 5).

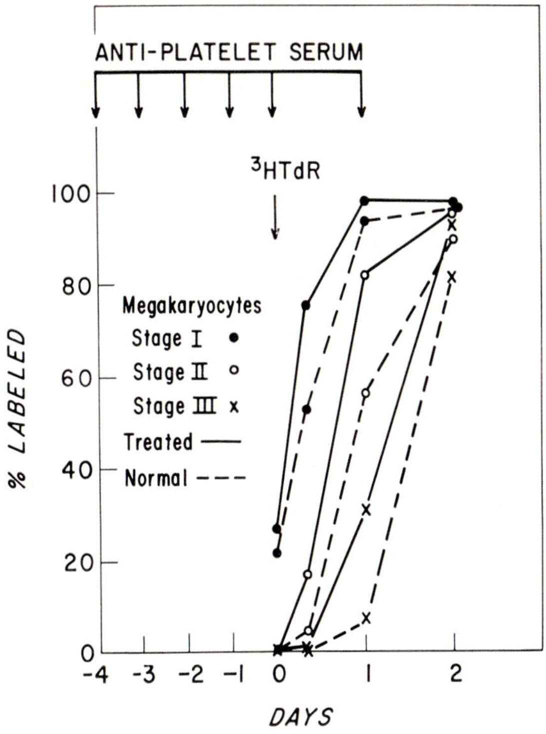

FIG. 5—Labeling of rat megakaryocytes by ^{3}HTdR (1 μc./Gr., I. V.) given after 4 daily doses of rabbit anti-rat platelet serum (APS). Each value for treated rats is the average for 4 animals; for normal, each point is the average for 3 rats; 250 megakaryocytes were evaluated in each rat. Platelet counts after the first dose of APS (day —4) averaged 3 per cent of normal, those on days 0, 1, and 2 were 12 per cent, 5 per cent, and 21 per cent of normal respectively.

Irradiation and Chronic Platelet Depletion

Rolovic et al.[26] reported the results of labeling of megakaryocytes with ^{3}HTdR in rats in which chronic immunothrombocytopenia was produced by repeated injections of antiplatelet serum. They concluded that the rate of megakaryocyte maturation in experimentally induced immune thrombocytopenia was depressed. Their findings suggested a different pattern of megakaryocytopoiesis in rats with chronic thrombocytopenia than had been seen in animals acutely stimulated by a single does of antiplatelet serum.[13] In similar studies, however, we were not able to demonstrate a depression of megakaryocytopoiesis in chronically stimulated animals. Figure 5 shows labeling indices for rat megakaryocytes when the ^{3}HTdR was administered after thrombocytopenia had been maintained for 4 days by daily doses of antiplatelet serum. The rate of megakaryocyte turnover was accelerated in these rats, as it had been in the acutely stimulated ones. However, megakaryocyte maturation was less rapid than it had appeared to be during the second day after acute platelet depletion. Total maturation time for recognizable megakaryocytes was estimated to be ~32 hours, or 44 per cent of controls, after acute stimulation.[13] In the more chronically thrombocytopenic rats it was ~42 hours, or 67 per cent of controls. This difference tended to corroborate the conclusion of Odell et al.[14] that a part of the apparent shortening of megakaryocyte maturation during the early response to thrombocytopenia was due to influx of increased numbers of cells labeled with ^{3}HTdR.

In spite of similarities of megakaryocyte maturation in acutely and chronically thrombocytopenic rats, chronically stimulated mice did not respond to sublethal radiation in the same way as did the the acutely stimulated ones.[5] The platelet counts shown in Figure 6 were from mice which were exposed to x-ray after they had been kept in a thrombocytopenic state for 4 days by daily injections of antiplatelet serum. The initial reactive thrombocytosis was not prevented by radiation, indicating that its magnitude was, in part, determined by compensatory changes already established in the pool of radioresistant megakaryocytic cells. However, its duration became shorter with increasing doses of radiation. This probably meant that persistence of post-thrombocytopenic thrombocytosis was dependent on continued replacement of the rapidly maturing megakaryocytes from a pool of proliferating precursor cells which, in turn, were damaged by the radiation.

The severity of radiation-induced thrombocytopenia was not modified by chronic immunothrombocytopenia present at the time of exposure to radiation. However, recovery from the lower doses of radiation (200 and 300 r) occurred somewhat more slowly than in the irradiated controls. The megakaryocytic system with this degree of chronic stimulation was at least as radiosensitive

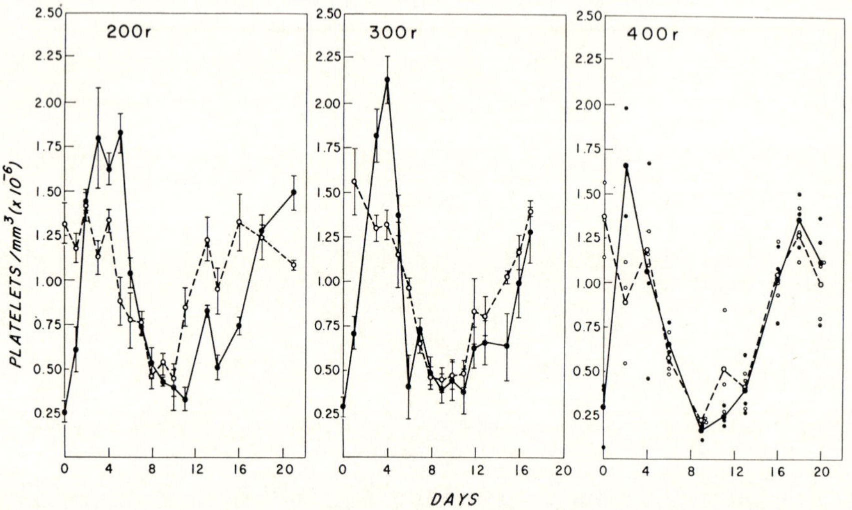

FIG. 6—Platelet counts of mice after exposure to sublethal radiation. Closed circles and solid lines represent mice in which thrombocytopenia was maintained for 4 days by daily injections of antiplatelet serum; the final injection was given 1 day prior to irradiation. Open circles and dashed lines are irradiated controls. Each point for 200 and 300 r is the average of 3–6 mice and vertical bars are ± 1 S.E.M. Each point on the lines for 400 r is the average of 3 mice with the other points being individual values. (From Reference 5).

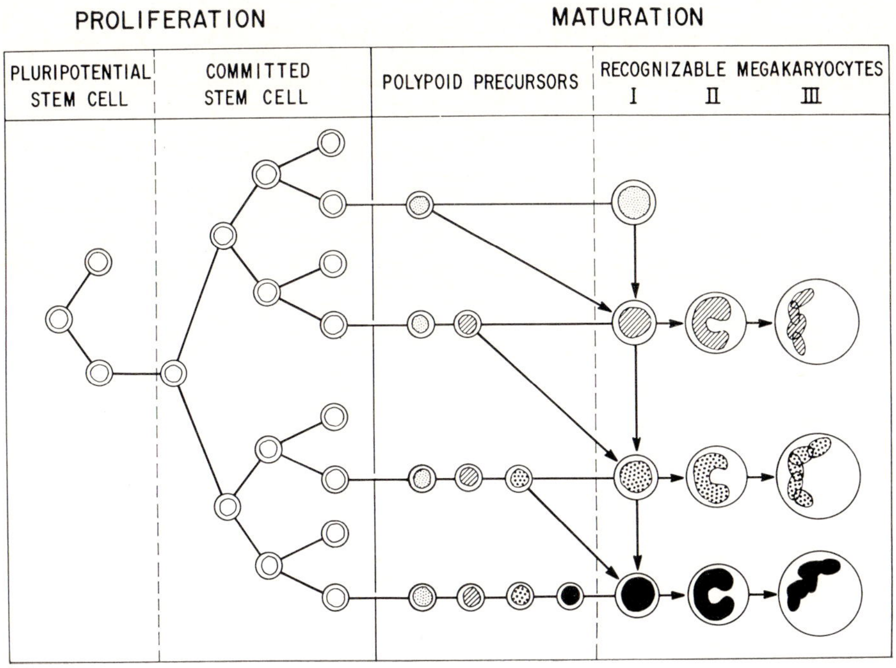

FIG. 7—Model for megakaryocytopoiesis in the normal rat. Different contents of nuclear DNA are indicated by different degrees of nuclear stippling: open nuclei are 2N, 4N, 8N, 16N, and 32N are represented from top to bottom in the compartment of recognizable megakaryocytes. (From Reference 29).

as normal and, perhaps, somewhat more so. Nakamura et al.[27] have reported that the 30 day survival of mice after exposure to lethal doses of x-ray was improved if the mice were made thrombocytopenic by pretreatment with antiplatelet serum 2 hours, 1 day or 2 days before exposure. They also found that such pretreatment given 2 hours or 1 day before exposure to sublethal radiation resulted in improvement of the platelet counts 10 days after exposure.

These results of the effects of radiation on the acutely and chronically stimulated thrombocytpietic system suggest that thrombopoietin may act by stimulating precursor cells which are radioresistant. Because of this radioresistance, a state of rest or prolonged cell cycle may be proposed. With chronic stimulation, a new steady state may be established in which this population of cells is expanded, but a normal or somewhat decreased proportion of them are in a resting state.

Models of Megakaryocytopiesis

We have previously proposed a model for megakaryocytopoiesis in the normal steady state.[28,29] In this model, the cell system originated from the pluripotential stem cell and terminated in the maturation pool of recognizable megakaryocytes (Fig. 7). One of the actions of thrombopoietin appears to be to hasten the progress of cells through all compartments committed to megakaryocytopoiesis. The immediate precursors of the recognizable megakaryocytes are probably cells which become polyploid before differentiating into the recognizable compartment, but the degree of polyploidy which these cells may develop before becoming large enough to be recognizable as megakaryocytes is not firmly established. This compartment could be considered, on the one hand, to be a multiplicative pool with regard to its nuclear replication or, on the other hand, to be a nonproliferating, maturation compartment since cell division probably does not occur. The compartment of polyploid precursors may be responsible for prolongation of normal thrombocytopoiesis in the early post-irradiation period for longer than anticipated from the maturation time of recognizable megakaryocytes. It could have some degree of radioresistance because of its polyploidy or because its cells may be almost continually in DNA synthesis. One could speculate that action of the thrombocytopoietic stimulus at this cell level would lead to increased DNA synthesis and production of macromegakaryocytosis.

In the absence of cell division within the compartment of recognizable megakaryocytes and, probably, also the polyploid precursors, maintenance of these compartments is dependent on continuous influx from precursor compartments. To supply this demand for cells, a multiplicative compartment of committed stem cells (2N megakaryocytes) between the pluripotential stem cell and the polyploid immediate precursors was incorporated into the model. Stimulation of the multiplicative compartment by thrombopoietin may provide the increased numbers of megakaryocytes to renew the rapidly maturing compartment and, with more prolonged stimulation, to expand it. These diverse responses (stimulation of cell proliferation, endomitosis, differentiation, and cytoplasmic maturation) result from a single event, namely, depletion of circulating platelets. They may all be mediated by the effects of a single thrombopoietin on cells at different stages of differentiation. Alternatively, there may be more than one thrombopoietin as suggested by the observations of Harker and Finch.[15,30]

Our model for megakaryocytopoiesis proposed three levels of cells which are not recognizable on the basis of morphologic appearance: the pluripotential stem cell, the committed stem cell, and the polyploid immediate precursors. Other than the pluripotential cell, it did not propose cells which were capable of proliferation but which were normally in a state of rest or prolonged cell cycle. The postirradiation stimulation of thrombocytopoiesis by acute platelet depletion strongly suggests that thrombopoietin acts on a cell population which is radioresistant and, hence, may be presumed to be resting (G_0) or in prolonged G_1. If thrombopoietin, like erythropoietin, acts at the committed stem cell level, then a portion of the 2N megakaryocytes would appear to be in G_0 or prolonged G_1. In that case, some committed stem cells would be multiplying to renew the maturation compartments, others would be resting. While this hypothesis is the most attractive, the available information does not rule out a separate committed, but resting, cell population between the pluripotential stem cells and the proliferating 2N megakaryocytes or, even less likely, a direct effect of thrombopoietin on the pluripotential stem cell. Until there are more experimental data which define the kinetic properties of megakaryocyte precursors, it is probably prudent to resist the temptation to propose yet another model for their behavior.

ACKNOWLEDGMENTS

I would like to thank Miss Elizabeth Phalen, Mr. Donald Howard, Mrs. Joan Overcash, and Mrs. Janet Donovan Horita for their excellent technical contributions to the work which was summarized in this report.

REFERENCES

1. Feinendegen, L. E., Odartchenko, N., Cottier, H., and Bond, V. P.: Kinetics of megakaryocyte proliferation. Proc. Soc. Exp. Biol. Med. 111:177, 1962.

2. Ebbe, S., and Stohlman, F., Jr.: Megakaryocytopoiesis in the rat. Blood 26:20, 1965.

3. Brecher, G., and Cronkite, E. P.: Morphology and enumeration of human blood platelets. J. Appl. Physiol. 3:365, 1950.

4. Ebbe, S., Stohlman, F., Jr., Overcash, J., Donovan, J., and Howard, D.: Megakaryocyte size in thrombocytopenic and normal rats. Blood 32:383, 1968.

5. Ebbe, S., and Stohlman, F., Jr.: Stimulation of thrombocytopoiesis in irradiated mice. Blood. In press.

6. Odell, T. T., Jr., and Anderson, B.: Production and life span of platelets. *In* Stohlman, F., Jr. (Ed.): The Kinetics of Cellular Proliferation. New York, Grune & Stratton, 1959, p. 278.

7. Ebbe, S., Stohlman, F., Jr., Donovan, J., and Howard, D.: Platelet survival in the rat as measured with tritium-labeled diisopropylfluorophosphate. J. Lab. Clin. Med. 68:233, 1966.

8. Patt, H. M., Maloney, M. A., and Jackson, E. M.: Recovery of blood neutrophils after acute peripheral depletion. Amer. J. Physiol. 188:585, 1957.

9. Matter, M., Hartmann, J. R., Kautz, J., DeMarsh, Q. B., and Finch, C. A.: A study of thrombopoiesis in induced acute thrombocytopenia. Blood 15:174, 1960.

10. Odell, T. T., Jr., McDonald, T. P., and Asano, M.: Response of rat megakaryocytes and platelets to bleeding. Acta Haemat. (Basel) 27:171, 1962.

11. Craddock, C. G., Jr., Adams, W. S., Perry, S., and Lawrence, J. S.: The dynamics of platelet production as studied by a depletion technique in normal and irradiated dogs. J. Lab. Clin. Med. 45:906, 1955.

12. Bessis, M.: Cytology of the Blood and Blood-forming Organs. New York, Grune & Stratton, 1956.

13. Ebbe, S., Stohlman, F., Jr., Donovan, J., and Overcash, J.: Megakaryocyte maturation rate in thrombocytopenic rats. Blood 32: 787, 1968.

14. Odell, T. T., Jr., Jackson, C. W., Friday, T. J., and Charsha, D. E.: Effects of thrombocytopenia on megakaryocytopoiesis. Brit. J. Haemat. 17:91, 1969.

15. Harker, L. A.: Kinetics of thrombopoiesis. J. Clin. Invest. 47:458, 1968.

16. Odell, T. T., Jr., and Jackson, C. W.: Polyploidy and maturation of rat megakaryocytes. Blood 32:102, 1968.

17. deLeval, M., and Paulus, J. M.: Nuclear maturation and cytoplasmic organelles and antigens in guinea pig and human megakaryocytes. (Abstract.) Blood 34:529, 1969.

18. McDonald, T. P., Odell, T. T., Jr., and Gosslee, D. G.: Platelet size in relation to platelet age. Proc. Soc. Exp. Biol. Med. 115:684, 1964.

19. Minter, N. and Ingram, M.: Density distribution of platelets. (Abstract.) Blood 30:551, 1967.

20. Karpatkin, S.: Heterogeneity of human platelets. I. Metabolic and kinetic evidence suggestive of young and old platelets. J. Clin. Invest. 48:1073, 1969.

21. Harker, L. A.: Significance of platelet volume alterations. (Abstract.) Abstracts of Simultaneous Sessions, XII Congress Int. Soc. Hemat., New York, 1968, p. 202.

22. Karpatkin, S., and Siskind, G. W.: In

vitro detection of platelet antibody in patients with idiopathic thrombocytopenic purpura and systemic lupus erythematosus. Blood 33:795, 1969.

23. Ebbe, S.: The megakaryocyte: Maturation and self-renewal. *In* The Platelet. Baltimore, Williams & Wilkins. In press.

24. Stohlman, F., Jr., Cronkite, E. P., and Brecher, G.: Stimulation of erythropoiesis in irradiated dogs and rats. Proc. Soc. Exp. Biol. Med. 88:402, 1955.

25. Stohlman, F., Jr., and Brecher, G.: Stimulation of erythropoiesis in sublethally irradiated rats by a plasma factor. Proc. Soc. Exp. Biol. Med. 91:1, 1956.

26. Rolovic, Z., Baldini, M., and Dameshek, W.: Megakaryocyte kinetics in experimentally induced immune thrombocytopenia. (Abstract.) Abstracts of Simultaneous Sessions, XII Congress Int. Soc. Hemat., New York, 1968, p. 206.

27. Nakamura, W., Kojima, E., Minamisawa, H., Kankura, T., Kobayaski, S., and Eto, H.: Role of thrombopoietic system in radiation induced hematopoietic death. *In* Bond, V. P., and Sugahara, T. (Eds.): Comparative Cellular and Species Radiosensitivity. Tokyo, Igaku Shoin Ltd., 1969, p. 202.

28. Ebbe, S.: Megakaryocytopoiesis and platelet turnover. Ser. Haemat. 1(2):65, 1968.

29. Ebbe, S.: Megakaryocytopoiesis. *In* Gordon, A. S. (Ed.): Regulation of Hematopoiesis. New York, Appleton-Century-Crofts. In press.

30. Harker, L. A., and Finch, C. A.: Thrombokinetics in man. J. Clin. Invest. 48: 963, 1969.

Thrombopoiesis and Thrombopoietin

By Jack Levin

The existence of thrombopoietin, a humoral factor analogous to erythropoietin for the control of thrombopoiesis, has been suggested by many studies. The evidence for a humoral control mechanism for platelet production has been reviewed recently.[1,2] Thrombocytopenia produced by exchange transfusion or platelet antiserum is followed by transient thrombocytosis, which occurs after a delayed increase in platelet production.[3–6] Stimulation of thrombopoiesis following induction of thrombocytopenia results in increased entry of precursor cells into the megakaryocyte pool,[2,5,7,8] an increase in the rate of maturation of megakaryocytes,[2,5,7–9] and an increase in megakaryocyte number and size.[2,6,8,10] In contrast, thrombocytosis secondary to platelet hypertransfusion is followed by transient thrombocytopenia.[11–13] These phenomena are compatible with a feedback mechanism, which has been suggested by Ebbe et al.[6] and Odell et al.[8] to be effective at the level of the megakaryocytic precursor cells.

Thrombocytopenia in some members of 2 families was attributed to deficiency of a plasma component with thrombopoietic stimulating activity.[14,15] Sullivan recently reported elevation of platelet levels in a patient following infusion with his own plasma, obtained during recovery from alcohol induced thrombocytopenia.[16] Whether or not these plasma factors are concerned with regulation of platelet production is unknown.

The presence of thrombopoietic stimulating activity in plasma or serum of normal or thrombocytopenic donors has been reported. Infusions of normal human plasma elevated the platelet counts of rats,[17,18] but other investigators have been unable to demonstrate thrombopoietic activity in the blood of normal animals when the recipient and donor animals were of the same species.[19–21] Plasma from thrombocytopenic animals caused a temporary rise in circulating platelet levels in normal recipients,[1,19,20,22] but such results have not been regularly produced.[19,23] Thrombopoietic activity has been detected in the plasma of patients with myeloproliferative disorders[24,25] and idiopathic thrombocytopenic purpura.[26,27] All of these studies used changes in levels of circulating platelets as a measure of activity. However, platelet counts do not necessarily reflect alterations in rate of production since the concentration of circulating cells in the blood is the resultant of the rate of production and of the rate of destruction. Furthermore, there is considerable fluctuation of platelet counts of rodents[11,26,28] which have been used in most

---

From the Department of Medicine, The Johns Hopkins University School of Medicine and Hospital, Baltimore, Maryland.

These investigations were carried out under Contract NYO–1208–103 between the United States Atomic Energy Commission and The Johns Hopkins University, and were supported in part by a Graduate Training Grant, TI–AM–5260, from the National Institute of Arthritis and Metabolic Diseases, U.S. Public Health Service.

studies. Odell, McDonald, and Howsden [29] have emphasized the possible misinterpretation of variations in platelet counts.

A few investigators have measured platelet production using isotopic labels. Induced thrombocytopenia and thrombocytosis altered the levels of the isotope, $Na_2{}^{35}SO_4$, in circulating platelets, suggesting a feedback control mechanism.[10,12] However, Ebbe, Stohlman, Donovan, and Howard [30] were not able to demonstrate alterations in platelet production after platelet hypertransfusion in rats, by studying the rate of fall of platelet radioactivity after labeling with diisopropylfluorophosphate-^{3}H.

In our studies, thrombopoiesis was investigated using ^{75}selenomethionine (^{75}SeM).[31] ^{75}SeM follows the metabolic pathway of methionine and is incorporated into proteins.[32–34] It appears in circulating platelets after its initial incorporation into megakaryocytes,[31,35] but is not incorporated into platelets already in the circulation.[31]

Incorporation of ^{75}SeM into platelets, in vivo. The rate of appearance of ^{75}SeM in the circulating platelets of rabbits following the intravenous administration of ^{75}SeM is shown in Figure 1. Maximum levels of ^{75}SeM in circulating platelets occurred approximately 3 days after injection. Radioactivity of the platelets began to decrease after 3 days, and by 8 days had fallen to

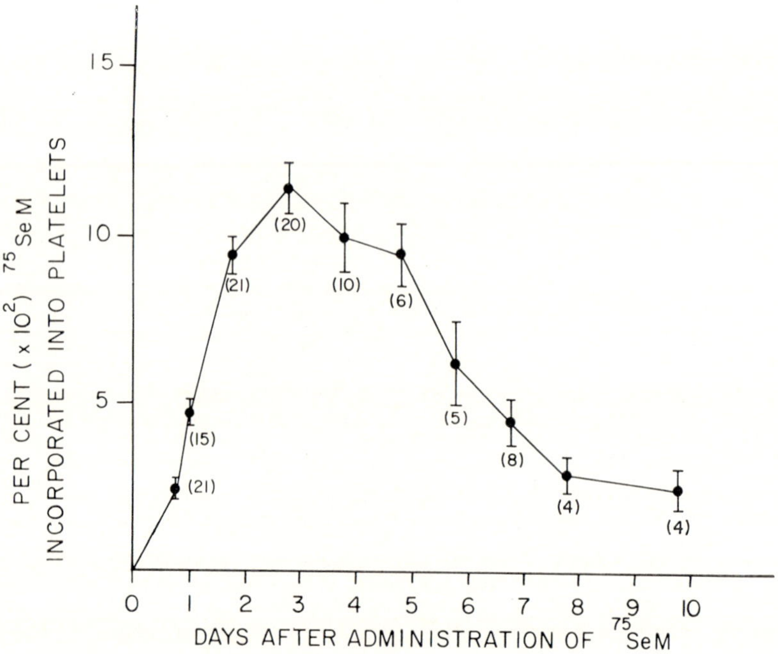

FIG. 1—Per cent incorporation of ^{75}SeM into circulating platelets in normal rabbits. Each rabbit received 15 μc. of ^{75}SeC intravenously. The Mean ± 1 SE is shown, and the numbers in parentheses indicate the number of animals studied. Maximum ^{75}SeM in circulating platelets occurred 3 days after administration of isotope. (Reproduced from Evatt and Levin, J. Clin. Invest. 48:1615, 1969, with permission of The Rockefeller University Press.)

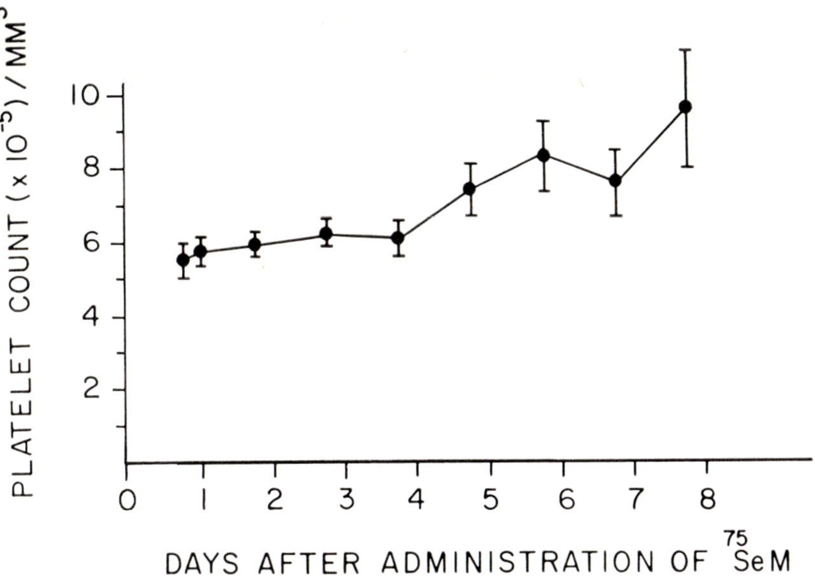

FIG. 2—Platelet counts in rabbits during study of incorporation of ^{75}SeM. Approximately 5 ml. of blood was obtained at each venipuncture. Platelet counts remained stable for 4 days, and then tended to rise, probably secondary to blood loss. The Mean $\pm$ 1 SE is shown. (Reproduced from Evatt and Levin, J. Clin. Invest. 48:1615, 1969, with permission of The Rockefeller University Press.)

the level observed 18 hours after administration of ^{75}SeM. Platelet counts (Fig. 2) of rabbits subjected to serial withdrawal of blood remained stable for 4 days but subsequently tended to rise, probably secondary to the effects of hemorrhage.

Effects of acute hemorrhage. Rabbits were bled 25 ml. by intracardiac puncture, and 24 hours later were given ^{75}SeM. Significantly more ^{75}SeM was incorporated into the platelets of animals that had been bled than in the control group during the 42 hour period after administration of the isotope (Fig. 3). However, the acute hemorrhage was not sufficient to produce either thrombocytopenia or a subsequent significant increase in the platelet counts (Fig. 4). These results suggested that incorporation of the isotope is a more sensitive indicator of altered platelet production than is the platelet count.

Effects of hypertransfusion of platelets. Thrombocytosis was produced and maintained by serial transfusions of fresh platelet concentrates. After 3 days of sustained thrombocytosis (platelet levels 2–4$\times$normal), ^{75}SeM was administered, and platelet concentrates were administered for an additional 3 days. The per cent of ^{75}SeM incorporated into platelets was markedly decreased in rabbits with sustained thrombocytosis produced by hypertransfusion of platelets (Fig. 5). Relative thrombocytopenia occurred after the period of transfusion-induced thrombocytosis (Fig. 6), further suggesting suppression of platelet production.

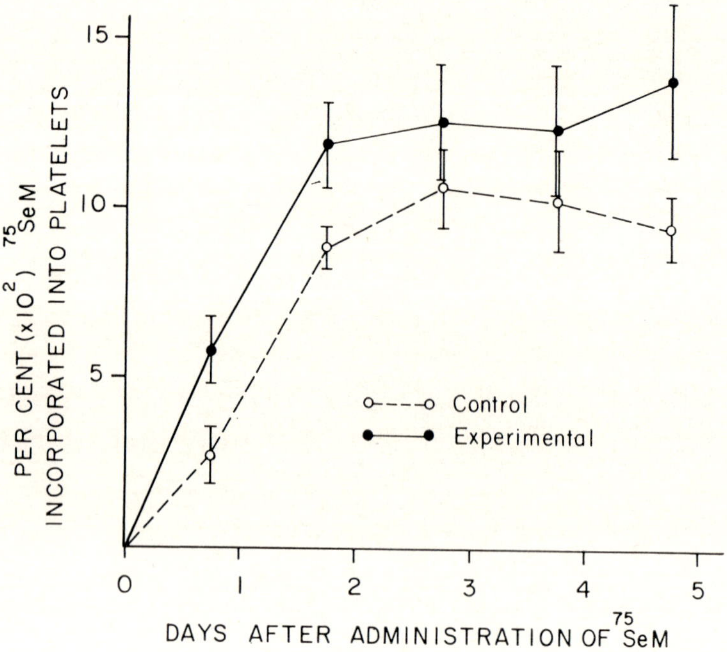

FIG. 3—Effect of hemorrhage upon incorporation of ^{75}SeM into platelets. Experimental animals were bled 25 ml., 1 day before administration of ^{75}SeM. There were 6 animals in each group. The mean ± 1 SE is shown. The rabbits which had been bled incorporated more ^{75}SeM into their platelets than did controls. (Reproduced from Evatt and Levin, J. Clin. Invest. 48:1615, 1969, with permission of The Rockefeller University Press.)

Detection of thrombopoietic activity using normal recipients. Twenty normal rabbits received from 75 to 225 ml. of platelet-free plasma from normal donors, and 18 rabbits received similar volumes of platelet-free plasma from rabbits rendered acutely thrombocytopenic by administration of antirabbit platelet serum. ^{75}SeM was administered 6 hours after the last of three equally divided doses of plasma. There was no significant alteration in the rate of appearance or levels of ^{75}SeM in the circulating platelets of normal recipients that were given plasma from normal donors (Fig. 7). In contrast, similar volumes of plasma from thrombocytopenic donors produced significant increases in both the rate of appearance and levels of ^{75}SeM in the platelets of normal recipients. There was no measurable dose response in the range tested, and the data were pooled (Fig. 8). When calculated as per cent change from control values, the increases at 18 and 24 hours were 39 per cent and 32 per cent respectively; P was less than 0.005 during the 42 hour period following the administration of ^{75}SeM. The results suggested that platelet-free plasma from thrombocytopenic animals contained a substance which increased the uptake of ^{75}SeM into the platelets of normal recipients. The platelet counts of rabbits given plasma from normal or thrombocytopenic donors did not increase or differ from each other.

Detection of thrombopoietic activity using recipients with suppressed thrombopoiesis. Thrombocytosis produced by platelet hypertransfusion suppressed thrombopoiesis (Fig. 5). This observation suggested that animals in which endogenous thrombopoiesis had been suppressed might provide a more sensitive assay model for the detection of thrombopoietic stimulating activity. Thrombopoietic stimulating activity of plasma was evaluated by its effect upon the levels of ^{75}SeM in the circulating platelet mass of rabbits that had received platelet transfusions for 5 days. Plasma was administered on the third and fourth days of a 5 day period of transfusion of platelet concentrates; ^{75}SeM was administered on the fourth day. Total doses of 150, 75, 37.5, or 20 ml. of plasma from thrombocytopenic donors were given. Maximum incorporation of ^{75}SeM was increased 75, 60, 41, and 36 per cent, respectively ($P<0.005$–0.01), as compared to hypertransfused controls.[36] The results demonstrated a dose-response relationship to thrombopoietic stimulating factor(s) in plasma of thrombocytopenic donors. Plasma from normal donors also increased the incorporation of ^{75}SeM, but produced a lesser effect than plasma from thrombocytopenic donors at each dose level tested.

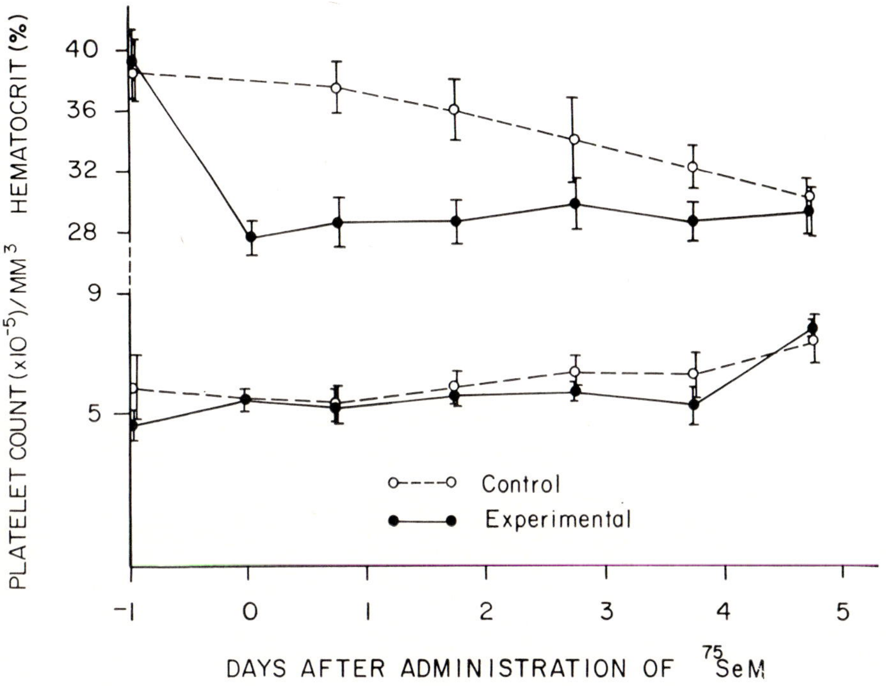

FIG. 4—Effect of hemorrhage upon peripheral platelet counts and hematocrit values. See legend for Figure 3. Removal of 25 ml. of blood reduced the average hematocrit value from 39.3 per cent to 27.5 per cent, but did not produce thrombocytopenia. There was no significant difference between the platelet counts of control animals and experimental animals which had been bled. The Mean ± 1 SE is shown.

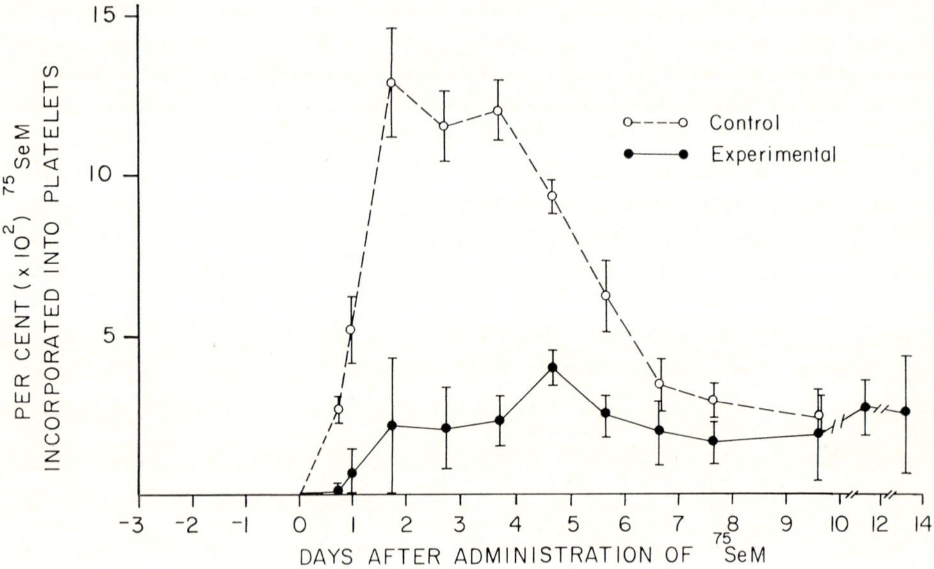

FIG. 5—Effect of hypertransfusion of platelets upon incorporation of ^{75}SeM into platelets. Two animals were hypertransfused with platelets, starting 3 days before administration of ^{75}SeM. Four control animals received equal volumes of fresh, platelet-free plasma. Thrombocytosis was maintained by daily platelet transfusions for a total period of 6 days (Fig. 6). The Mean ± 1 SE is shown. Hypertransfusion of platelets produced a marked decrease of incorporation of ^{75}SeM into platelets. (Reproduced from Evatt and Levin, J. Clin. Invest. 48:1615, 1969, with permission of The Rockefeller University Press.)

DISCUSSION

Maintenance of homeostasis requires various feedback mechanisms by which the body is able to regulate production of proteins. The existence of a humoral factor, erythropoietin, which helps regulate red cell production, is well established. Detection of thrombopoietic activity utilizing changes in platelet counts has been unsatisfactory. Studies by Odell et al.[12] and Harker,[10] using $Na_2{}^{35}SO_4$, and by Evatt and Levin,[31] using ^{75}SeM, have indicated that the use of an isotope that is incorporated into megakaryocytes provides a more satisfactory method for the measurement of the rate of thrombopoiesis. Evatt and Levin have shown [31] that the blood of thrombocytopenic donors contains a substance that increases the incorporation of ^{75}SeM into the platelets of normal recipients. Plasma from normal donors did not increase the incorporation of ^{75}SeM into the platelets of normal recipients. Additional studies, by Shreiner and Levin,[36] have indicated that rabbits in which endogenous thrombopoiesis has been suppressed by platelet hypertransfusion provide a more sensitive assay for the detection of thrombopoietic stimulating factor(s) in plasma, just as polycythemic animals provide a sensitive bioassay for the detection of erythropoietin. Using the latter model, it was possible to detect thrombopoietic activity in plasma from thrombocytopenic and normal donors and to demonstrate a dose response relationship to the administration of plasma from either type of donor. However, less thrombopoietic activity was detected in normal plasma than in the plasma of thrombocytopenic donors. Harker recently reported measurement of thrombopoietic activity in rat plasma using a similar model.[37]

Maximum changes in the rate and levels of incorporation of ^{75}SeM in normal and thrombocytotic rabbits were manifested during the 3 day period following administration of the isotope, or approximately 2–4 days after initiation of the experimental stimulation. This interval of time is consistent with previous estimations of the transit time of megakaryocytes in rabbits[38] and rats.[39] It is compatible with increasing entry of megakaryocytic precursors from an unidentified pool of stem cells into the compartment of recognizable megakaryocytes, their subsequent maturation, and release of platelets. ^{75}SeM probably is incorporated into the cytoplasm of precursor cells and megakaryoblasts, as well as later stages.

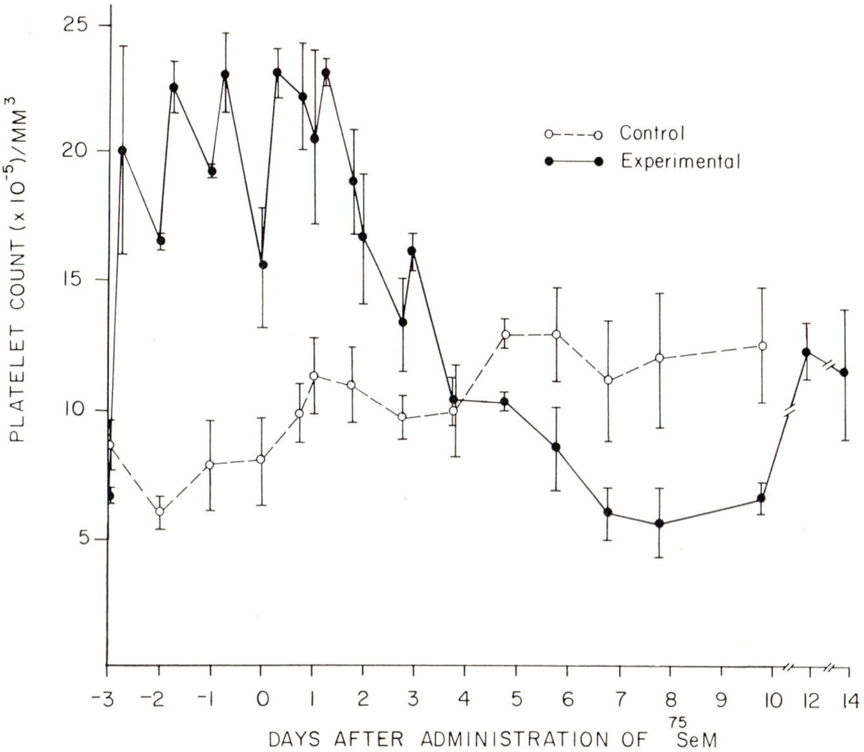

FIG. 6—Effect of hypertransfusion of platelets upon peripheral platelet counts. The Mean ± 1 SE is shown. Peripheral platelet levels were maintained from 2–4 × normal for a period of 6 days. Relative thrombocytopenia was observed after platelet transfusions were discontinued (days 7–10), at a time when control rabbits had elevated platelet counts secondary to repeated withdrawal of blood samples. A rebound above the original control level was seen following this period in the experimental animals. (Reproduced from Evatt and Levin, J. Clin. Invest. 48:1615, 1969, with permission of The Rockefeller University Press.)

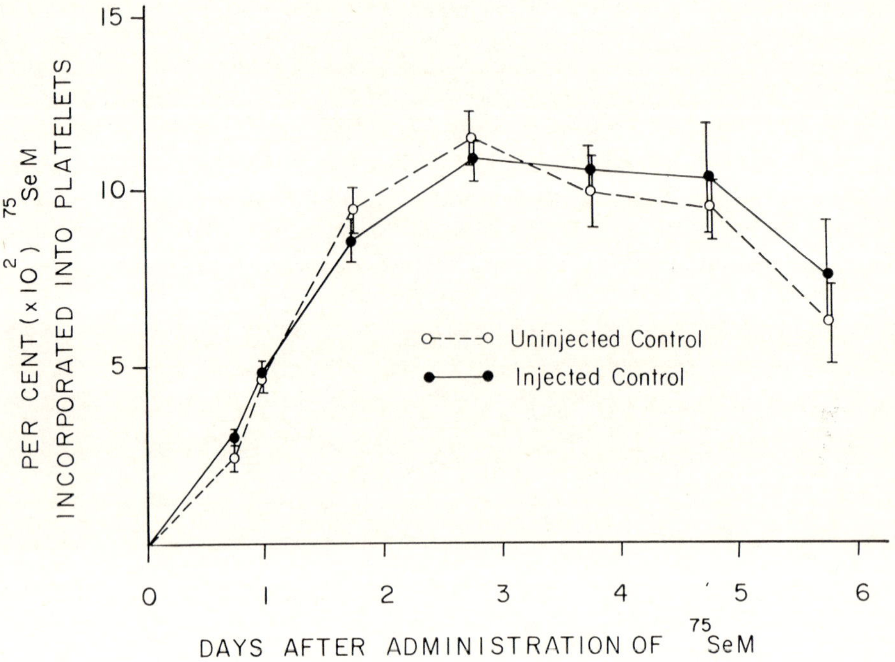

FIG. 7—Effect of normal plasma on incorporation of ^{75}SeM into platelets. Each of 41 rabbits received 15 μc. of ^{75}SeM intravenously. Twenty-one of this group (uninjected controls) received either no material or 1 ml. of normal saline daily before administration of isotope. The remaining 20 (injected controls) received from 75–225 ml. of platelet-free plasma from normal donors, in three equally divided doses, during the 32 hour period before isotope administration. The Mean $\pm$ 1 SE is shown. There was no significant difference in per cent incorporation of ^{75}SeM into platelets between the 2 groups. (Reproduced from Evatt and Levin, J. Clin. Invest. 48:1615, 1969, with permission of The Rockefeller University Press.)

The data are compatible with an increased release of platelets and/or an increase in platelet size. It is tempting to speculate that the latter was the case, since increases in platelet levels were not detected in animals in which thrombopoiesis was stimulated. Young platelets are larger and more dense than old platelets, and presumably contain more protein.[40–42] Changes in the rate and levels of incorporation of ^{75}SeM may primarily indicate changes in platelet protein, and may be more sensitive indicators of the state of thrombopoiesis than are alterations in the numbers of circulating platelets.

These results are compatible with a feedback mechanism by which circulating platelets affect the rate of thrombopoiesis. The observations strongly support the concept of a humoral agent, i.e., thrombopoietin, that acts on megakaryocytes to regulate platelet production. It is likely that such humoral factors act at least in part upon megakaryocytic precursors and immature megakaryocytes. The site of production, chemical nature, and mode of action of thrombopoietin(s) remain unknown, along with the location and nature of the mechanism that senses platelet levels or total platelet mass.

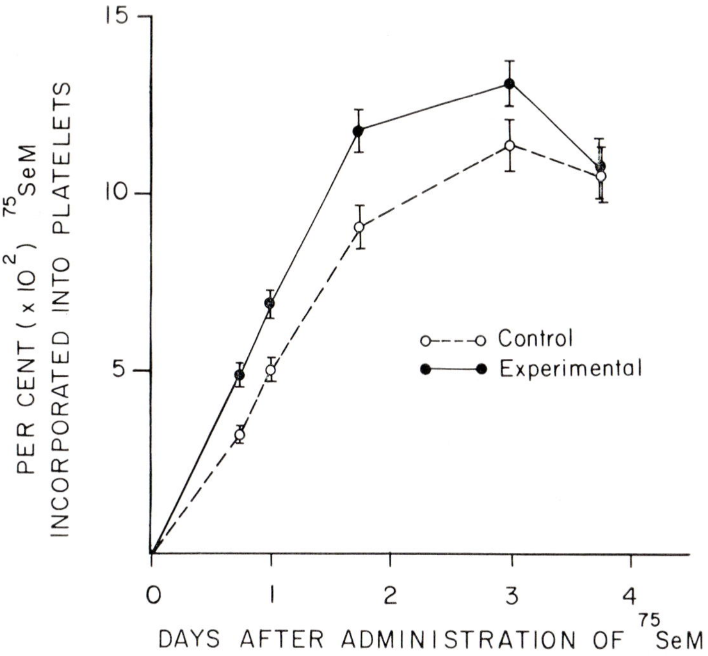

FIG. 8—Effect of plasma from thrombocytopenic donors upon incorporation of ^{75}SeM into platelets. 75–225 ml. of platelet-free plasma from either normal or thrombocytopenic donors was injected into normal recipients in three equally divided doses during the 32 hour period preceding the administration of ^{75}SeM. There were 18 animals in each group. The Mean ± 1 SE is shown. There was a significant increase in incorporation of isotope into platelets of animals that received plasma from thrombocytopenic donors (P less than 0.005 at 18, 24, and 42 hours). (Reproduced from Evatt and Levin, J. Clin. Invest. 48:1615, 1969, with permission of The Rockefeller University Press.)

REFERENCES

1. Abildgaard, C. F., and Simone, J. V.: Thrombopoiesis. Seminars Hemat. 4:424, 1967.

2. Ebbe, S.: Megakaryocytopoiesis and platelet turnover. Ser. Haemat. 1:65, 1968.

3. Craddock, C. G., Adams, W. S., Perry, S., and Lawrence, J. S.: The dynamics of platelet production as studied by a depletion technique in normal and irradiated dogs. J. Lab. Clin. Med. 45:906, 1955.

4. Matter, M., Hartmann, J. R., Kautz, J., DeMarsh, Q. B., and Finch, C. A.: A study of thrombopoiesis in induced acute thrombocytopenia. Blood 15:174, 1960.

5. Odell, T. T., Jr., McDonald, T. P., and Asano, M.: Response of rat megakaryocytes and platelets to bleeding. Acta Haemat. (Basel) 27:171, 1962.

6. Ebbe, S., Stohlman, F., Jr., Overcash, J., Donovan, J., and Howard, D.: Megakaryocyte size in thrombocytopenic and normal rats. Blood 32:383, 1968.

7. Odell, T. T., Jr., and Kniseley, R. M.: The origin, life span, regulation and fate of blood platelets. *In* Tocantins, L. M. (Ed.): Progress in Hematology. Vol. III. New York, Grune & Stratton, 1962, p. 203.

8. Odell, T. T., Jr., Jackson, C. W., Friday, T. J., and Charsha, D. E.: Effects of thrombocytopenia on megakaryocytopoiesis. Brit. J. Haemat. 17:91, 1969.

9. Ebbe, S., Stohlman, F., Jr., Donovan, J., and Overcash, J.: Megakaryocyte maturation rate in thrombocytopenic rats. Blood 32:787, 1968.

10. Harker, L. A.: Kinetics of thrombopoiesis. J. Clin. Invest. 47:458, 1968.

11. Cronkite, E. P., Bond, V. P., Fliedner, T. M., Paglia, D. A., and Adamik, E. R.: Studies on the origin, production and destruction of platelets. *In* Johnson, S. A., Monto, R. W., Rebuck, J. W., and Horn, R. C., Jr. (Eds.): Blood Platelets. Boston, Little, Brown, 1961, p. 595.

12. Odell, T. T., Jr., Jackson, C. W., and Reiter, R. S.: Depression of the megakaryocyte-platelet system in rats by transfusion of platelets. Acta Haemat. (Basel) 38:34, 1967.

13. de Gabriele, G., and Penington, D. G.: Physiology of the regulation of platelet production. Brit. J. Haemat. 13:202, 1967.

14. Schulman, I., Pierce, M., Lukens, A., and Currimbhoy, Z.: Studies on thrombopoiesis. I. A factor in normal human plasma required for platelet production; chronic thrombocytopenia due to its deficiency. Blood 16:943, 1960.

15. Vildosola, J., and Emparanza, E.: Hereditary familial thrombocytopenia. (Abstract.) Proceedings of the Xth Int. Cong. Paediat., Lisbon, 1962, p. 36.

16. Sullivan, L. W.: Studies of a humoral factor affecting platelet production in man. (Abstract.) Clin. Res. 17:345, 1969.

17. Linman, J. W., and Pierre, R. V.: Thrombocytosis-promoting activity of normal plasma. Proc. Soc. Exp. Biol. Med. 110: 463, 1962.

18. Steinberg, B., Cheng, F. H. F., Martin, R. A., and Deitz, A. A.: Thrombocytic and thrombopenic factors in normal human serum. Acta Haemat. (Basel) 33:341, 1965.

19. Spector, B.: *In vivo* transfer of a thrombopoietic factor. Proc. Soc. Exp. Biol. Med. 108:146, 1961.

20. Odell, T. T., Jr., McDonald, T. P., and Detweiler, T. C.: Stimulation of platelet production by serum of platelet-depleted rats. Proc. Soc. Exp. Biol. Med. 108:428, 1961.

21. Kelemen, E., Lehoczky, D., Cserháti, I., Krizsa, F., and Rák, K.: Demonstrability of a serum factor inducing thrombocytosis prior to acute rises of platelets in mice and men. Acta Haemat. (Basel) 29:16, 1963.

22. de Gabriele, G., and Penington, D. G.: Regulation of platelet production: "Thrombopoietin." Brit. J. Haemat. 13:210, 1967.

23. Abildgaard, C. F., Simone, J. V., and Schulman, I.: Nonreproducibility of thrombopoietin assays in the rat—a common experience? (Abstract.) Blood 30:547, 1967.

24. Keleman, E., Cserháti, I., and Tanos, B.: Demonstration and some properties of human thrombopoietin in thrombocythaemic sera. Acta Haemat. (Basel) 20:350, 1958.

25. Linman, J. W., and Pierre, R. V.: Studies on thrombopoiesis. III. Thrombocytosis-promoting effects of "thrombocythemic" and "polycythemic" plasmas. J. Lab. Clin. Med. 62:374, 1963.

26. Chat, L. X., Ebbe, S.: Assay of plasma thrombopoietic activity in rats. Med. Exp. (Basel) 7:317, 1962.

27. Schulman, I., Abildgaard, C. F., Cornet, J., Simone, J. A., and Currimbhoy, Z.: Studies on thrombopoiesis. II. Assay of human plasma thrombopoietic activity. J. Lab. Clin. Med. 66:604, 1965.

28. Penington, D. G.: Studies of platelet production and sequestration in the experimental animal. Proc. Roy. Soc. Med. 61: 601, 1968.

29. Odell, T. T., Jr., McDonald, T. P., and Howsden, F. L.: Native and foreign stimulators of platelet production. J. Lab. Clin. Med. 64:418, 1964.

30. Ebbe, S., Stohlman, F., Jr., Donovan, J., and Howard, D.: Platelet survival in rats with transfusion-induced thrombocytosis. J. Lab. Clin. Med. 68:813, 1966.

31. Evatt, B. L., and Levin, J.: Measurement of thrombopoiesis in rabbits using ^{75}selenomethionine. J. Clin. Invest. 48:1615, 1969.

32. Spencer, R. P., and Blau, M.: Intestinal transport of Selenium-75 selenomethionine. Science 136:155, 1962.

33. Hansson, E., and Blau, M.: Incorporation of Se^{75} Se-methionine into pancreatic juice proteins in vivo. Biochem. Biophys. Res. Commun. 13:71, 1963.

34. McConnell, K. P., and Cho, G. J.: Transmucosal movement of selenium. Amer. J. Physiol. 208:1191, 1965.

35. Cohen, P., Cooley, M. H., and Gardner, F. H.: The use of selenomethionine (Se^{75}) as a label for canine and human platelets. (Abstract.) J. Clin. Invest. 44:1036, 1965.

36. Shreiner, D. P., and Levin, J.: Detection of thrombopoietic activity in plasma by

stimulation of suppressed thrombopoiesis. (Abstract.) Blood 34:850, 1969.

37. Harker, L. A.: The regulation of thrombopoiesis (Abstract.) J. Clin. Invest. 48:35a, 1969.

38. Cooney, D. P., and Smith, B. A.: Maturation time of rabbit megakaryocytes. Brit. J. Haemat. 11:484, 1965.

39. Ebbe, S., and Stohlman, F., Jr.: Megkaryocytopoiesis in the rat. Blood 26:20, 1965.

40. McDonald, T. P., Odell, T. T., Jr., and Gosslee, D. G.: Platelet size in relation to platelet age. Proc. Soc. Exp. Biol. Med. 115:684, 1964.

41. Minter, N., and Ingram, M.: Density distribution of platelets. (Abstract.) Blood 30:551, 1967.

42. Karpatkin, S.: Heterogeneity of human platelets. I. Metabolic and kinetic evidence suggestive of young and old platelets. J. Clin. Invest. 48:1073, 1969.

Abnormal Patterns of Platelet Turnover

By Mario G. Baldini

The technic for the measure of platelet survival in human subjects by the use of radioactive sodium chromate [1] has been improved in recent years [2,3] so that the rate of platelet destruction and, indirectly, the platelet turnover can now be quantitated with relative accuracy. In normal subjects, the rate of platelet disappearance from the circulation is a finite function influenced mainly by senescence. The survival time is, on the average, 9 days.[1–4] In the steady state, the platelet turnover can be calculated by the formula proposed by Harker and Finch [4]:

$$\frac{\text{platelet count}/\mu\text{l}}{\text{survival time, days}} \times \frac{90 \text{ per cent}}{\text{per cent recovery}}$$

The recovery is expressed by the percentage of labeled platelets remaining in the circulation after infusion and is 65 per cent in normal subjects. Since splenectomized patients have a recovery of 90 per cent,[5,6] the lower value obtained in normal subjects is attributed to pooling of the platelets in the spleen. Any smaller value can be due to either abnormally increased platelet sequestration in the spleen or to abrupt destruction of part of the platelets immediately after infusion. In normal subjects with a platelet count of 220,000 per microliter ± 38,000 (S.D.), the platelet turnover equals 33,000 platelets per microliter per day ± 3,800 (S.D.), a value similar to that reported by others.[4]

Several abnormal patterns of turnover in thrombocytopenic states have recently been recognized. On most occasions, precise evaluation of megakaryocytopoiesis was not necessary to define the pattern involved, since gross evaluation of megakaryocyte number from the bone marrow study (whether increased, normal or decreased), together with estimation of platelet survival, provided sufficient evidence.

Thrombocytopenia due to lack of platelet production has long been recognized,[7] and is characterized by a reduction in the bone marrow megakaryocytes and a normal platelet survival time. It is illustrated by a variety of hypoplastic and infiltrative marrow lesions.

More recently, thrombocytopenia due to ineffective platelet production has also been defined. It is characterized by normal or increased numbers of megakaryocytes in the bone marrow and a normal platelet survival time indicating a decreased platelet turnover. It is presumably due to an incapacity of the megakaryocytes to reach the stage of platelet production. Ineffective thrombocytopoiesis has been described in Vitamin B_{12} deficiency, in the DiGuglielmo syndrome and in a not well defined form of congenital thrombocytopenia.[4]

From Brown University, Providence, Rhode Island, and The Memorial Hospital, Pawtucket, Rhode Island.

This research was supported by the U.S. Atomic Energy Commission Contract #AT (30–1) 3590.

We have observed ineffective platelet production in 2 patients with preleukemia. A 16-year-old girl and a 62-year-old man had unexplained thrombocytopenia (60,000 and 22,000 platelets per μl respectively) associated with increased megakaryocytes in the bone marrow and a normal survival of autologous platelets. There were no other abnormal findings and their spleens were not enlarged. The thrombocytopenia persisted until they developed acute myeloblastic leukemia 6 months and 4 years thereafter, respectively.

"Hypersplenic" thrombocytopenia is mainly a disorder of platelet distribution as demonstrated by Aster.[5] In patients with hypersplenic thrombocytopenia, a nearly normal platelet survival time has often been found associated with a markedly low recovery value. In the spleen of these patients, the number of platelets was far in excess of the theoretic figure considering the spleen as a "lake of blood." [8] The splenic platelet pool may be in a dynamic equilibrium with the platelets in the general circulation to maintain a constant ratio in each subject.

Recent investigations were done on megakaryocytopoiesis in hypersplenic rats by the use of bone marrow autoradiography after ^{3}HTdR injection.[9] These demonstrated that megakaryocytopoiesis was slightly stimulated in these animals and proved that thrombocytopenia in the splenomegalic rats could not be ascribed to the functional inhibition of megakaryocyte maturation. This result also is consistent with the idea that in "hypersplenism" the thrombocytopenia occurs mainly because of excessive pooling of the platelets in the enlarged spleen, while the bone marrow is providing a somewhat larger than normal number of platelets.

Although sequestration in the enlarged spleen may predominate as a mechanism for the thrombocytopenia in hypersplenism, we have seen, however, that an increase in platelet destruction may at times be associated with it.[6] Increased thrombocytolysis could be clearly demonstrated in rats with experimental hypersplenism produced by repeated intraperitoneal injections of methylcellulose.[10]

In these animals, not only the recovery value was lower, but the platelet survival time became progressively shorter when the spleen became larger, indicating that in severe hypersplenism the thrombocytopenia is due to a double mechanism, sequestration and destruction, although the former prevails on most occasions.

Thrombocytolytic thrombocytopenia is represented by various disorders in which increased platelet destruction is the common denominator. By analogy with red cell kinetics, one would expect a compensatory increase in platelet production in every case of thrombocytolytic thrombocytopenia. Idiopathic thrombocytopenic purpura (ITP) has been particularly investigated in this respect. In this disease, the rapid platelet destruction is due, we believe, to an autoantibody. Recent investigations by Najean [11] and by Harker and Finch [4,12] indicate that in all patients with ITP, the bone marrow megakaryocytes are indeed capable of increased platelet production which can reach 6 to 8 times the normal rate and is in direct relationship with the degree of the thrombocytopenia.[4,12] This result is in contrast to previous findings by others.[13–15] Since the technic for the measurement of platelet survival in human subjects has been recently improved,[2,3] we have reexamined this problem in 11 patients with acute or chronic ITP. Fifteen studies were done in these patients. Autologous platelets were used when the platelet count was above 100,000 per microliter, and allogeneic platelets when it was below this value. The studies performed with donor platelets were done only in patients who had not been previously transfused and had had no pregnancies. Figure 1 indicates no clear relationship between platelet count and degree of platelet production as measured by the platelet turnover. Only when one considers the 6 studies performed in patients with platelet counts above 50,000 per microliter, a direct relationship becomes probably apparent. Although these studies are still too few to allow a definitive conclusion, it appears that in severe ITP, with platelet counts below 50,000 per microliter, the thrombocytopenia is associated with only a normal or moderately increased rate of platelet production (1 to 3.5 times the normal) with no relationship to the degree of the thrombocytopenia. We interpret these findings as indicating a relative depression of the bone marrow in the severe cases of this disorder. In fact, in no patient with ITP could we ever find a normal platelet count in the presence of an increased platelet turnover. Only in occasional, splenectomized patients with ITP, examined 1 to 3 months after splenectomy we found increased thrombocytolysis associated with a normal platelet count. In all cases, however, this was only a transitional stage and, later, the platelet survival became normal while the platelet count increased further becoming stable at a higher

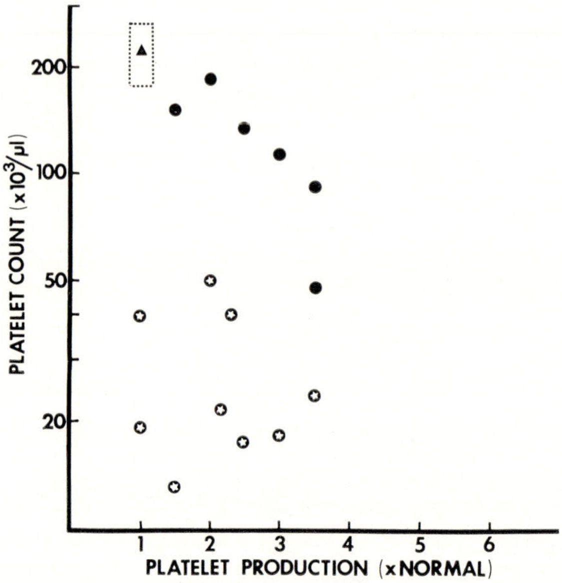

FIG. 1—In idiopathic thrombocytopenic purpura, rates of platelet production appeared to have no reciprocal relationship with the circulating platelet counts. Such relationship was probably apparent only in patients with platelet counts above 50,000 per microliter (●). Normal values (± 1 S.D.) are indicated by the rectangle.

than normal level. In conclusion, we have so far been unable to demonstrate the existence of a "compensated thrombocytolytic disease" analogous to the well known "compensated hemolytic disease," although we do not exclude that it may exist with disorders different from ITP.

These findings seem to indicate that in ITP the bone marrow megakaryocytes are inhibited to various degrees. This conclusion correlates well with the observations by Rolovic et al.[16] on megakaryocytopoiesis in experimentally induced immune thrombocytopenia in rats. In these animals, sustained thrombocytopenia was produced by titrated injections of a potent heteroimmune antiplatelet serum and megakaryocytopoiesis was studied by the use of ^{3}HTdR and bone marrow autoradiography. Data in the rats indicated a severely impaired and depressed megakaryocyte maturation, whereas in control rats in which the platelet level was maintained low by repeated thrombocytophereses, the pattern of megakaryocytopoiesis was that of accelerated maturation.

It seems clear, therefore, that in ITP the platelet autoantibody may also produce an injurious effect on the megakaryocytes in the bone marrow thereby depressing platelet production so that the thrombocytopenia in this disorder is the result of both increased platelet destruction and defective platelet production.

By analogy with the red cells, thrombocytopenias due to increased platelet destruction should be caused either by an extracorpuscular factor (an antibody, a microangiopathic disorder) or by an intracorpuscular platelet defect. The proof that in a given patient the thrombocytopenia is due to an intracorpuscular platelet defect is classically based on three findings: (1) a shortened survival of autologous platelets in the patient; (2) a normal survival of donor platelets in the patient; (3) a shortened survival of patient's platelets in a normal recipient.

Although in the past a few cases had been suspected or incompletely demonstrated as having thrombocytolytic thrombocytopenia of the intracorpuscular variety, clear-cut proof has only recently been obtained when we demonstrated this mechanism of platelet destruction in patients with the Wiskott-Aldrich syndrome.[17] This sex-linked hereditary disorder is known for its triad of eczema, thrombocytopenia, and infections. Studies in the past have primarily dealt with the immune deficiency aspects of this disease. Allogeneic platelets had been shown on several occasions to have a normal survival in these patients[18,19] and we confirmed this result.[14] Since autologous platelets had not been used in these experiments, the problem of the platelet turnover in the Wiskott-Aldrich syndrome needed reevaluation. We recently investigated three boys, 5, 7, and 9 years old with the syndrome, one normal 11-year-old sibling and their mother. The 3 children had had all the typical features of the syndrome with hemorrhage, recurrent infections, and eczema since early in life. Isohemagglutinins were low-normal or reduced. At the time of study, their platelet counts varied between 9000 and 56,000 per microliter and the bone marrow showed increased numbers of megakaryocytes. Survival of autologous platelets collected by pheresis and labeled with radioactive sodium chromate was measured in the children. Results demonstrated (Fig. 2) that

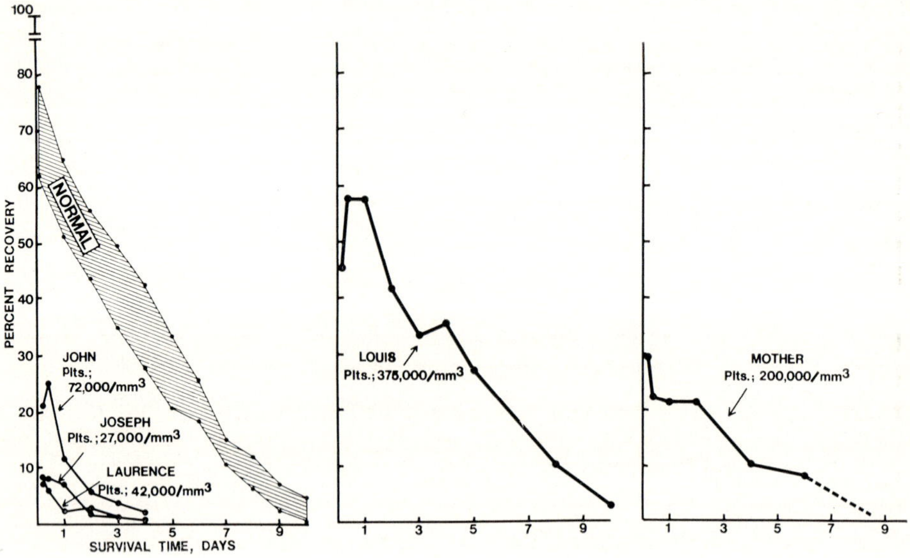

FIG. 2—Autologous platelet survival studies in 3 brothers with the Wiskott-Aldrich syndrome (left) show rapid platelet destruction. The recovery values were 8 to 25 per cent, the survival curves were exponential with a *T½* of less than 24 hours. The 10 per cent values in the curves were reached in 3 to 4 days. (Shaded area represents average normal values.) In the nonaffected sibling (middle), the autologous platelet survival was normal. It was moderately reduced in the clinically normal mother (right).

the autologous radioactive platelets disappeared rapidly from the circulation with a *T½* of less than 24 hours and that platelet turnover was increased 6.5 to 8 times the normal. The autologous platelet survival was normal in the nonaffected sibling whereas in the mother, who had a platelet count of 200,000 per microliter and no history of hemorrhage, infections or eczema, the autologous platelet survival was subnormal. In other experiments, the patients' platelets were infused in normal, group-compatible recipients and survival was shown again to be drastically reduced with no gross difference between the diminutive curves obtained in the 3 affected children and those observed in the normal recipients. These data clearly demonstrated that the increased thrombocytolysis in the affected children was due to an intracorpuscular platelet defect.

A number of studies of platelet physiology and metabolism were then performed in an attempt to delineate the platelet defect. Studies of platelet function disclosed that platelet aggregation with ADP and collagen was grossly reduced, but the result was most striking with epinephrine. With this agent, no second wave of aggregation could be elicited and the result was grossly abnormal also with platelets from the clinically normal mother.[20] Platelet factor III activity and clot retraction, on the other hand, were normal in all the children.

Studies of platelet biochemistry revealed that glucose consumption and lactate production, ATP level, glycogen content, the uptake of ^{14}C-serotonin in vitro and protein synthesis were moderately reduced in all the affected children with values ranging between 30 and 64 per cent of normal. They were, however, normal in the platelets of the mother and in those of the normal sibling. A number of glycolytic intermediates were determined. These were glucose-6-phosphate, fructose-diphosphate, triose-phosphate, 3-phospho-glycerate, phospho-enol-pyruvate, and pyruvate. No specific block in the glycolytic pathway could be shown, but there was an over-all reduction to a level of 50–60 per cent of normal in all intermediates.

Several key enzymes of glycolysis were repeatedly measured. These included hexokinase, glucose-6-phosphate dehydrogenase, phospho-fructo-kinase, lactic dehydrogenase, and pyruvate kinase. The enzyme which was found consistently reduced was hexokinase which in the affected children was 38 to 55 per cent of normal, while it was normal in the platelets of the mother and of the normal sibling. No evidence for a hexokinase inhibitor could be ascertained in mixture experiments employing combinations of patients' and normal platelet extracts. When allowance was made for platelet size which is known to be reduced in the Wiskott-Aldrich syndrome,[21] the above values were still significantly reduced. Platelet size in these experiments was expressed by the platelet protein content which, in the affected children, was reduced by an average of 35 per cent and was normal in the platelets of the mother (1.6 mg. 1×10^9 platelets).

A functional parameter which revealed a striking alteration was the citric acid cycle activity during incubation of the platelets at 37 C. with polystyrene-latex particles.[20] When normal platelets are incubated in a plasma medium to which polystyrene-latex particles of small diameter (0.234 μ) are added, aerobic glycolysis as measured by the $^{14}CO_2$ production from glucose-^{14}C labeled in the –6– position becomes stimulated 8 to 12 times the control level.[22] With the platelets from the affected children, no stimulation of the citric acid cycle was obtained (Fig. 3). Similar results were observed using epinephrine or collagen. It was of interest to see that platelets of the mother also had a severe lack of this reactive function. ^{14}C-lactic acid production from glucose-6-^{14}C in the presence of the latex particles was only moderately reduced in the platelets of the 3 children and of the mother.

We do not know, as yet, the primary defect responsible for the multiple abnormalities that we have demonstrated in the platelets of the Wiskott-Aldrich syndrome. Additional experiments with iodoacetate, which is a stimulant of aerobic glycolysis in the mitochondria, did not demonstrate an intrinsic functional block within the citric acid cycle in the platelets of the affected children.[20] However, the fact that physiologic stimulation of this pathway by the use of latex particles, epinephrine or collagen was missing in the platelets of the affected children may still be of important significance. These agents (phagocytosis, epinephrine, and collagen) are all known to cause platelet degranulation.[23] It is therefore concluded that the primary defect in the platelets of the Wiskott-Aldrich syndrome may be in the platelet α-granules which would not contain (or not release in sufficient quantities) factors which physiologically regulate or stimulate the activity of the citric acid cycle.

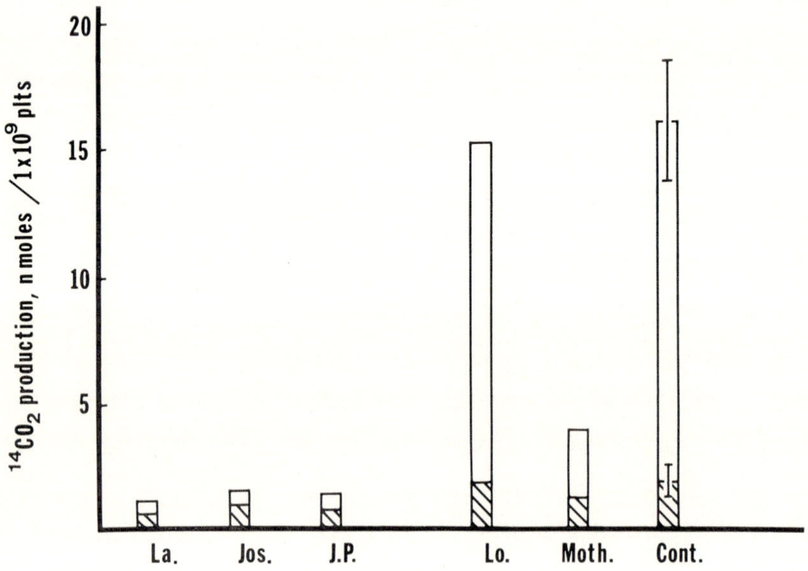

FIG. 3—Citric acid cycle activity in platelets before (hatched bars) and after (solid bars) stimulation with polystyrene-latex particles. In normal controls (right) and in the normal sibling (Lo) there was an 8 to 12 times increase of activity after particle stimulation (brackets express ± 1 S.D.). In the 3 brothers with the Wiskott-Aldrich syndrome, practically no stimulation of the citric acid cycle was obtained. The mother's platelets also had a severe lack of this reactive function. The activity of the citric acid cycle was measured by the $^{14}CO_2$ production from glucose –6– ^{14}C.

The relationship between the above functional, biochemical, and metabolic defects and the shortened platelet survival is not as yet clarified, but it is logical to assume that a causal relationship exists. In any case, these defects confirm the intracorpuscular nature of the thrombocytolytic thrombocytopenia in this disorder.

In conclusion, it is clear that a pathophysiologic classification of thrombocytopenic disorders should recognize:

1. Disorders of platelet production, which can be:
   (a) amegakaryocytic, i.e., hypoproliferative, or
   (b) megakaryocytic, i.e., due to ineffective thrombocytopoiesis.
2. Disorders of platelet distribution, due predominantly to pooling of the platelets in the spleen.
3. Disorders of platelet destruction, which can be:
   (a) extracorpuscular, i.e., due to an immune process or to a microangiopathic disorder, or
   (b) intracorpuscular, due to an acquired or inborn error of platelet metabolism.

REFERENCES

1. Aas, K. A., and Gardner, F. H.: Survival of blood platelets labeled with Chromium[51]. J. Clin. Invest. 37:1257, 1958.

2. Aster, R. H., and Jandl, J. H.: Platelet sequestration in man. I. Methods. J. Clin. Invest. 43:843, 1964.

3. Morrison, F. S., and Baldini, M.: The favorable effect of ACD on the viability of fresh and stored platelets. Vox Sang. 12:90, 1967.

4. Harker, L. A., and Finch, C. A.: Thrombokinetics in man. J. Clin. Invest. 48: 963, 1969.

5. Aster, R. H.: Pooling of platelets in the spleen: Role in the pathogenesis of "hypersplenic" thrombocytopenia. J. Clin. Invest. 45:645, 1966.

6. Baldini, M.: Platelet survival. Proceedings of the Xth Congr. Europ. Soc. Haemat, 1965; part II, p. 1091, Basel, Karger, 1967.

7. Stefanini, M., and Dameshek, W.: Collection, preservation and transfusion of platelets. New Eng. J. Med. 248:797, 1953.

8. Penny, R., Rozenberg, M. C., and Firkin, B. G.: The splenic platelet pool. Blood 27:1, 1966.

9. Rolovic, Z., and Baldini, M.: Megakaryocytopoiesis in splenectomized and "hypersplenic" rats. Brit. J. Haemat. In press.

10. Ciucci, A., and Baldini, M.: Platelet kinetics in experimental "hypersplenism." In preparation.

11. Najean, Y., Ardaillou, N., Caen, J., Larrieu, M. J., and Bernard, J.: Survival of radiochromium-labeled platelets in thrombocytopenias. Blood, 22:718, 1963.

12. Harker, L. A., and Finch, C. A.: Mechanisms of human thrombokinetics. (Abstract.) J. Clin. Invest. 47:45a, 1968.

13. Ebbe, S., and Baldini, M.: Studies of platelet kinetics in idiopathic thrombocytopenic purpura. (Abstract.) Clin. Res. 11: 193, 1963.

14. Baldini, M.: Idiopathic thrombocytopenic purpura. New Eng. J. Med. 274:1245, 1302, 1360, 1966.

15. Aster, R. H.: Observation on survival time, sites of sequestration, and production rate of platelets in idiopathic thrombocytopenic purpura. (Abstract.) Blood 28:1014, 1966.

16. Rolovic Z., Baldini, M., and Dameshek, W.: Megakaryocytopoiesis in experimentally induced immune thrombocytopenia. Blood. In press.

17. Baldini, M., Kim, B., Steiner, M., Kuramoto, A., Okuma, M., and Otridge, B.: Metabolic platelet defect in the Wiskott-Aldrich syndrome. (Abstract.) Proc. Amer. Soc. Pediat. Res., 1969, p. 48.

18. Pearson, H. A., Shulman, N. R., Oski, F. A., and Eitzman, D. V.: Platelet survival in Wiskott-Aldrich syndrome. J. Pediat. 68: 754, 1966.

19. Krivit, W., Yunis, E., and White, J. C.: Platelet survival studies in Aldrich syndrome. Pediatrics 37:339, 1966.

20. Kuramoto, A., Steiner, M., and Baldini, M.: Lack of platelet response to stimulation in the Wiskott-Aldrich syndrome. New Eng. J. Med. In press.

21. Gelzer, J., and Gasser, C.: Wiskott-Aldrich syndrome. Helv. Paediat. Acta 16: 17, 1961.

22. Kuramoto, A., Steiner, M., and Baldini, M. G.: Metabolic basis of platelet phagocytosis. Biochim. Biophys. Acta In press.

23. Movat, H. Z., Weiser, W. G., Glynn, M. F., and Mustard, J. F.: Platelet phagocytosis and aggregation. J. Cell. Biol. 27: 531, 1965.

Conference Summary

Frederick Stohlman, Jr.

I should like to thank all of you for joining us in this symposium which has been held in conjunction with the Centennial Celebration of St. Elizabeth's Hospital. It has been for me a most enjoyable symposium up until this point but I find it rather difficult, indeed impossible, to do justice to the many fine presentations and discussions of the past two days. Undoubtedly my remarks will demonstrate a measure of personal bias, but I hope that you will forgive me if I appear to neglect what to you may seem to be more important considerations.

I should like to thank the Chairmen, Drs. E. P. Cronkite, V. P. Bond, W. B. Castle, W. C. Moloney, and G. Brecher. They have done a remarkable job in keeping the conference on schedule and to the point. I would also like to thank the participants for their excellent and lucid presentations and for remaining within the allotted time thus providing us with adequate chance for discussion and reflection on the points which they raised.

Dr. Morley opened the meeting with a discussion of the oscillatory nature of hemopoiesis. The model which he proposed and the experimental data on which it is based appears to provide a new approach not only to the nature of the steady state but to an understanding of certain clinical diseases in man. It is clear that the implications of such a model are far-reaching and involve not only such diseases as cyclical neutropenia but also are relevant to the chemotherapeutic approach to diseases of the bone marrow and, perhaps, understanding of compensated hemolytic syndromes. Dr. Okunewick indicated that he had developed a similar model and had data which would substantiate the observations of Dr. Morley.

Dr. McCulloch of Toronto reported on the recent studies of his group on what he terms the short-range regulation of hemopoiesis. He emphasized that our attention has been directed over the past several years to "long-range regulation" by hormonal mechanisms such as erythropoietin. However, the growth of the pluripotential cell in his view is controlled by short-range regulation through a cell-cell interaction, perhaps through the membrane H_2 antigen as suggested by Cudkowicz. Documentation for this hypothesis derived from parental-hybrid transplants in which the administration of cytoxan influenced the rate of pluripotential cell growth. Dr. Joan Wright Goodman, whose initial observations on the effect of thymus cells in a parental-hybrid system prompted the Toronto group to use such a model, reported accelerated recovery of myeloid and megakaryocytic colonies as well as erythropoiesis in parental-hybrid transplants given thymic cells. The ineffectiveness of syngeneic thymic cells or dead parental cells together with the localization of the response to the spleen led her to suggest that the mechanism by which thymus cells affect recovery is not "humoral" in nature but more likely involves growth and recognition of "self" at the cellular level, i.e., a cell-cell interaction. If one can eliminate immunologic considerations in such a system, then it would appear that this model provides a most useful tool for the further study of cell interaction on the regulation of compartment size. In some respects this notion might be considered reminiscent of the sterile and fertile cell concepts advanced by Lajtha and ourselves a decade ago. The differentiation of sterile but viable cells still capable of exerting a negative feedback on the remaining fertile cells was invoked as an explanation for the early erythroid recovery in animals receiving erythropoietin or bled immediately after sublethal irradiation. Observations not dissimilar from these were reported in this conference by Dr. Ebbe who observed that thrombocytopenia and consequent differentiation of precursor cells shortly after sublethal irradiation accelerated recovery. It is clear that further work along the lines suggested by the observations of Drs. Goodman, McCulloch, and Ebbe should provide insight into the importance of compartment size, both of the committed and pluripotential, in the regulation of cell growth in normal and perturbed hemopoiesis.

The notion that there are functionally if not morphologically two different precursor cells appears to have been finally accepted; further convincing documentation was provided by Drs. Hellman, Morse, and Rickard. The question of "stem-cell competition" was raised by Dr. Hellman's observations on the relative recovery of myeloid and erythroid committed cells after exposure to radiation. Immediately after irradiation the pluripotential stem cell preferentially differentiates along granulocytic lines. Thereafter there appears to be a shift toward erythroid differentiation with an ultimate swingback toward normal. Clearly these changes are of importance in considering experimental results obtained with split-dose irradiation or radiation followed by chemotherapeutic agents such as those described by both Dr. Smith and Dr. Hellman. The mechanism responsible for stem-cell differentiation along a given line awaits further study. Exploration of these areas appear to be a most promising line of investigation and should yield important insight into interrelationship between the committed and the pluripotential stem-cell compartments. It would seem most likely that some form of cell-cell interaction will prove to be the mechanism by which this feedback control is maintained. It should also be noted that these relationships might provide a better understanding of the differential effects of certain chemotherapeutic agents which are currently in use for malignant disease of the bone marrow.

Ecology has been the subject of widespread interest in many fields of endeavor but only recently has it received attention at the cellular level in hemopoiesis. In addition to the previous considerations on cell-cell interaction, the problem of interrelationship of bone marrow structure to cell production evoked a most provocative discussion. Much of the experimental work on hemopoiesis, particularly in respect to the red cell, has been done in mice in which the spleen is an erythropoietic organ. Dr. Hoogstraten questioned the relevance of this to man where splenic hemopoiesis is seen only under pathologic circumstances. The difference appears to lie primarily in bone marrow structure. Thus, in adult mice the bone marrow is almost exclusively red and little increase in production can be accomplished within the rigid structure of the bone; the spleen serves as an alternative and expansible site of hemopoiesis. In adult man, in contrast, much of the marrow reverts to an inactive, yellow or fatty form and, after increased demand, hemopoiesis can expand into these areas. Noteworthy in this respect are the observations reported by Dr. Crosby's group on the growth characteristics of bone marrow after subcutaneous implantation of bone. These observations are an extension of those made by Dr. Moloney more than a decade ago. Dr. Crosby's group observed that both red and yellow marrow, after subcutaneous implantation, passed through an initial period in which there was proliferation of primitive cells, cancellous bone formation, the appearance of sinusoids, and finally active marrow over a period of a couple of weeks. Thereafter, marrow derived from a fatty source reverted to yellow marrow; in that derived from a red source, hemopoiesis continued. The mechanism responsible for such programming is yet to be defined, but clearly it is of clinical as well as biologic significance.

Dr. Fliedner proposed that the fine regulation of hemopoiesis is accomplished by a neural regulation of a pulsatile flow through the microcirculation of the marrow, pointing out that the rigid characteristics of the bone are admirably suited for such regulation. He further presented evidence, using his technic for continuous labeling, that the small round cell, usually referred to as the lymphocyte, appears to be the stem cell. Dr. McCulloch challenged the latter on the basis of work with velocity sedimentation which suggested that the transplantable cell was a large rather than a small cell. Distortion of the cell during velocity sedimentation or differences in density may account for these two apparently divergent views. One might further comment that, based on morphologic grounds and projected numbers of CFU's, there are not enough unidentifiable "large cells" to satisfy the requirements for both pluripotential and committed stem cells.

Regeneration of the marrow after physical destruction has been studied most extensively by Drs. Patt and Maloney using both cytokinetic and histologic technics. These elegant studies have led to the conclusion that regeneration stems from mesenchymal cells of the connective tissue of the haversian canals. Trabeculization appears to be an important factor in this process, perhaps because it produces a favorable microenvironment. It is of particular interest that under these conditions transplanted pluripotential cells did not restore hemopoiesis in the irradiated, depopulated marrow but that hemopoiesis regenerated from local cells. Perhaps the failure of transplanted cells to colonize in the depopulated limb reflected a failure of trapping of such cells due to the loss of structure.

These studies, while pointing out the importance of bone marrow structure and microenvironment in the support of hemopoiesis, do not provide an answer as to why in rodents hemopoiesis proceeds with relative ease in the spleen but does not do so in higher order animals. Presumably the answer lies in the microenvironment, but Crosby's experiments suggest that it is more complex Certainly it is an area in which further study is urgently needed.

There was lively speculation on the relationship of the above observations to the clinical diseases associated with bone marrow failure. Dr. Patt has observed myelofibrosis in an occasional animal after mechanical depopulation. Dr. Crosby postulated that the normal involutional process of the yellow marrow may in some people extend beyond that to the red marrow and perhaps due to inadequacies of the microvasculature result in aplasia. It appears clear, however, that two forms of aplasia occur. One of these is due to failure of the pluripotential stem cell as is seen following irradiation, cytotoxic agents, and the like. Presumably these would be amenable to transplantation should histocompatability permit. The other type of aplasia is associated with changes in bone marrow structure with or without the eventual evolution of fibrosis. Under such circumstances the microenvironment would appear to be such that transplantation of pluripotential cells would be inadequate for restoration of hemopoiesis.

After the discussion of these more general aspects of hemopoiesis, the symposium turned to consideration of more specific areas in discussing regulation of each of the three differentiated cell lines, erythropoiesis, myelopoiesis, and megakaryocytopoiesis. Dr. A. S. Gordon reported on the current status of the renal-erythropoietic factor, which he now terms erythrogenin. The interaction of a plasma factor with erythrogenin leads to the production of erythropoietin. The existence of this renal precursor of erythropoietin has now been documented by a number of laboratories and appears to be well established. Dr. Lange described his immunologic technic for measurement of erythropoietin which holds promise of providing a significant breakthrough on the clinical level. Replacement of the time-consuming and expensive bioassay technic with an immunoassay should accelerate the evolution of information on the relationship of erythropoietin to clinical disease states. Since the production of appropriate antigen for both biologic and immunologic studies has presented a problem in a number of laboratories, we were all delighted to hear of Dr. Lange's generous offer to provide appropriate antisera for immunologic studies. Our attentions were next turned to the importance of the shape of the oxygen dissociation curve to erythropoietin production both experimentally and in clinical disease states. Dr. J. W. Adamson discussed both high and low oxygen-affinity hemoglobins and their etiologic role in polycythemia and anemia. One would project that within the next few years the number of abnormal hemoglobins which will be demonstrated to be of causal relationship to abnormal erythropoiesis will grow perhaps geometrically.

The mechanism by which erythropoietin exerts its effect on the bone marrow was the subject of extensive discussion. Dr. S. H. Boyer described the work which he and his associates have done on the A-C Switch in sheep which have been bled or given erythropoietin. His data clearly indicates that erythropoietin is responsible for this switch and his model provides an opportunity to explore on a molecular level the action of erythropoietin. Dr. A. Kretchmar presented his model for the relationship of erythropoietin effectiveness to the generative cycle. The projected time parameters derived from his model were remarkably compatible with experimental data presented by Dr. B. S. Morse, as well as information derived from Dr. Kretchmar and Lange's laboratory. Both have suggested a variable G_1 for the committed cell, prolongation of G_1 being observed in the suppressed state. Erythropoietin shortens the G_1. This in effect leads to what Dr. Schooley has termed recruitment. This he discussed together with the importance of the spleen in the latter phenomenon. Final proof for Dr. Kretchmar's model would appear to require data indicating the necessity that erythropoietin be present during a part of S as well as during G_1. Dr. Stephen Robinson extended his studies on the fate of "stressed" red cells. It is clear that under conditions of severe erythroid stimulation there is a loss of hemoglobin either through fragmentation of the cell, cell death, or both in the marrow and the peripheral blood. Dr. Robinson's data does not exclude the possibility that the peripheral loss was due to fragmentation rather than cell death but Dr. Cronkite pointed out in discussion that the data of the Brookhaven group using single-cell autoradiography was incompatible with fragmentation and clearly indicated that "stressed" erythrocytes are short-lived.

During the past several years there has been much apparently conflicting evidence produced on the role of testosterone in red cell production. Significant light on this problem was shed by studies which Dr. Clifford Gurney reported utilizing a testosterone analogue, cyproterone. His studies indicate that androgens have a stimulatory effect on erythropoiesis which can be explained in significant measure by increased production of erythropoietin. The studies with cyproterone, however, indicate that in addition there is a small direct effect of androgens which may be on the bone marrow, the nature of which remains to be established. In the discussion Dr. Gurney also raised the importance from a clinical standpoint of inhibitors of the erythropoietin response which have been described in some cases of red cell aplasia by Dr. Sanford Krantz. He briefly summarized some of the previously published work of Dr. Krantz, who unfortunately was unable to attend.

Differences in the regulation of fetal and adult erythropoiesis have been the subject of interest in a number of laboratories including our own over the past several years. There is evidence suggesting that there are significant differences in the kinetics of fetal and adult erythropoiesis. Dr. J. F. Duplan of Paris presented further, most cogent evidence pointing to this fact. His studies demonstrate a clear-cut difference in the radiosensitivity of fetal as compared with adult pluripotential cells. Not only is the over-all radiosensitivity for adult cells lower but it is particularly prominent during G_1 and G_2. Of equal and perhaps greater significance was his demonstration that in vivo pluripotential cells in G_1 were less radiosensitive than cells in S, unlike the situation in many synchronized in vitro cell systems. Dr. Kubanek, in discussion, pointed out differences in the pattern of colonization of fetal cells after transplantation and also in both the plating efficiency and growth rate. Finally, Dr. G. Lucarelli described in vivo and in vitro studies which demonstrate that erythropoietin does in fact play a role in the regulation of red cell production in the newborn rat, although significant differences exist between this and adult animals.

The session on myelopoiesis was initiated by a review of the kinetics of normal granulopoiesis by Dr. E. P. Cronkite. In so doing he speculated on the mechanisms by which differentiation in the committed stem cell is switched on. Of particular interest to me were his ideas on the mechanism by which division occurs. In his view this relates to the development of a critical cell mass rather than intra-nuclear "programming." Although some evidence for this thesis exists in other systems, that it is the case for the hemopoietic cell remains to be established. Obligatory changes in the generation time would accompany any change in the rate of maturation. There are some studies to suggest a shortening of generation time in the erythroid cell in the face of severe stimulation and accelerated maturation, but the order of magnitude is small. Moreover, the data of Alpen and Cranmore indicate that there is no change at least in the pronormoblast. Obviously further experimental data is needed before a final judgment can be reached on the most provocative notion that the critical mass governs division of the hemopoietic cells.

The most lively interchange of the symposium centered about the interpretation of data derived from DFP labeling in myelopoiesis. Dr. Carter Bishop reported studies on double labeling of myeloid precursors using both cytoplasmic and nuclear labels. The controversy centered mainly about the interpretation of the "die-away" curve of the peripheral blood which appeared to be similar for both labels. If one assumes sequential entry of labeled cells into the peripheral blood, then the number of myelocyte divisions is greater than had been estimated from studies of labeled mitoses and grain count halving time in the bone marrow myelocytes by a number of workers including Drs. Cronkite and Patt. As an outsider, it appeared to me that the resolution to this problem rests in the arguments advanced by Drs. Boggs and Patt that there is random rather than strictly sequential entry of labeled granulocytes into the periphery. Final adjudication of this controversy will undoubtedly come with time, but it appears to be important in assessing the role of early release in granulocytosis.

As previously mentioned, it appears certain that the cell responsible for in vitro myeloid cloning with the Metcalf technic is a committed rather than a pluripotential stem cell. The data presented by Dr. Rickard presented cogent support for this thesis. One of the most exciting reports was by Dr. W. Robinson who has succeeded in cloning granulocytic elements from human bone marrow. His technic, which was presented in detail, relies on granulocytic feeder layers to initiate growth. This is in contrast to the murine system in which Sachs has reported that mature granulocytes inhibit colony growth—another illustration of the difference between mice and men. Dr. Robinson has not only succeeded in growing normal cells in culture, but also reports the growth of leukemic cells. Normal granulocytes in feeder layers support growth of cells but leukemic cells do not. Identification of the factor(s) responsible for initiation of cell growth in these cultures may well provide insight into the important problem of the failure of leukemic cells to differentiate. It is indeed an exciting prospect.

The extent to which the basic information derived from the study of cell proliferation has been applied to human disease was further shown by the reports of Drs. Killmann and Mauer. Dr. Killmann presented in a most lucid fashion his notion of clonal succession or "sleepers and feeders" and its etiologic role in leukemia. The hypothesis provides insight not only into the evolution of the disease itself, but also into the varied patterns of response to chemotherapy. Dr. Mauer approached the therapy of leukemia by utilizing kinetic technics and attempting to trigger dormant cells into cycle. His methods may not be applicable on a routine clinical basis because of time factors, as pointed out by Dr. Carbone. But, from an experimental standpoint, his approach holds great promise for providing a truly scientific approach to the rational development and application of chemotherapy.

Dr. Odell reviewed the concepts of megakaryocytopoiesis which have evolved over the past several years through work in his, as well as several other, laboratories. In his view the committed compartment is in large measure self-maintaining under normal conditions but, under conditions of increased demand, may be fed from the pluripotential compartment. The most immature of the recognizable class of megakaryocyte consists of a subclass of cells which are synthesizing DNA, the largest proportion of which are 4N but all ploidy classes are represented. The mechanism by which nucleic acid synthesis is discontinued is unknown. The possibility that it relates to the concentration of a unique protein such as thrombasthenin was considered, but the production of this protein appears to occur too late in maturation to affect DNA synthesis. He suggested that the size of the megakaryocyte and perhaps the eventual numbers of platelets produced relates to the ploidy class. Dr. S. Ebbe presented data on changes in megakaryocytopoiesis which result from altered peripheral blood platelet demands. These consist of changes of input into the differentiated compartment both in number and in eventual size. Macromegakaryocytes appear to be programmed chiefly in the committed compartment. With severe stimulation, shortening of maturation or transit time also is seen. She has also observed that acute thrombocytopenia when present at the time of irradiation will modify the irradiation effect on megakaryocytopoiesis and platelet production. These studies have broad implications beyond those of modification of irradiation injury alone and perhaps may provide an assay system for thrombopoietin. Dr. Levin reviewed the evidence for the existence of a hormone thrombopoietin analogous to erythropoietin. The data which he presented as well as that presented earlier by Odell and Ebbe appear to implicate the presence of such a hormone which is operative at the committed stem-cell level. The technics suggested for measuring thrombopoietin are many and varied. Perhaps one of the more promising is the use of ^{75}selenomethionine which Dr. Levin described. The technic requires repetitive massive infusions with thrombocytopenic plasma. Although it permits one to conclude that there is indeed a humoral feedback, the amounts of plasma required are such that refinement of the technic will be necessary before this assay can be widely applied to physiologic and clinical studies. It is perhaps noteworthy in this respect that the art of measuring erythropoietin started much the same way when Erslev used massive infusions of anemic rabbit plasma. Perhaps at some future conference we can look forward to measurments of thrombopoietin with more refined technics, perhaps immunologic, such as those described by Lange for erythropoietin.

Finally Dr. Mario Baldini presented a classification for clinical thrombocytopenia which reflects the kinetic studies previously presented. These related not only to megakaryocyte kinetics but also to platelet destruction. Implicit in his concept is that of ineffective megakaryocytopoiesis as well as decreased production due to reduced input from the committed compartment. His observations on the Wiskott-Aldrich syndrome provide strong evidence for intracorpuscular defects as a cause of thrombocytolysis and consequent platelet depletion. Presumably, as was the case with the red cell, other metabolic defects leading to shortening of life span will be described and will lead to better understanding and treatment of clinical disorders of the platelet.

As one listened to this conference over the past two days, it was clear that the past decade has seen the accumulation of much significant and important information relative to hemopoiesis. Our understanding of the complicated cellular interrelationships has provided important insight into the mechanism of control of cell production under normal conditions and after increased physiologic demand. Perhaps most importantly we have reached a point where clinical application of these basic observations is already a reality. One may, I think, look forward to the future with confidence that further studies built on the solid base that you have presented in this conference will provide greater understanding and a better therapeutic approach to diseases of hemopoiesis.

Finally I would like to thank all of you for joining us in this Centennial Celebration of St. Elizabeth's Hospital, for your participation in the conference, and for your willingness and cooperation in providing manuscripts. I would be remiss if I did not extend a special thanks to the chairmen of the various sessions without whose help we could not have developed as meaningful a discussion of the varied subjects which have been presented.

Index